EXPONENTIAL STABILITY OF STOCHASTIC DIFFERENTIAL EQUATIONS

PURE AND APPLIED MATHEMATICS

A Program of Monographs, Textbooks, and Lecture Notes

MONOGRAPHS AND TEXTBOOKS IN PURE AND APPLIED MATHEMATICS

1. *K. Yano,* Integral Formulas in Riemannian Geometry (1970)
2. *S. Kobayashi,* Hyperbolic Manifolds and Holomorphic Mappings (1970)
3. *V. S. Vladimirov,* Equations of Mathematical Physics (A. Jeffrey, ed.; A. Littlewood, trans.) (1970)
4. *B. N. Pshenichnyi,* Necessary Conditions for an Extremum (L. Neustadt, translation ed.; K. Makowski, trans.) (1971)
5. *L. Narici et al.,* Functional Analysis and Valuation Theory (1971)
6. *S. S. Passman,* Infinite Group Rings (1971)
7. *L. Dornhoff,* Group Representation Theory. Part A: Ordinary Representation Theory. Part B: Modular Representation Theory (1971, 1972)
8. *W. Boothby and G. L. Weiss, eds.,* Symmetric Spaces (1972)
9. *Y. Matsushima,* Differentiable Manifolds (E. T. Kobayashi, trans.) (1972)
10. *L. E. Ward, Jr.,* Topology (1972)
11. *A. Babakhanian,* Cohomological Methods in Group Theory (1972)
12. *R. Gilmer,* Multiplicative Ideal Theory (1972)
13. *J. Yeh,* Stochastic Processes and the Wiener Integral (1973)
14. *J. Barros-Neto,* Introduction to the Theory of Distributions (1973)
15. *R. Larsen,* Functional Analysis (1973)
16. *K. Yano and S. Ishihara,* Tangent and Cotangent Bundles (1973)
17. *C. Procesi,* Rings with Polynomial Identities (1973)
18. *R. Hermann,* Geometry, Physics, and Systems (1973)
19. *N. R. Wallach,* Harmonic Analysis on Homogeneous Spaces (1973)
20. *J. Dieudonné,* Introduction to the Theory of Formal Groups (1973)
21. *I. Vaisman,* Cohomology and Differential Forms (1973)
22. *B.-Y. Chen,* Geometry of Submanifolds (1973)
23. *M. Marcus,* Finite Dimensional Multilinear Algebra (in two parts) (1973, 1975)
24. *R. Larsen,* Banach Algebras (1973)
25. *R. O. Kujala and A. L. Vitter, eds.,* Value Distribution Theory: Part A; Part B: Deficit and Bezout Estimates by Wilhelm Stoll (1973)
26. *K. B. Stolarsky,* Algebraic Numbers and Diophantine Approximation (1974)
27. *A. R. Magid,* The Separable Galois Theory of Commutative Rings (1974)
28. *B. R. McDonald,* Finite Rings with Identity (1974)
29. *J. Satake,* Linear Algebra (S. Koh et al., trans.) (1975)
30. *J. S. Golan,* Localization of Noncommutative Rings (1975)
31. *G. Klambauer,* Mathematical Analysis (1975)
32. *M. K. Agoston,* Algebraic Topology (1976)
33. *K. R. Goodearl,* Ring Theory (1976)
34. *L. E. Mansfield,* Linear Algebra with Geometric Applications (1976)
35. *N. J. Pullman,* Matrix Theory and Its Applications (1976)
36. *B. R. McDonald,* Geometric Algebra Over Local Rings (1976)
37. *C. W. Groetsch,* Generalized Inverses of Linear Operators (1977)
38. *J. E. Kuczkowski and J. L. Gersting,* Abstract Algebra (1977)
39. *C. O. Christenson and W. L. Voxman,* Aspects of Topology (1977)
40. *M. Nagata,* Field Theory (1977)
41. *R. L. Long,* Algebraic Number Theory (1977)
42. *W. F. Pfeffer,* Integrals and Measures (1977)
43. *R. L. Wheeden and A. Zygmund,* Measure and Integral (1977)
44. *J. H. Curtiss,* Introduction to Functions of a Complex Variable (1978)
45. *K. Hrbacek and T. Jech,* Introduction to Set Theory (1978)
46. *W. S. Massey,* Homology and Cohomology Theory (1978)
47. *M. Marcus,* Introduction to Modern Algebra (1978)
48. *E. C. Young,* Vector and Tensor Analysis (1978)
49. *S. B. Nadler, Jr.,* Hyperspaces of Sets (1978)
50. *S. K. Segal,* Topics in Group Kings (1978)
51. *A. C. M. van Rooij,* Non-Archimedean Functional Analysis (1978)
52. *L. Corwin and R. Szczarba,* Calculus in Vector Spaces (1979)

53. *C. Sadosky,* Interpolation of Operators and Singular Integrals (1979)
54. *J. Cronin,* Differential Equations (1980)
55. *C. W. Groetsch,* Elements of Applicable Functional Analysis (1980)
56. *I. Vaisman,* Foundations of Three-Dimensional Euclidean Geometry (1980)
57. *H. I. Freedan,* Deterministic Mathematical Models in Population Ecology (1980)
58. *S. B. Chae,* Lebesgue Integration (1980)
59. *C. S. Rees et al.,* Theory and Applications of Fourier Analysis (1981)
60. *L. Nachbin,* Introduction to Functional Analysis (R. M. Aron, trans.) (1981)
61. *G. Orzech and M. Orzech,* Plane Algebraic Curves (1981)
62. *R. Johnsonbaugh and W. E. Pfaffenberger,* Foundations of Mathematical Analysis (1981)
63. *W. L. Voxman and R. H. Goetschel,* Advanced Calculus (1981)
64. *L. J. Corwin and R. H. Szcarba,* Multivariable Calculus (1982)
65. *V. I. Istrăţescu,* Introduction to Linear Operator Theory (1981)
66. *R. D. Järvinen,* Finite and Infinite Dimensional Linear Spaces (1981)
67. *J. K. Beem and P. E. Ehrlich,* Global Lorentzian Geometry (1981)
68. *D. L. Armacost,* The Structure of Locally Compact Abelian Groups (1981)
69. *J. W. Brewer and M. K. Smith, eds.,* Emily Noether: A Tribute (1981)
70. *K. H. Kim,* Boolean Matrix Theory and Applications (1982)
71. *T. W. Wieting,* The Mathematical Theory of Chromatic Plane Ornaments (1982)
72. *D. B. Gauld,* Differential Topology (1982)
73. *R. L. Faber,* Foundations of Euclidean and Non-Euclidean Geometry (1983)
74. *M. Carmeli,* Statistical Theory and Random Matrices (1983)
75. *J. H. Carruth et al.,* The Theory of Topological Semigroups (1983)
76. *R. L. Faber,* Differential Geometry and Relativity Theory (1983)
77. *S. Barnett,* Polynomials and Linear Control Systems (1983)
78. *G. Karpilovsky,* Commutative Group Algebras (1983)
79. *F. Van Oystaeyen and A. Verschoren,* Relative Invariants of Rings (1983)
80. *I. Vaisman,* A First Course in Differential Geometry (1984)
81. *G. W. Swan,* Applications of Optimal Control Theory in Biomedicine (1984)
82. *T. Petrie and J. D. Randall,* Transformation Groups on Manifolds (1984)
83. *K. Goebel and S. Reich,* Uniform Convexity, Hyperbolic Geometry, and Nonexpansive Mappings (1984)
84. *T. Albu and C. Năstăsescu,* Relative Finiteness in Module Theory (1984)
85. *K. Hrbacek and T. Jech,* Introduction to Set Theory: Second Edition (1984)
86. *F. Van Oystaeyen and A. Verschoren,* Relative Invariants of Rings (1984)
87. *B. R. McDonald,* Linear Algebra Over Commutative Rings (1984)
88. *M. Namba,* Geometry of Projective Algebraic Curves (1984)
89. *G. F. Webb,* Theory of Nonlinear Age-Dependent Population Dynamics (1985)
90. *M. R. Bremner et al.,* Tables of Dominant Weight Multiplicities for Representations of Simple Lie Algebras (1985)
91. *A. E. Fekete,* Real Linear Algebra (1985)
92. *S. B. Chae,* Holomorphy and Calculus in Normed Spaces (1985)
93. *A. J. Jerri,* Introduction to Integral Equations with Applications (1985)
94. *G. Karpilovsky,* Projective Representations of Finite Groups (1985)
95. *L. Narici and E. Beckenstein,* Topological Vector Spaces (1985)
96. *J. Weeks,* The Shape of Space (1985)
97. *P. R. Gribik and K. O. Kortanek,* Extremal Methods of Operations Research (1985)
98. *J.-A. Chao and W. A. Woyczynski, eds.,* Probability Theory and Harmonic Analysis (1986)
99. *G. D. Crown et al.,* Abstract Algebra (1986)
100. *J. H. Carruth et al.,* The Theory of Topological Semigroups, Volume 2 (1986)
101. *R. S. Doran and V. A. Belfi,* Characterizations of C*-Algebras (1986)
102. *M. W. Jeter,* Mathematical Programming (1986)
103. *M. Altman,* A Unified Theory of Nonlinear Operator and Evolution Equations with Applications (1986)
104. *A. Verschoren,* Relative Invariants of Sheaves (1987)
105. *R. A. Usmani,* Applied Linear Algebra (1987)
106. *P. Blass and J. Lang,* Zariski Surfaces and Differential Equations in Characteristic $p > 0$ (1987)
107. *J. A. Reneke et al.,* Structured Hereditary Systems (1987)

108. *H. Busemann and B. B. Phadke,* Spaces with Distinguished Geodesics (1987)
109. *R. Harte,* Invertibility and Singularity for Bounded Linear Operators (1988)
110. *G. S. Ladde et al.,* Oscillation Theory of Differential Equations with Deviating Arguments (1987)
111. *L. Dudkin et al.,* Iterative Aggregation Theory (1987)
112. T. Okubo, *Differential Geometry* (1987)
113. *D. L. Stancl and M. L. Stancl,* Real Analysis with Point-Set Topology (1987)
114. *T. C. Gard,* Introduction to Stochastic Differential Equations (1988)
115. *S. S. Abhyankar,* Enumerative Combinatorics of Young Tableaux (1988)
116. *H. Strade and R. Farnsteiner,* Modular Lie Algebras and Their Representations (1988)
117. *J. A. Huckaba,* Commutative Rings with Zero Divisors (1988)
118. *W. D. Wallis,* Combinatorial Designs (1988)
119. *W. Więsław,* Topological Fields (1988)
120. *G. Karpilovsky,* Field Theory (1988)
121. *S. Caenepeel and F. Van Oystaeyen,* Brauer Groups and the Cohomology of Graded Rings (1989)
122. *W. Kozlowski,* Modular Function Spaces (1988)
123. *E. Lowen-Colebunders,* Function Classes of Cauchy Continuous Maps (1989)
124. *M. Pavel,* Fundamentals of Pattern Recognition (1989)
125. *V. Lakshmikantham et al.,* Stability Analysis of Nonlinear Systems (1989)
126. *R. Sivaramakrishnan,* The Classical Theory of Arithmetic Functions (1989)
127. *N. A. Watson,* Parabolic Equations on an Infinite Strip (1989)
128. *K. J. Hastings,* Introduction to the Mathematics of Operations Research (1989)
129. *B. Fine,* Algebraic Theory of the Bianchi Groups (1989)
130. *D. N. Dikranjan et al.,* Topological Groups (1989)
131. *J. C. Morgan II,* Point Set Theory (1990)
132. *P. Biler and A. Witkowski,* Problems in Mathematical Analysis (1990)
133. *H. J. Sussmann,* Nonlinear Controllability and Optimal Control (1990)
134. *J.-P. Florens et al.,* Elements of Bayesian Statistics (1990)
135. *N. Shell,* Topological Fields and Near Valuations (1990)
136. *B. F. Doolin and C. F. Martin,* Introduction to Differential Geometry for Engineers (1990)
137. *S. S. Holland, Jr.,* Applied Analysis by the Hilbert Space Method (1990)
138. *J. Okniński,* Semigroup Algebras (1990)
139. *K. Zhu,* Operator Theory in Function Spaces (1990)
140. *G. B. Price,* An Introduction to Multicomplex Spaces and Functions (1991)
141. *R. B. Darst,* Introduction to Linear Programming (1991)
142. *P. L. Sachdev,* Nonlinear Ordinary Differential Equations and Their Applications (1991)
143. *T. Husain,* Orthogonal Schauder Bases (1991)
144. *J. Foran,* Fundamentals of Real Analysis (1991)
145. *W. C. Brown,* Matrices and Vector Spaces (1991)
146. *M. M. Rao and Z. D. Ren,* Theory of Orlicz Spaces (1991)
147. *J. S. Golan and T. Head,* Modules and the Structures of Rings (1991)
148. *C. Small,* Arithmetic of Finite Fields (1991)
149. *K. Yang,* Complex Algebraic Geometry (1991)
150. *D. G. Hoffman et al.,* Coding Theory (1991)
151. *M. O. González,* Classical Complex Analysis (1992)
152. *M. O. González,* Complex Analysis (1992)
153. *L. W. Baggett,* Functional Analysis (1992)
154. *M. Sniedovich,* Dynamic Programming (1992)
155. *R. P. Agarwal,* Difference Equations and Inequalities (1992)
156. *C. Brezinski,* Biorthogonality and Its Applications to Numerical Analysis (1992)
157. *C. Swartz,* An Introduction to Functional Analysis (1992)
158. *S. B. Nadler, Jr.,* Continuum Theory (1992)
159. *M. A. Al-Gwaiz,* Theory of Distributions (1992)
160. *E. Perry,* Geometry: Axiomatic Developments with Problem Solving (1992)
161. *E. Castillo and M. R. Ruiz-Cobo,* Functional Equations and Modelling in Science and Engineering (1992)
162. *A. J. Jerri,* Integral and Discrete Transforms with Applications and Error Analysis (1992)
163. *A. Charlier et al.,* Tensors and the Clifford Algebra (1992)

164. *P. Biler and T. Nadzieja*, Problems and Examples in Differential Equations (1992)
165. *E. Hansen*, Global Optimization Using Interval Analysis (1992)
166. *S. Guerre-Delabrière*, Classical Sequences in Banach Spaces (1992)
167. *Y. C. Wong*, Introductory Theory of Topological Vector Spaces (1992)
168. *S. H. Kulkarni and B. V. Limaye*, Real Function Algebras (1992)
169. *W. C. Brown*, Matrices Over Commutative Rings (1993)
170. *J. Loustau and M. Dillon*, Linear Geometry with Computer Graphics (1993)
171. *W. V. Petryshyn*, Approximation-Solvability of Nonlinear Functional and Differential Equations (1993)
172. *E. C. Young*, Vector and Tensor Analysis: Second Edition (1993)
173. *T. A. Bick*, Elementary Boundary Value Problems (1993)
174. *M. Pavel*, Fundamentals of Pattern Recognition: Second Edition (1993)
175. *S. A. Albeverio et al.*, Noncommutative Distributions (1993)
176. *W. Fulks*, Complex Variables (1993)
177. *M. M. Rao*, Conditional Measures and Applications (1993)
178. *A. Janicki and A. Weron*, Simulation and Chaotic Behavior of α-Stable Stochastic Processes (1994)
179. *P. Neittaanmäki and D. Tiba*, Optimal Control of Nonlinear Parabolic Systems (1994)
180. *J. Cronin*, Differential Equations: Introduction and Qualitative Theory, Second Edition (1994)
181. *S. Heikkilä and V. Lakshmikantham*, Montotone Iterative Techniques for Discontinuous Nonlinear Differential Equations (1994)
182. *X. Mao*, Exponential Stability of Stochastic Differential Equations (1994)

Additional Volumes in Preparation

EXPONENTIAL STABILITY OF STOCHASTIC DIFFERENTIAL EQUATIONS

Xuerong Mao

Department of Statistics and Modelling Science
University of Strathclyde
Glasgow, Scotland

Marcel Dekker, Inc. New York • Basel • Hong Kong

Library of Congress Cataloging-in-Publication Data

Mao, Xuerong.
Exponential stability of stochastic differential equations / Xuerong Mao.
p. cm. — (Monographs and textbooks in pure and applied mathematics ; 182)
Includes bibliographical references and index.
ISBN 0-8247-9080-4
1. Stochastic differential equations. 2. Delay differential equations. I. Title. II. Series.
QA274.23.M35 1994
519.2—dc20 94-6019
CIP

The publisher offers discounts on this book when ordered in bulk quantities. For more information, write to Special Sales/Professional Marketing at the address below.

This book is printed on acid-free paper.

MARCEL DEKKER, INC.
270 Madison Avenue, New York, New York 10016

Current printing (last digit):
10 9 8 7 6 5 4 3 2 1

PRINTED IN THE UNITED STATES OF AMERICA

To my grandmother and parents
and
to my wife Weihong
with love

Preface

It was only recently observed that the most natural examples of stochastic flows (namely the isotropic ones) could not be described by stochastic differential equations involving a finite number of real-valued Brownian motions (cf. Baxendale (1984) and Le Jan (1982)). For example, Brownian flows, i.e., the flows with independent increments, are described by stochastic differential equations driven by vector-field-valued Brownian motions. In other words, in applications to various stochastic problems concerning stochastic differential equations and stochastic flows, it is necessary to define stochastic integrals with nonlinear integrators of the form

$$\int F(x(s), ds),$$

where $F(x, t)$ is a continuous C-semimartingale with spatial parameter $x \in R^d$. This modern stochastic integral has been introduced recently by several leading mathematicians e.g. Carmona, Kunita, Le Jan, Naulart, and Watanabe. It is defined in such a way that if $F(x, t) = x \cdot B(t)$ with B a Brownian motion, then the above integral coincides with the classical Itô integral. For an intensive treatment on this subject we refer to Kunita (1990) and Carmona & Nualart (1990).

On the other hand, numerous problems in the sciences, engineering, economy, biology and so on lead to the study of stochastic differential (delay) equations driven by nonlinear integrators and their exponential stability. So far there has been no book dealing with the exponential stability for such advanced stochastic differential equations, although there exist several books on stability of stochastic differential equations based on Brownian motions or (usual) semimartingales, for example, Kushner (1967), Has'minskii (1969), Arnold (1974), and Mao (1991). The theory of the exponential stability for such equations has been developed very quickly in the past few years, and they require monographs to systematize the results.

Usually the existence and uniqueness of the solution for a stochastic differential equation driven by a nonlinear integrator are obtained under the Lipschitz condition. It is only recently proved by the author that the existence and uniqueness of the solution for such an equation under non-Lipschitz conditions-e.g. Hölder's continuity-can be established by successive approximation procedure, and therefore the approximate solutions, which are useful in applications, are also obtained.

A more general concept of stochastic bounded integral contractors has been introduced to study the existence and uniqueness of the solution for a stochastic differential delay equation driven by a nonlinear integrator. In particular, the stochastic bounded integral contractors include the Lipschitz condition as a special case. This new method seems to possess great potential.

Despite the apparent diversity of problems of modern society, often tools developed in one context are adopted for an entirely different situation. For example, consider Lyapunov's second method. This interesting and fruitful technique has gained increasing significance and has given decisive impetus for modern development of stability theory of stochastic differential delay equations. A manifest advantage of this method is that it does not require the knowledge of solutions and therefore has great power in applications.

It is now recognized that the Lyapunov function and the theory of Lebesgue–Stieltjes integral inequalities can be utilized to study qualitative and quantitative properties of stochastic differential equations driven by nonlinear integrators. The Lyapunov function serves as a vehicle to transform a given complicated system into a relatively simpler system and therefore it is enough to investigate the properties of this simpler system. It is also being realized that the same versatile tools are adopted to discuss entirely different problems (e.g. the existence of global solution) and the almost surely asymptotic bounds of solutions and various stabilities, in particular, exponential stability.

It is in this spirit that we see the importance of this book. It is the aim of the book to systematize the recent developments, in particular, our own research on stochastic differential (delay) equations driven by nonlinear integrators and their exponential stability.

Some important features of the book are as follows. This is the first book that:

(i) presents a systematic study of stochastic differential (delay) equations driven by nonlinear integrators and their exponential stability;

(ii) establishes the theory of existence and uniqueness of the solution under a weaker condition than the Lipschitz one via a successive approximation procedure;

(iii) demonstrates the manifestations of the general Lyapunov method by showing how this effective technique can be adapted to study entirely differently qualitative and quantitative properties of the stochastic equations, e.g. asymptotic bounds, exponential stability;

(iv) introduces the concept of stochastic bounded integral contractors in the context of stochastic differential delay equations – this method seems to possess great potential in the study of the existence and uniqueness of solutions to delay equations;

(v) presents many applications, e.g. stochastic stabilization and destabilization, stochastic flows and stochastic oscillators to illustrate the important uses of the theory in real situations.

This book consists of eight chapters. Chapter 1 introduces the theory of C-semimartingales with spatial parameters and stochastic integrals based on them. It therefore forms a basis for the book. We start with a very brief account on the general theory of stochastic processes, in particular, semimartingales. We then introduce C-semimartingales with spatial parameters and give the definition of stochastic integrals based on the C-semimartingales (which are also called nonlinear integrators) along with a discussion of properties of the integrals including the generalized Itô's formula and the time change formula of the integrals. This chapter ends up with a discussion on Lebesgue-Stieltjes integral inequalities, e.g. the generalized Gronwall inequality and Bihari inequality, which will be needed very frequently throughout this book.

In Chapter 2 we establish the existence and uniqueness of the solution for a stochastic differential equation driven by a nonlinear integrator under a weaker condition than Lipschitz's, for instance, Hölder's continuity, via successive approximation procedure. We also obtain L^p-estimates and almost surely asymptotic bounds for the solutions.

Chapter 3 is devoted to the discussion of stochastic differential delay equations driven by nonlinear integrators. The local solutions are first studied under the local Lipschitz conditions and the global solutions are obtained under the additional linear growth conditions as usual. But a more general concept of stochastic bounded integral contractors is then introduced to study the existence and uniqueness of the global solutions. In particular, the stochastic bounded integral contractors include the Lipschitz condition as a special case. Approximate solutions to the delay equations not only are very useful in applications, in particular, in obtaining numerical solutions, but also represent an alternative to the procedure for establishing the existence and uniqueness of the solutions. The Cauchy-Maruyama and Carathéodory approximation procedures will be discussed. This chapter closes with a study of the regularities of the solutions, e.g. L^p-bounds and almost surely asymptotic bounds.

From Chapter 4 on we shall start to study exponential stability of stochastic differential (delay) equations driven by nonlinear integrators. In Chapter 4, we first apply the Lyapunov functions to study the almost sure exponential stability and moment exponential stability for the stochastic differential equations. As applications, we discuss the stochastic stabilization and destabilization, the Lyapunov exponents of stochastic flows and stochastic oscillators.

Chapter 5 forms the study of almost sure exponential stability for the stochastic differential delay equations. In order to reduce complications and to make the theory more understandable, we start with the discussion of almost sure exponential stability for the equations with constant delay and then carry on to deal with the equations with variable delay. We furthermore generalize these results by studying the almost sure exponential stability for stochastic differential-functional equations. We shall close this chapter by giving several examples to illustrate the theory of almost sure exponential stability.

Chapter 6 consists of the discussions of the moment exponential stability for stochastic differential delay equations. The Lyapunov functions are first used to establish sufficient criteria for the moment exponential stability of the delay equations. We then investigate the robustness problem of the stability, i.e. how much stochastic perturbation can be added to a delay equation, which is moment exponentially stable, so that the stochastically perturbed system remains stable in the same sense. The absolutely exponential stability for delay equations are also studied and a number of examples are given to illustrate the results.

In Chapter 7 we investigate the moment exponential stability for a kind of stochastic differential equations with small time lag. More precisely we shall study equations of the type

$$dx(t) = F(x(t), dt) + G(x(t) - x(t-\tau), dt)$$

with small τ. The aim of this chapter is to show under various hypotheses that if the non-delay equation $dx(t) = F(x(t), dt)$ is moment exponentially stable, then the above delay equations remains stable provided τ small enough. Moreover, we also give estimates for the bounds of such τ in various cases, and obviously such bounds are useful and important in applications.

The last chapter of this book is devoted to the study of exponential stability for large-scale stochastic differential delay systems. We obtain a sufficient and necessary condition for the exponential stability of a large-scale system in hierarchical form. We also obtain a sufficient condition for the exponential stability of a general large-scale system. The exponential stability for linear large-scale systems will be discussed in more detail. We shall close this chapter by discussing several applications.

Acknowledgements

I wish to thank Professors K. D. Elworthy and L. Markus for their kind assistance and constant encouragements. I am particularly indebted to the Faculty of Science at the University of Strathclyde, the London Mathematical Society, the Royal Society for their financial support during the writing of the book. I also wish to thank my family for their constant support, in particular, I should thank my beloved wife Weihong.

Xuerong Mao

Contents

Preface v

General Notation xi

1. Semimartingales with Spatial Parameters and Stochastic Integrals
   - 1.1. Introduction 1
   - 1.2. Preliminaries on stochastic processes 2
   - 1.3. C-semimartingales with spatial parameters 12
   - 1.4. Stochastic integrals with nonlinear integrators 17
   - 1.5. Generalized Itô's formula and others 20
   - 1.6. Time change formula of stochastic integrals 22
   - 1.7. Inequalities 24

2. Stochastic Differential Equations
   - 2.1. Introduction 33
   - 2.2. One-dimensional stochastic differential equations 36
   - 2.3. Special d-dimensional stochastic differential equations 55
   - 2.4. General d-dimensional stochastic differential equations 58
   - 2.5. L^p-estimates of solutions 72
   - 2.6. Almost surely asymptotic bounds of solutions 76
   - 2.7. Stochastic flows and stochastic differential equations 89

3. Stochastic Differential Delay Equations
   - 3.1. Introduction 93
   - 3.2. Existence and uniqueness of solutions 94
   - 3.3. Bounded stochastic integral contractors and solutions 101
   - 3.4. Cauchy-Maruyama's approximate solutions 117
   - 3.5. Carathéodory's approximate solutions 124
   - 3.6. L^p-estimates 132
   - 3.7. Almost surely asymptotic bounds of solutions 136

4. Exponential Stability of Stochastic Differential Equations
4.1. Introduction 147
4.2. Almost sure exponential stability 148
4.3. Pth moment exponential stability 163
4.4. Stochastic stabilization and destabilization 167
4.5. Lyapunov exponents of stochastic flows 172
4.6. Stochastic oscillators 176

5. Almost Sure Exponential Stability of Stochastic Differential Delay Equations
5.1. Introduction 183
5.2. Equations with constant delay 184
5.3. Equations with variable delay 195
5.4. Stochastic differential functional equations 201
5.5. Examples 215

6. Moment Exponential Stability of Stochastic Differential Delay Equations
6.1. Introduction 223
6.2. Pth moment exponential stability of delay equations 224
6.3. Generalizations 229
6.4. Pth moment exponential stability of functional equations 238
6.5. Pth moment exponential stability of functional equations with perturbation 243
6.6. Examples 251

7. Exponential Stability of Stochastic Differential Equations with Small Time Lag
7.1. Introduction 259
7.2. Linear equations with small time lag 260
7.3. Semi-linear equations with small time lag 265
7.4. Non-linear equations with small time lag 266
7.5. Further remarks and examples 273

8. Exponential Stability of Large-Scale Stochastic Differential Delay Systems
8.1. Introduction 277
8.2. Definitions and relationships between various exponential stabilities 279
8.3. Large-scale stochastic delay systems in hierarchical form 283
8.4. General large-scale stochastic delay systems 288
8.5. Applications 290

References 295

Index 305

General Notation

Theorem 4–3.2, for example, means Theorem 3.2 (the second theorem in Section 3) in Chapter 4. If this theorem is quoted in Chapter 4, it is written as Theorem 3.2 only.

positive	: > 0.
nonnegative	: ≥ 0.
a.s.	: almost surely or P–almost surely.
$A := B$	: A is defined by B or A is denoted by B.
$A(x) \equiv B(x)$	: A(x) and B(x) are identically equal, i.e. $A(x) = B(x)$ for all x.
$\mathbb{1}_A$	: the indicator function of A, i.e. $\mathbb{1}_A(x) = 1$ or 0 accordingly as $x \in A$ or $x \notin A$.
$a \vee b$	: the maximum of a and b.
$a \wedge b$	: the minimum of a and b.
$f : A \to B$	: the mapping f from A to B.
$R = R^1$	: the real line.
R_+	: the set of all nonnegative real numbers, i.e. $[0, \infty)$.
R^d	: the d–dimensional Euclidean space.
S_h	: $S_h := \{x \in R^d : \|x\| \leq h\}$.
$\|x\|$	: the Euclidean norm of a vector x.
A^T	: the transpose of a vector or matrix A.
(x, y)	: $(x, y) = x^T y$, i.e. the scale product of vectors x and y.
$\|A\|$	: the operator norm of A, i.e. $\|A\| := \sup\{\|Ax\| : \|x\| = 1\}$.
$\operatorname{trace} A$	: $\operatorname{trace} A := \sum_{1 \leq i \leq d} a_{ii}$ for a square matrix $A = (a_{ij})_{d \times d}$.
$C(D; R^d)$	: the family of continuous R^d–valued functions defined on D.
$C^m(D; R^d)$	: the family of continuously m–times differentiable R^d–valued functions defined on D.
$C([-\tau, 0]; R^d)$	: the space of all continuous R^d–valued functions defined on $[-\tau, 0]$ with a norm $\|u\|_\tau = \sup\{\|u(t)\| : -\tau \leq t \leq 0\}$.
$C^{2,1}(D \times R_+; R)$	: the family of all real–valued functions $V(x, t)$ defined on $D \times R_+$

		which are continuously twice differentiable in $x \in D$ and once differentiable in $t \in R_+$.
$\|\xi\|_{L^p}$	:	$\|\xi\|_{L^p} := (E\|\xi\|^p)^{1/p}$.
$L^p(\Omega; R^d)$	:	the family of R^d–valued random variables ξ such that $\|\xi\|_{L^p} < \infty$.
$\hat{x}(t)$	:	$\hat{x}(t) := \{x(t+s) : -\tau \le s \le 0\}$.

Other notation will be explained where it first appears.

CHAPTER 1

Semimartingales with Spatial Parameters and Stochastic Integrals

1.1. INTRODUCTION

This chapter introduces the theory of semimartingales with spatial parameters and stochastic integrals, and therefore forms a basis for the book.

Stochastic integrals were first introduced by Itô (1942) to formulate the stochastic differential equation that determined Kolmogorov's diffusion process (cf. Kolmogorov (1931)). It was Doob (1953) who pointed out the martingale character of stochastic integrals and suggested that a unified theory of stochastic integrals should be established in the framework of martingale theory. His program was accomplished by Courrège (1963), Fisk (1963), Kunita and Watanabe (1976), Meyer (1972). We shall give a very brief account of this modern theory of stochastic integrals in Section 2 in order to make the book self-contained. For the details we refer the readers to Dellacherie and Meyer (1980), Doob (1953), Elworthy (1982), Jacod (1979), Métivier (1982), Yan (1980) among others.

On the other hand, it was only recently observed that the most natural examples of stochastic flows (namely the isotropic ones) could not be described by stochastic differential equations involving a finite number of real-valued Brownian motions (cf. Baxendale (1984) and Le Jan (1982)). For example, Brownian flows, i.e. the flows with independent increments, are described by stochastic differential equations driven by vector-field-valued Brownian motions. In other words, in applications to various stochastic problems concerning stochastic differential equations and stochastic flows, it is necessary to define stochastic integrals with nonlinear integrators of the form

$$\int F(x(s), ds),$$

where $F(x, t)$ is a continuous C-semimartingale with spatial parameter $x \in R^d$. This modern stochastic integral has been introduced recently by several leading mathematicians e.g. Carmona, Kunita, Le Jan, Naulart, Watanabe. It is defined in such a way that if $F(x, t) = x \cdot B(t)$ with B a Brownian motion, then the above integral coincides with the classical Itô integral. For an intensive treatment on this subject we mention two new books by Kunita (1990) and Carmona & Nualart (1990).

In this chapter we start with a very brief account on the general theory of stochastic processes, in particular, semimartingales. We then introduce C-semimartingales with spatial parameters and give the definition of stochastic integrals based on the C-semimartingales (which are also called nonlinear integrators) along with a discussion on the properties of the integrals including the generalized Itô formula and the time change formula of the integrals. This chapter ends up with a discussion on the Lebesgue-Stieltjes integral inequalities, in particular, the generalized Gronwall inequality, which will be needed very frequently throughout this book. All results in this chapter are stated without proofs.

1.2. PRELIMINARIES ON STOCHASTIC PROCESSES

Throughout this book we let $(\Omega, \mathcal{F}, \{\mathcal{F}_t\}_{t\geq 0}, P)$ be a complete probability space which satisfies the usual conditions, i.e. $\{\mathcal{F}_t\}_{t\geq 0}$ is an increasing and right continuous family of sub-σ-algebras of $\mathcal{F}$ and $\mathcal{F}_0$ contains all P-null sets. We also define $\mathcal{F}_\infty = \bigvee_{t\geq 0}\mathcal{F}_t$, the σ-algebra generated by $\bigcup_{t\geq 0}\mathcal{F}_t$.

A stochastic process $X = (X_t(\omega))_{t\geq 0}$ is said to be *continuous* (or *left continuous*, or *right continuous*) if for almost all $\omega \in \Omega$ function $X_t(\omega)$ is continuous on $t \in R_+$ (or left continuous on $t > 0$, or right continuous on $t \geq 0$). It is said to be *cadlag* (*right continuous and left limit*) if it is right continuous and for almost all $\omega \in \Omega$ the left limit $X_{t-}(\omega) = \lim_{s\uparrow t}X_s(\omega)$ exists and is finite for all $t > 0$. It is said to be *adapted* (to $\{\mathcal{F}_t\}$) if, for every t, $X_t(\omega)$ is $\mathcal{F}_t$-measurable in ω. It is said to be *measurable* if $X_t(\omega)$ is

$\mathcal{B}(R_+) \times \mathcal{F}$-measurable in $(t, \omega) \in R_+ \times \Omega$, where $\mathcal{B}(R_+)$ stands for the all Borel subsets of R_+. The process is said to be *progressively measurable* or *progressive* if $X_t(\omega)$ is $\mathcal{B}([0, T]) \times \mathcal{F}$-measurable in $(t, \omega) \in [0, T] \times \Omega$ for any $T > 0$. Let $\mathcal{O}$ denote the smallest σ-algebra on $R_+ \times \Omega$ with respect to which every cadlag adapted process is measurable in (t, ω). Let $\mathcal{P}$ denote the smallest σ-algebra on $R_+ \times \Omega$ with respect to which every left continuous adapted process is measurable in (t, ω). A stochastic process $X = (X_t(\omega))_{t\geq 0}$ is said to be *optional* or *predictable* if $X_t(\omega)$ is $\mathcal{O}$-measurable or $\mathcal{P}$-measurable, respectively, in (t, ω). We have the following relationship among the various processes:

predictable processes $\subset$ optional processes $\subset$
$\subset$ progressive processes $\subset$ adapted processes.

A stochastic process $A = (A_t(\omega))_{t\geq 0}$ is said to be *increasing* if for almost all $\omega \in \Omega$, $A_t(\omega)$ is nonnegative nondecreasing right continuous on $t \geq 0$. It is called *a process of finite variation* if $A = A^1 - A^2$ with A^1 and A^2 both increasing processes. It is obvious that processes of finite variation are cadlag. Hence adapted processes of finite variation are optional. An increasing process (A_t) is said to be *integrable* if $A_\infty = \lim_{t\uparrow\infty} A_t$ is an integrable random variable.

Let (X_t) be a stochastic process. Another stochastic process (Y_t) is called a *version* or *modification* of (X_t) if $X_t = Y_t$ a.s. for all $t \geq 0$. Two stochastic processes (X_t) and (Y_t) are said to be *indistinguishable* if for almost all $\omega \in \Omega$, $X_t(\omega) = Y_t(\omega)$ on $t \geq 0$.

A random variable $\tau(\omega) : \Omega \to [0, \infty]$ (it may take the value ∞) is called an $\{\mathcal{F}_t\}$-*stopping time* (or simply a *stopping time*) if $\{\omega : \tau(\omega) \leq t\} \in \mathcal{F}_t$ for any $t \geq 0$. Let τ and σ be two stopping times with $\tau \leq \sigma$ a.s. We define $[[\tau, \sigma]] = \{(t, \omega) \in R_+ \times \Omega : \tau(\omega) \leq t \leq \sigma(\omega)\}$ and call it a *stochastic interval.* Similarly, we can define stochastic intervals $[[\tau, \sigma[[$, $]]\tau, \sigma]]$ and $]]\tau, \sigma[[$, the meaning being evident. If τ is a stopping time, define $\mathcal{F}_\tau = \{A \in \mathcal{F} : A \cap \{\tau \leq t\} \in \mathcal{F}_t \text{ for all } t \geq 0\}$ which is a σ-algebra. If τ, σ are two stopping times with $\tau \leq \sigma$ a.s., then $\mathcal{F}_\tau \subset \mathcal{F}_\sigma$. If $(X_t)_{t\geq 0}$ is a progressive process and τ is a stopping time, then $X_\tau \mathbb{1}_{\{\tau<\infty\}}$ is $\mathcal{F}_\tau$-measurable, where $\mathbb{1}_{\{\tau<\infty\}}$ is the indicator function of $\{\tau<\infty\}$.

An adapted process $M = (M_t)_{t\geq 0}$ is called a *martingale* if

$$E(M_t|\mathcal{F}_s) = M_s \quad \text{a.s. for } 0 \leq s < t < \infty.$$

It should be pointed out that every martingale has a cadlag modification since we always assume the filtration $\{\mathcal{F}_t\}$ is right continuous. Therefore we can always assume that any martingale is cadlag in the sequel. The following is the well-known Doob stopping theorem.

Theorem 2.1 Let $M = (M_t)_{t\geq 0}$ be a martingale and τ, σ two stopping times. Then

$$E(M_\tau|\mathcal{F}_\sigma) = M_{\tau\wedge\sigma} \quad \text{a.s.}$$

In particular, if $\sigma \leq \tau$ a.s., then

$$E(M_\tau|\mathcal{F}_\sigma) = M_\sigma \quad \text{a.s.}$$

A martingale $M = (M_t)_{t\geq 0}$ is said to be *uniformly integrable* if $\{M_t : t \geq 0\}$ is a family of uniformly integrable random variables.

Theorem 2.2 Let $M = (M_t)_{t\geq 0}$ be a martingale. Then the following properties are equivalent:

(1) $(M_t)_{t\geq 0}$ is uniformly integrable.

(2) There exists a random variable ξ such that $M_t = E(\xi|\mathcal{F}_t)$ a.s. for all $t \geq 0$.

(3) $(M_t)_{t\geq 0}$ converges in $L^1 = L^1(\Omega; R)$ to a random variable M_∞ which is $\mathcal{F}_\infty$-measurable.

The following is the well-known Doob martingale inequality.

Theorem 2.3 Let $p \in (1, \infty)$ and $M = (M_t)_{t\geq 0}$ be a martingale with property

$$\sup_{t\geq 0} \|M_t\|_{L^p} < \infty.$$

Then $M_\infty = \lim_{t\to\infty} M_t$ exists almost surely, and

$$\| \sup_{t\geq 0} |M_t| \|_{L^p} \leq \frac{p}{p-1} \|M_\infty\|_{L^p} \leq \frac{p}{p-1} \sup_{t\geq 0} \|M_t\|_{L^p} .$$

Here $\|\xi\|_{L^p} = (E|\xi|^p)^{1/p}$.

A martingale $M = (M_t)_{t\geq 0}$ is said to be *square integrable* if

$$\sup_{t\geq 0} E|M_t|^2 < \infty.$$

Let $\mathcal{M}^2$ denote the family of all square integrable martingales. If $M \in \mathcal{M}^2$, then there exists a unique predictable integrable increasing process denoted by $\langle M, M\rangle$ such that $M^2 - \langle M, M\rangle$ is a uniformly integrable martingale which vanishes at $t = 0$. The process $\langle M, M\rangle$ is called the *predictable quadratic variation* or the *Meyer process of* M. In particular, for any finite stopping time τ,

$$E(M_\tau)^2 = E\langle M, M\rangle_\tau.$$

If $M, N \in \mathcal{M}^2$, we define

$$\langle M, N\rangle = \frac{1}{2}\Big(\langle M + N, M + N\rangle - \langle M, M\rangle - \langle N, N\rangle\Big)$$

which is called the *joint predictable quadratic variation* of M and N. It is useful to know that $\langle M, N\rangle$ is a unique predictable integrable process of finite variation such that $MN - \langle M, N\rangle$ is a uniformly integrable martingale that vanishes at $t = 0$. We now introduce a norm for $M \in \mathcal{M}^2$

$$\|M\|_{\mathcal{M}^2} = \| \sup_{t\geq 0}|M_t| \|_{L^2}$$

under which $\mathcal{M}^2$ is a Hilbert space. Let $\mathcal{M}^{2,c}$ denote the space of continuous square integrable martingales, and it is a closed subspace of $\mathcal{M}^2$. Two martingales M, N in $\mathcal{M}^2$ are *orthogonal* if $M_0N_0 = 0$ and $E(M_\tau N_\tau) = 0$ for any finite stopping time τ, and we shall then write $M \perp N$. Let $\mathcal{M}^{2,d} = (\mathcal{M}^{2,c})^\perp$ which is the subspace orthogonal to $\mathcal{M}^{2,c}$. Martingales in $\mathcal{M}^{2,d}$ are said to be *purely discontinuous.* Any $M \in \mathcal{M}^2$ has a unique decomposition $M = M^c + M^d$ such that $M^c \in \mathcal{M}^{2,c}$ and $M^d \in \mathcal{M}^{2,d}$. We shall then call M^c the *continuous martingale part* and M^d the *purely discontinuous martingale part.* Let $M \in \mathcal{M}^2$. Define

$$[M, M]_t = \langle M^c, M^c\rangle_t + \sum_{0<s\leq t} (\Delta M_s)^2 \quad \text{for } t \geq 0$$

which is an optional increasing process and is called the *optional quadratic variation* of M, where $\Delta M_t = M_t - M_{t-}$. Particularly, if $M \in \mathcal{M}^{2,c}$, then $[M, M] = \langle M, M\rangle$ and we shall simply call it the *quadratic variation* of M. If $M, N \in \mathcal{M}^2$, define

$$[M, N]_t = \langle M^c, N^c\rangle_t + \sum_{0<s\leq t} \Delta M_s \Delta N_s \quad \text{for } t \geq 0$$

which is called the *joint optional quadratic variation* of M and N. It should be pointed out that $MN - [M, N]$ is a uniformly integrable martingale which vanishes at $t = 0$. If $M, N \in \mathcal{M}^{2,c}$, then $[M, N] = \langle M, N\rangle$ and we shall simply call it the *joint quadratic variation* of M and N.

If $X = (X_t)_{t\geq 0}$ is a progressive process and τ is a stopping time, we denote

$$X^\tau = (X_{\tau\wedge t})_{t\geq 0},$$

which will be called a stopped process of X. A right continuous adapted process $M = (M_t)_{t\geq 0}$ is called a *local martingale* if there exist stopping times τ_n, $n \geq 1$, with $\tau_n \uparrow \infty$ a.s. such that each $M^{\tau_n} - M_0$ is a uniformly integrable martingale. Similarly, we can define *local processes of finite variation, locally square integrable martingales* and so on. Moreover, an increasing process (A_t) is said to be *prelocally integrable* if there exist stopping times τ_n, $n \geq 1$, with $\tau_n \uparrow \infty$ a.s. such that $A_{\tau_n -}$ is integrable, where we use conventions $A_{0-} = A_0$ and $A_{\infty -} = A_\infty$. It is not difficult to see that any local martingale is cadlag and any martingale is a local martingale.

In particular, we can define in an obvious way the *predictable quadratic variation* $\langle M, M\rangle$ for a continuous local martingale and the *joint predictable quadratic variation* $\langle M, N\rangle$ for continuous local martingales M and N. It can be showed that $\langle M, M\rangle$ is a unique predictable increasing process such that $M^2 - \langle M, M\rangle$ is a local martingale vanishing at $t = 0$, and $\langle M, N\rangle$ is a unique predictable process of finite variation such that $MN - \langle M, N\rangle$ is a local martingale vanishing at $t = 0$.

Denote by $\mathcal{M}_{loc}$ the space of all local martingales. $M \in \mathcal{M}_{loc}$ is called a *purely discontinuous local martingale* if $M_0 = 0$ and it has the decomposition $M = U+V$, where U is a purely discontinuous local square integrable martingale, V a local martingale of finite variation and both vanish at $t = 0$. Every local martingale M has a unique decomposition $M = M^c + M^d$ such that M^c is a continuous local martingale and M^d a purely discontinuous local martingale. We shall call M^c the *continuous local martingale part* and M^d the *purely discontinuous local martingale part.* If $M \in \mathcal{M}_{loc}$, define

$$[M, M]_t = \langle M^c, M^c\rangle_t + \sum_{0<s\leq t} (\Delta M_s)^2 \quad \text{for } t \geq 0,$$

which is called the *optional quadratic variation* of M. Similarly, if $M, N \in \mathcal{M}_{loc}$, define the *joint optional quadratic variation* of M and N by

$$[M, N] = \frac{1}{2}\Big([M + N, M + N] - [M, M] - [N, N] \Big).$$

Obviously, if M is a continuous local martingale, then $\langle M, M\rangle = [M, M]$ which is simply called the *quadratic variation* of M. Moreover, if M and N are both continuous local martingales, then $\langle M, N\rangle = [M, N]$ which is simply called the *joint quadratic variation* of M and N.

We now state four well-known theorems on continuous local martingales which will be used in the sequel of this book very frequently. Let us first give the maximal inequality for the exponential of a local martingale, which is usually called the exponential martingale inequality.

Theorem 2.4 Let $M = (M_t)$ be a continuous local martingale. Then for any positive constants τ, γ, δ, we have

$$P\Big(\sup_{t\le\tau} \Big[M_t - \frac{\gamma}{2}\langle M, M\rangle_t \Big] > \delta \Big) \le e^{-\gamma\delta}.$$

The next one is the famous Lévy Theorem which gives the martingale character of a Brownian motion.

Theorem 2.5 Let (B_t) be a continuous adapted process vanishing at zero. Then (B_t) is a Brownian motion if and only if (B_t) is a local martingale and $\langle B, B\rangle_t = t$ for all $t \ge 0$.

The following is the well-known Burkholder-Davis-Gundy inequality.

Theorem 2.6 There exist universal constants c_p and C_p for any $0 < p < \infty$ such that for every continuous local martingale M vanishing at zero and any stopping time τ,

$$c_p\, E\Big(\langle M, M\rangle_\tau^{p/2} \Big) \le E\Big(\sup_{0\le s\le\tau} |M_s|^p \Big) \le C_p\, E\Big(\langle M, M\rangle_\tau^{p/2} \Big).$$

In particular, one may have

$$c_p = (p/2)^p, \qquad C_p = (32/p)^{p/2} \qquad \text{if } 0 < p < 2;$$

$$c_p = 1, \qquad C_p = 4 \qquad \text{if } p = 2;$$

$$c_p = (2p)^{-p/2}, \qquad C_p = \Big(\frac{p^{p+1}}{2(p-1)^{p-1}} \Big)^{p/2} \qquad \text{if } p > 2,$$

although they may not be optimal, for example, one could have $C_1 = 2\sqrt{2}$.

Theorem 2.7 Let $\{M^n\}$ be a sequence of continuous local martingales. It converges uniformly in any finite time interval in probability to a continuous local martingale M if and only if $\langle M^n - M, M^n - M\rangle_t$ converges to 0 in probability for any $t \ge 0$.

We now introduce the concept of semimartingales. A right continuous adapted process $X = (X_t)$ is called a *semimartingale* if X has a decomposition $X = M + A$ with M a local martingale and A an adapted process of finite variation. If we require $A_0 = 0$ in this decomposition, it is easy to see that the continuous local martingale part M^c of M is uniquely determined by X. Hence we can set $X^c = M^c$ and call it the *continuous local martingale part* of X. Similarly, we introduce the *optional quadratic variation* of X

$$[X, X]_t = \langle X^c, X^c \rangle_t + \sum_{0<s\le t} (\Delta X_s)^2 \qquad \text{for } t \ge 0.$$

If X, Y are two semimartingales, define

$$[X, Y]_t = \langle X^c, Y^c \rangle_t + \sum_{0<s\le t} \Delta X_s \Delta Y_s \quad \text{for } t \ge 0,$$

which is called the *joint optional quadratic variation* of X and Y.

We can now introduce stochastic integrals with respect to martingales. Let us start with a general modern definition of stochastic integral.

Definition 2.1 Let M be a local martingale with $M_0 = 0$ and H a measurable process such that

$$\left(\int_0^\cdot H_s^2 \, d[M, M]_s\right)^{1/2}$$

is a locally integrable increasing process and

$$\int_0^\cdot H_s^2 \, d[M, M]_s$$

is a prelocally integrable increasing process. Then there exists a unique local martingale L with $L_0 = 0$ such that for any stopping time τ satisfying

$$E\left(\int_{[[0, \tau]]} H_s^2 \, d[M, M]_s\right)^{1/2} < \infty$$

and

$$E\left(\int_{[[0, \tau[[} H_s^2 \, d[M, M]_s\right) < \infty,$$

and any bounded martingale N, we have

$$E(L_\tau N_\tau) = E\int_{[[0, \tau]]} H_s \, d[M, N]_s.$$

Such an L is called the *stochastic integral of* H *with respect to* M, and is denoted by H.M or

$$\int_0^\cdot H_s dM_s.$$

For a general local martingale M (without zero initial condition), we define

$$H.M = E(H_0 M_0 | \mathcal{F}_0) + H.(M - M_0).$$

Remark 2.1 Let M be a local martingale and H a progressive process such that $(H^2.[M, M])^{1/2}$ is a locally integrable increasing process, where . denotes the Lebesgue-Stieltjes integral. Hence $H^2.[M, M]$ is an adapted increasing process and so prelocally integrable increasing process. Thus H.M is well defined. In this case H.M is a unique local martingale L such that $L_0 = H_0M_0$ and for any bounded martingale N, LN - H.[M, N] is a local martingale.

Remark 2.2 Let M and H be the same as in Definition 2.1. Then there exists an optional process K such that $(K^2.[M, M])^{1/2}$ is locally integrable and H.M = K.M. Hence, in order to study the properties of stochastic integrals one can always assume that the integrands are optional processes.

Remark 2.3 Let M be a continuous local martingale and H a progressive process such that H^2 is integrable with respect to [M, M]. Then there exists a predictable process K such that K.M = H.M.

Remark 2.4 In order to show that the modern general definition of stochastic integral is reasonable, let us give here a simple fact. Let M be a local martingale with $M_0 = 0$. Let τ, σ be two stopping times with $\tau \leq \sigma$ a.s. Let ξ be an $\mathcal{F}_\tau$-measurable random variable. Set $H = \xi 1\!\!1_{]]\tau, \sigma]]}$. Then H is a predictable process and $H.M = \xi(M^\sigma - M^\tau)$.

Remark 2.5 The definition of stochastic integral here looks much different from the classical definition of Itô's integral. By the classical definition, we mean the procedure where stochastic integrals are first defined for simple predictable processes and then are extended to general processes by limit in probability. As a matter of fact, the definition here coincides with the classical one. This fact follows from Remark 2.4 and the following stochastic integral convergence theorem.

Theorem 2.8 Let M be a local square integrable martingale with $M_0 = 0$. Let $\{H, H^n, n \geq 0\}$ be a sequence of predictable processes such that they are integrable with respect to M, and both H.M and $H^n.M$ are all local square integrable martingales. Let $D \in \mathcal{F}$ and τ be a finite stopping time. If

$$1\!\!1_D \int_{[[0, \tau]]} (H_s^n - H_s)^2 \, d\langle M, M\rangle_s \to 0 \quad \text{in probability,}$$

then

$$1\!\!1_D \sup_{t \leq \tau} |(H^n.M)_t - (H.M)_t| \to 0 \quad \text{in probability.}$$

The following theorem gives some important properties of stochastic integrals.

Theorem 2.9 (a) Let M be a local martingale and H a measurable process which is integrable with respect to M. Then for any stopping time τ,

$$(H.M)^{\tau} = H.M^{\tau} = (H\mathbb{1}_{[[0,\,\tau]]}).M.$$

(b) If H and K are two measurable processes which are integrable with respect to M, then for any real numbers a, b, aH + bK is integrable with respect to M and

$$(aH + bK).M = aH.M + bK.M.$$

(c) If M and N are two local martingales and H is a measurable process which is integrable with respect to both M and N, then for any real numbers a, b, the process H is integrable with respect to aM + bN and

$$H.(aM + bN) = aH.M + bH.N.$$

(d) Let M be a local martingale and H a predictable process which is integrable with respect to M. Then $\Delta(H.M)$ and $H\Delta M$ are indistinguishable, and for any local martingale N,

$$[H.M, N] = H.[M, N].$$

(e) Let M be a local martingale and H, K predictable processes. If H is integrable with respect to M, then K is integrable with respect to H.M if and only if KH is integrable with respect to M. In this case, we have

$$(KH).M = K.(H.M).$$

We now introduce stochastic integrals with respect to semimartingales.

Definition 2.2 Let X be a semimartingale and H a predictable process. If there exists a decomposition X = M + A, in which M is a local martingale and A is a process of finite variation, such that H is integrable with respect to M and H is also Lebesgue–Stieltjes integrable with respect to A, then we say H is *integrable with respect to* X (or X-*integrable*) and X = M + A is called an H-decomposition. If X = N + B is another H-decomposition, we have H.M + H.A = H.N + H.B, where H.A and H.B are the Lebesgue-Stieltjes integrals. Therefore, we can write H.X = H.M + H.A which is independent of the H-decomposition and shall be called the *stochastic integral of* H *with respect to* X.

The following is the dominated convergence theorem of stochastic integrals.

Theorem 2.10 Let X be a semimartingale and H a predictable process which is integrable with respect to X. Let K, K^n ($n \geq 1$) be predictable processes such that $|K| \leq |H|$ and $|K^n| \leq |H|$. Then the K and K^n are all integrable with respect to X. Moreover, let $D \in \mathcal{F}$ and let τ be a finite stopping time. If for almost all $\omega \in D$,

$$\lim_{n\to\infty} K_t^n(\omega) = K_t(\omega) \quad \text{on } t \in [0, \tau(\omega)],$$

then

$$\mathbb{1}_D \sup_{t\leq\tau} |\,(K^n.M)_t - (K.M)_t\,| \to 0 \quad \text{in probability as } n \to \infty.$$

A stochastic process $X = (X^1, ..., X^d)$ is called a *d-dimensional semimartingale* if $X^1, ..., X^d$ are all (1-dimensional) semimartingales. Similarly, we can define other d-dimensional processes. We finally introduce the very important Itô formula to close this section.

Theorem 2.11 Let $F(x_1, ..., x_d; y_1, ..., y_m)$ be a real-valued function defined on R^{d+m} which is continuously once differentiable in x_j and continuously twice differentiable in y_k. Let $X = (X^1, ..., X^d)$ be a d-dimensional adapted process of finite variation and $Y = (Y^1, ..., Y^m)$ an m-dimensional semimartingale. Then

$$F(X, Y) = F(X^1, ..., X^d; Y^1, ..., Y^m)$$

is a semimartingale, and for any $t \geq 0$ we have the following well-known Itô formula

$$\begin{aligned} &F(X_t, Y_t) \\ &= F(X_0, Y_0) + \sum_{i=1}^{d} \int_0^t \frac{\partial}{\partial x_i} F(X_{s-}, Y_{s-})\, dX_s^i \\ &+ \sum_{j=1}^{m} \int_0^t \frac{\partial}{\partial y_j} F(X_{s-}, Y_{s-})\, dY_s^i \\ &+ \frac{1}{2} \sum_{j,k=1}^{m} \int_0^t \frac{\partial^2}{\partial y_j\, \partial y_k} F(X_{s-}, Y_{s-})\, d\langle (Y^j)^c, (Y^k)^c \rangle_s \end{aligned}$$

$$+ \sum_{0<s\le t} \Big[\Delta F(X_s, Y_s) - \sum_{i=1}^{d} \frac{\partial}{\partial x_i} F(X_{s-}, Y_{s-}) \Delta X_s^i$$

$$- \sum_{j=1}^{m} \frac{\partial}{\partial y_j} F(X_{s-}, Y_{s-}) \Delta Y_s^j \Big] \qquad \text{a.s.}$$

1.3. C-SEMIMARTINGALES WITH SPATIAL PARAMETERS

Let $F(x, t)$, $t \ge 0$ be a family of real valued predictable processes with spatial parameter $x \in D$, where D is a domain in R^d. We can regard it as a random field with double parameters x and t. If $F(x, t, \omega)$ is a continuous function of x for almost all ω for any t, we can regard $F(., t)$ as a predictable process with values in $C := C(D; R)$ or C-valued predictable process.

For the multi-index of nonnegative integers $\alpha = (\alpha_1, ..., \alpha_d)$ we define the differential operator

$$D_x^\alpha = \frac{\partial^{\alpha_1}}{\partial x_1^{\alpha_1}} \cdots \frac{\partial^{\alpha_d}}{\partial x_d^{\alpha_d}}.$$

Set $|\alpha| = \sum_i \alpha_i$. In the sequel of this section, m is a nonnegative integer.

Definition 3.1 A continuous random field $F(x, t)$, $x \in D$, $t \ge 0$ is called a *continuous process with value in* C^m or *continuous* C^m*-process* if it is m-times differentiable in x for any t a.s. and the derivative

$$D_x^\alpha F(x, t), \quad \text{for any } \alpha \text{ with } |\alpha| \le m,$$

is continuous in (x, t) a.s. Furthermore, let $\delta \in (0, 1]$. If for any α with $|\alpha| = m$,

$$\sup_{\substack{x,y\in K,\ x\ne y \\ 0\le t\le T}} \frac{|D_x^\alpha F(x, t) - D_x^\alpha F(y, t)|}{|x - y|^\delta} < \infty \qquad \text{a.s.}$$

holds for any compact convex subset K of D and $T > 0$, the random field $F(x, t)$ is then called a *continuous process with value in* $C^{m,\delta}$ or *continuous* $C^{m,\delta}$*-process.*

Definition 3.2 Let $M(x, t)$, $x \in D$ be a family of continuous local martingales. It is called a *continuous local martingale with values in* C^m or a *continuous* C^m*-local martingale* if it has a modification of a continuous C^m-process. In the case $m = 0$, it is called a *continuous* C*-local martingale.* Furthermore, it is called a *continuous* *local martingale with values in* $C^{m,\delta}$ or a *continuous* $C^{m,\delta}$*-local martingale* if it has a modification of a continuous $C^{m,\delta}$-process, where $\delta \in (0, 1]$.

Now let $M(x, t)$, $x \in D$ and $N(y, t)$, $y \in D$ be families of continuous local martingales with spatial parameters x, y respectively and let $A^{MN}(x, y, t)$ be the joint quadratic variation of $M(x, t)$ and $N(y, t)$. We shall discuss the regularity of $M(x, t)$ and $N(y, t)$ with respect to the spatial parameter in connection with the regularity of the joint quadratic variation. We first classify the joint quadratic variation according to its regularity.

Definition 3.3 A continuous random field $A(x, y, t)$, $x, y \in D$, $t \geq 0$ is said to belong to the *class* $\tilde{C}^m$ if it is m-times differentiable with respect to each of variables x and y, and the derivatives

$$D_x^\alpha D_y^\beta A(x, y, t), \qquad |\alpha| \leq m, \ |\beta| \leq m$$

are continuous in (x, y, t) a.s. Furthermore, let $\delta \in (0, 1]$. If for any α, β with $|\alpha| = |\beta| = m$,

$$\tilde{A}(x, y, t) := D_x^\alpha D_y^\beta A(x, y, t)$$

satisfies

$$\sup_{\substack{x,y,\bar{x},\bar{y} \in K, \ x \neq \bar{x}, \ y \neq \bar{y} \\ 0 \leq t \leq T}} \frac{|\tilde{A}(x, y, t) - \tilde{A}(\bar{x}, y, t) - \tilde{A}(x, \bar{y}, t) + \tilde{A}(\bar{x}, \bar{y}, t)|}{|x - \bar{x}|^\delta \, |y - \bar{y}|^\delta} < \infty$$

almost surely for any compact convex subset K of D and $T > 0$, then the random field $A(x, y, t)$ is said to belong to the *class* $\tilde{C}^{m,\delta}$.

The following theorem describes the regularity of $M(x, t)$ in connection with the regularity of its joint quadratic variation.

Theorem 3.1 Let $M(x, t)$, $x \in D$ be a family of continuous local martingales. Suppose that the joint quadratic variation $\langle M(x, t), M(y, t)\rangle$ of $M(x, t)$ and $M(y, t)$ has a modification denoted by $A(x, y, t)$ belonging to the class $\tilde{C}^{m,\delta}$ for some nonnegative integer m and $\delta \in (0, 1]$. Then for any $0 < \gamma < \delta$, $M(x, t)$ has a modification of a continuous $C^{m,\gamma}$-local martingale. Moreover, for any α with $|\alpha| \leq m$,

$$D_x^\alpha M(x, t), \ x \in D$$

is a family of continuous local martingales with the joint quadratic variation

$$D_x^\alpha D_y^\alpha A(x, y, t).$$

The following is the inverse of Theorem 3.1 which shows the regularity of a joint quadratic variation of continuous $C^{m,\delta}$-local martingales.

Theorem 3.2 Let both $M(x, t)$, $x \in D$ and $N(y, t)$, $y \in D$ be continuous $C^{m,\delta}$-local martingales, where m is a nonnegative integer and $\delta \in (0, 1]$. Then their joint quadratic variation $\langle M(x, t), N(y, t)\rangle$ has a modification belonging to the class $\tilde{C}^{m,\gamma}$ for any $\gamma < \delta$. Moreover, the modification denoted still by $\langle M(x, t), N(y, t)\rangle$ satisfies that

$$D_x^\alpha D_y^\beta \langle M(x, t), N(y, t)\rangle = \langle D_x^\alpha M(x, t), D_y^\beta N(y, t)\rangle \quad \text{a.s.}$$

for any $t \geq 0$ provided $|\alpha|, |\beta| \leq m$.

Now let $M(x, t)$, $x \in D$ be a family of continuous local martingales vanishing at zero. Let $A(x, y, t)$ denote the joint quadratic variation of $M(x, t)$ and $M(y, t)$. Then there exists a continuous increasing process A_t such that $A(x, y, t)$ is absolutely continuous with respect to process A_t a.s. for any x, y. Therefore there exists a predictable process $a(x, y, t)$ such that

$$A(x, y, t) = \int_0^t a(x, y, s)\, dA_s$$

holds for any $t \geq 0$ a.s. For each $T > 0$, introduce a measure μ_T on $([0, T]\times\Omega, \mathcal{P}_T)$ by

$$\mu_T(B) = E \int_0^T 1\!\!1_B(s, \omega)\, dA_s \qquad \text{for any } B \in \mathcal{P}_T,$$

where $\mathcal{P}_T$ is the predictable σ-field on $[0, T] \times \Omega$. It is easy to see that the density function $a(x, y, t, \omega)$ is uniquely determined up to $\mu_T = 0$ on $[0, T] \times \Omega$ for any x and y. The pair $(a(x, y, t), A_t)$ is called the *local characteristic* of $M(x, t)$. When A_t is identical to t, the random field $a(x, y, t)$ itself is often called the *local characteristic* of $M(x, t)$.

Definition 3.4 Let $m \geq 0$ and $\delta \in [0, 1]$. The local characteristic $(a(x, y, t), A_t)$ is said to belong to the *class* $B^{m,\delta}$ if $a(x, y, t)$ has a modification with respect to μ_T on $[0, T] \times \Omega$ for each $T > 0$ and the modification denoted still by $a(x, y, t)$ has the following two properties:

(i) $a(x, y, t)$ is a predictable process belonging to the class $\tilde{C}^m$.

(ii) Set

$$\tilde{a}(x, y, t) := D_x^\alpha D_y^\alpha\, a(x, y, t), \quad |\alpha| \leq m.$$

For any compact subset K of D there exists a predictable process H_t with

$$\int_0^T H_t \, dA_t < \infty \quad \text{a.s.}$$

such that both

$$|\tilde{a}(x, y, t)| \leq H_t (1 + |x|)^{0 \vee (1-|\alpha|)} (1 + |y|)^{0 \vee (1-|\dot{\alpha}|)}$$

and

$$|\tilde{a}(x, x, t) - 2\tilde{a}(x, y, t) + \tilde{a}(y, y, t)| \leq H_t |x - y|^{2\delta}$$

hold for any $x, y \in K$ and $0 \leq t \leq T$.

Observe that if $M(x, t)$, $x \in D$ is a family of continuous local martingales with the local characteristic belonging to the class $B^{m,\delta}$, where $\delta \in (0, 1]$, then for any $0 < \gamma < \delta$, M has a modification of a continuous $C^{m,\gamma}$–local martingale by Theorem 3.1.

Now let $M(x, t)$, $x \in D$ and $N(x, t)$, $x \in D$ be two continuous C–local martingales vanishing at $t = 0$ with the local characteristics $(a^M(x, y, t), A_t)$ and $(a^N(x, y, t), A_t)$, respectively, which belong to the class $B^{m,\delta}$. Observe that the joint quadratic variation of $M(x, t)$ and $N(y, t)$, denoted by $A^{MN}(x, y, t)$, is absolutely continuous with respect to A_t. Therefore, for each x, y, there exists a predictable process $a^{MN}(x, y, t)$ such that

$$A^{MN}(x, y, t) = \int_0^t a^{MN}(x, y, s) \, dA_s \qquad \text{a.s. for any } t \geq 0.$$

The pair $(a^{MN}(x, y, t), A_t)$ is called the *joint local characteristic* of M and N.

Theorem 3.3 The joint local characteristic of M and N has a modification such that $a^{MN}(x, y, t)$ is continuous with respect to (x, y) for μ_T–almost every (t, ω) on $[0, T] \times \Omega$ for any $T > 0$. Furthermore,

$$|a^{MN}(x, y, t)| \leq a^M(x, y, t)^{1/2} \, a^N(x, y, t)^{1/2} \quad \text{a.e. } (t, \omega).$$

We now introduce continuous semimartingales with spatial parameters. Let $F(x, t)$, $x \in D$ be a family of continuous semimartingales with initial value $F(x, 0) \equiv 0$ and the decomposition $F(x, t) = M(x, t) + B(x, t)$, where $M(x, t)$ is a continuous local martingale and $B(x, t)$ a continuous process of finite variation.

Definition 3.5 A family of continuous semimartingales $F(x, t)$, $x \in D$ is called a *continuous* C^m*-semimartingale* if $M(x, t)$ is a continuous C^m-local martingale and $B(x, t)$ is a continuous C^m-process such that

$$D_x^\alpha B(x, t), \quad x \in D, \ |\alpha| \leq m$$

are all processes of finite variation. In the case $m = 0$, it is called a *continuous C-semimartingale.* Furthermore, let $\delta \in (0, 1]$. The family $F(x, t)$, $x \in D$ is called a *continuous* $C^{m,\delta}$*-semimartingale* if $M(x, t)$ is a continuous $C^{m,\delta}$-local martingale and $B(x, t)$ is a continuous $C^{m,\delta}$-process such that

$$D_x^\alpha B(x, t), \quad x \in D, \ |\alpha| \leq m$$

are all processes of finite variation.

We shall assume that $M(x, t)$ has a local characteristic $(a(x, y, t), A_t)$ and $B(x, t)$ is written as

$$B(x, t) = \int_0^t b(x, s)\, dA_s,$$

where $b(x, t)$, $x \in D$ is a family of predictable processes. The triple $(a(x, y, t), b(x, t), A_t)$ is called the *local characteristic* of $F(x, t)$. In the case A_t is identical to t, the pair $(a(x, y, t), b(x, t))$ is often called the *local characteristic* of $F(x, t)$.

Definition 3.6 Let m' be a nonnegative integer and $\delta' \in [0, 1]$. The local characteristic $(b(x, t), A_t)$ is said to belong to the *class* $B^{m',\delta'}$ if $b(x, t)$ is a predictable process belonging to the class $C^{m'}$ and for any $T > 0$ and any compact subset K of D there exists a predictable process H_t with

$$\int_0^T H_t\, dA_t < \infty \quad \text{a.s.}$$

such that both

$$|D_x^\alpha b(x, t)| \leq H_t (1 + |x|)^{0 \vee (1-|\alpha|)}$$

and

$$|D_x^\alpha b(x, t) - D_x^\alpha b(y, t)| \leq H_t |x-y|^{\delta'}$$

hold for any $x, y \in K$, $0 \leq t \leq T$ and α with $|\alpha| \leq m'$.

Definition 3.7 Let m, m′ be both nonnegative integers and $\delta, \delta' \in [0, 1]$. The triple $(a(x, y, t), b(x, t), A_t)$ is said to belong to the *class* $(B^{m,\delta}, B^{m',\delta'})$ if (a, A) belongs to the class $B^{m,\delta}$ and (b, A) belongs to the class $B^{m',\delta'}$. When $m = m'$ and $\delta = \delta'$, the triple is simply said to belong to the *class* $B^{m,\delta}$.

1.4. STOCHASTIC INTEGRALS WITH NONLINEAR INTEGRATORS

Let F(x, t) be a continuous C–semimartingale with spatial parameter $x \in D$ such that $F(x, 0) \equiv 0$ and let f_t be a predictable process with values in D. In this section we shall introduce a stochastic integral of f_t based on the kernel F(x, dt) of the form $\int F(f_t, dt)$. This is a generalization of the classical Itô integral. As a matter of fact, if $F(x, t) = x X_t$, where X_t is a one–dimensional continuous semimartingale and $x \in R$, our definition in this section coincides with the stochastic integral $\int f_t \, dX_t$ defined in Section 2. However, the stochastic integral $\int F(f_t, dt)$ to be defined in this section will not have the linear property : $\int F(af_t+bf_t, dt) = a\int F(f_t, dt) + b\int F(f_t, dt)$, where a and b are constants, because F(x, t) is not a linear function of x in general. That is why F(x, t) is also called a *nonlinear integrator.*

We shall first define the integral in the case where the nonlinear integrator is a continuous C–local martingale. Let M(x, t), $x \in D$ be a continuous C–local martingale with the local characteristic $(a(x, y, t), A_t)$ belonging to the class $B^{0,\delta}$ for some $\delta \in [0, 1]$. Let f_t, $t \geq 0$ be a predictable process with values in D such that

$$\int_0^t a(f_s, f_s, s) \, dA_s < \infty \quad \text{a.s.} \quad \text{for all } t \geq 0. \tag{4.1}$$

We shall now start to define the stochastic integral of f_t based on the kernel M(x, dt) step be step. In the first step, we will treat the case where f_t is a simple process, i.e. f_t can be written as

$$f_t = \sum_{i=0}^{k} f_{t_i} \mathbb{1}_{[t_i, t_{i+1})}(t) \qquad \text{on } t \geq 0,$$

where t_i are all constants such that $0 \leq t_0 < t_1 < \ldots < t_k < \infty$ and k is a positive integer. The stochastic integral of f_t based on the kernel M(x, dt) is defined by the finite sum of the form:

$$M_t(f) = \sum_{i=0}^{k} \Big(M(f_{t_i \wedge t}, t_{i+1} \wedge t) - M(f_{t_i \wedge t}, t_i \wedge t) \Big).$$

It can be shown that $M_t(f)$ is a continuous local martingale. Moreover, let g_t be another simple process with values in D. Then $M_t(g)$ is also a continuous local martingale. In particular, the joint quadratic variation of $M_t(f)$ and $M_t(g)$ are given by

$$<M_t(f), M_t(g)> = \int_0^t a(f_s, g_s, s)\, dA_s \quad \text{a.s.}$$

We next let f_t be a predictable process with values in a compact subset K of D. Then there exists a sequence of simple processes $\{f^n\}$ with values in K such that

$$\int_0^t \Big(a(f_s^n, f_s^n, s) - 2a(f_s^n, f_s, s) + a(f_s, f_s, s) \Big)\, dA_s \to 0 \quad \text{a.s.} \tag{4.2}$$

for all $t \geq 0$ as $n \to \infty$. This implies that

$$\int_0^t \Big(a(f_s^n, f_s^n, s) - 2a(f_s^n, f_s^m, s) + a(f_s^m, f_s^m, s) \Big)\, dA_s \to 0 \quad \text{a.s.} \tag{4.3}$$

for all $t \geq 0$ as $n, m \to \infty$. So

$$<M(f^n) - M(f^m), M(f^n) - M(f^m)>_t \to 0 \quad \text{a.s.}$$

for all $t \geq 0$. Therefore, by Theorem 2.7, $\{M_t(f^n)\}$ converges uniformly in any finite time interval in probability to a continuous local martingale which is denoted by $M_t(f)$.

Finally let f_t be an arbitrary predictable process satisfying (4.1). Let $\{K_n\}$ be a sequence of compact subsets of D such that $K_n \uparrow D$. Let f^n be the truncation of f associated with the set K_n. The convergence (4.3) holds again. Therefore $\{M_t(f^n)\}$ converges uniformly in any finite time interval in probability and the limit denoted by $M_t(f)$ is a continuous local martingale.

Definition 4.1 The continuous local martingale $M_t(f)$ defined through the above is called the *Itô stochastic integral* (or simply, *Itô integral*) *of* f_t *based on the kernel* M(x, dt). It is also denoted by

$$\int_0^t M(f_s, ds).$$

Now let g_t, $t \geq 0$ be another predictable process with values in D such that

$$\int_0^t a(g_s, g_s, s)\, dA_s < \infty \quad \text{a.s.} \quad \text{for all } t \geq 0.$$

So the stochastic integral

$$M_t(g) = \int_0^t M(g_s, ds)$$

is well defined and it is another continuous local martingale. The following theorem describes the joint quadratic variation of these two continuous local martingales $M_t(f)$ and $M_t(g)$.

Theorem 4.1 The joint quadratic variation of $M_t(f)$ and $M_t(g)$ satisfies

$$\langle M_t(f), M_t(g)\rangle = \int_0^t a(f_s, g_s, s)\, dA_s \qquad \text{a.s.}$$

More generally, let $M(x, t)$, $x \in D$ and $N(x, t)$, $x \in D$ be two continuous C-local martingales vanishing at $t = 0$ with the local characteristics $(a^M(x, y, t), A_t)$ and $(a^N(x, y, t), A_t)$, respectively, which both belong to the class $B^{0,\delta}$ for some $\delta \in [0, 1]$. Assume that they have the joint local characteristic $(a^{MN}(x, y, t), A_t)$. Let f_t, $t \geq 0$ and g_t, $t \geq 0$ be two predictable processes with values in D such that

$$\int_0^t a^M(f_s, f_s, s)\, dA_s < \infty \qquad \text{a.s.,}$$

$$\int_0^t a^N(g_s, g_s, s)\, dA_s < \infty \qquad \text{a.s.}$$

for all $t \geq 0$. So the stochastic integrals

$$M_t(f) = \int_0^t M(f_s, ds),$$

$$N_t(g) = \int_0^t N(g_s, ds)$$

are both well defined as continuous local martingales.

Theorem 4.2 The joint quadratic variation of $M_t(f)$ and $N_t(g)$ satisfies

$$\langle M_t(f), N_t(g)\rangle = \int_0^t a^{MN}(f_s, g_s, s)\, dA_s \qquad \text{a.s.}$$

We shall now start to define the Itô stochastic integral based on a continuous C-semimartingale kernel F(x, t). It is quite easy to do so since we have already defined the integral based on a continuous C-local martingale kernel and the stochastic integral based on the absolutely continuous process kernel like $B(x, dt) = b(x, t)\, dA_t$ is immediate.

Let F(x, t), $x \in D$ be a continuous C-semimartingale vanishing at t = 0 with the decomposition $F(x, t) = M(x, t) + B(x, t)$, where M(x, t) is a continuous C-local martingale and B(x, t) is a continuous process of finite variation. Assume that F(x, t) has the local characteristic $(a(x, y, t), b(x, t), A_t)$ belonging to the class $B^{0,\delta}$ for some $\delta \in [0, 1]$. Let f_t, $t \geq 0$ be a predictable process with values in D such that

$$\int_0^t a(f_s, f_s, s)\, dA_s < \infty \qquad \text{a.s.} \tag{4.4}$$

and

$$\int_0^t |b(g_s\, s)|\, dA_s < \infty \qquad \text{a.s.} \tag{4.5}$$

for all $t \geq 0$. Then the Itô stochastic integral of f_t based on the kernel F(x, t) is defined by

$$\int_0^t F(f_s, ds) = \int_0^t M(f_s, ds) + \int_0^t b(f_s, s)\, dA_s$$

which is a continuous semimartingale obviously.

If f_t is a continuous predictable process, conditions (4.4) and (4.5) are always satisfied. The integral can be approximated in the usual way as

$$\int_0^t F(f_s, ds) = \lim_{|\Delta| \to 0} \sum_{i=0}^{\infty} \Big(F(f_{t_i \wedge t}, t_{i+1} \wedge t) - F(f_{t_i \wedge t}, t_i \wedge t) \Big),$$

where $0 = t_0 < t_1 < \ldots < t_i \uparrow \infty$ and

$$|\Delta| := \sup_{0 \leq i < \infty} (t_{i+1} - t_i).$$

1.5. GENERALIZED ITÔ'S FORMULA AND OTHERS

Let F(x, t), $x \in D$ be not only a continuous C^2-process but also a continuous C^1-semimartingale. Let f_t be a continuous semimartingale with values in D. Then the composite process $F(f_t, t)$ is a continuous semimartingale. We shall give the differential rule of $F(f_t, t)$ with respect to time t. If $F(x, t)$ is a deterministic function continuously twice differentiable in x and continuously once differentiable in t as well, the differential rule of $F(f_t, t)$ is the Itô formula introduced in Section 2. However, if $F(x, t)$ is a continuous C^1-semimartingale, the differential formula is not the same as this one. It requires an additional term or a correction form as described below.

Theorem 5.1 (Generalized Itô's formula) Let $F(x, t)$, $x \in D$ be not only a continuous C^2-process but also a continuous C^1-semimartingale with the local characteristic belonging to the class $B^{1,\delta}$. Let f_t be a continuous semimartingale with values in D. Then $F(f_t, t)$ is a continuous semimartingale and satisfies

$$F(f_t, t) - F(f_0, t)$$

$$= \int_0^t F(f_s, ds) + \sum_{i=1}^d \int_0^t \frac{\partial}{\partial x^i} F(f_s, s)\, df_s^i$$

$$+ \frac{1}{2} \sum_{i,j=1}^d \int_0^t \frac{\partial^2}{\partial x^i \partial x^j} F(f_s, s)\, d\langle f_s^i, f_s^j \rangle$$

$$+ \sum_{i=1}^d \left\langle \int_0^t \frac{\partial}{\partial x^i} F(f_s, ds), f_s^i \right\rangle.$$

Apart from the above Itô formula, we shall now give some other integral and differential formulae.

Theorem 5.2 (i) Let $F(x, t)$, $x \in D$ be a continuous $C^{m,\delta}$-semimartingale with the local characteristic belonging to the class $B^{m,\delta}$, where $\delta > 0$. Let $f(\lambda, t)$, $\lambda \in \Lambda$ be a continuous predictable process with values in $C^{k,\gamma} := C^{k,\gamma}(\Lambda; D)$, where $\gamma > 0$ and Λ is a domain in R^e. Set

$$L(\lambda, t) = \int_0^t F(f(\lambda, s), ds).$$

Then $L(\lambda, t)$ has a modification of a continuous $C^{m\wedge k,\varepsilon}(\Lambda; R)$-semimartingale with the local characteristic of the class $B^{m\wedge k,\gamma\delta}$, where $\varepsilon < \gamma\delta$. If $m \geq 1$ and $k \geq 1$, then

$$\frac{\partial}{\partial \lambda^i} L(\lambda, t) = \sum_{j=1}^d \int_0^t \frac{\partial}{\partial \lambda^i} f^j(\lambda, t) \frac{\partial}{\partial x^j} F(f(\lambda, s), ds),$$

where $f(\lambda, s) = (f^1(\lambda, s), \ldots, f^d(\lambda, s))$. Furthermore, if g_t is a continuous predictable process with values in Λ, then

$$\int_0^t L(g_s, ds) = \int_0^t F(f(g_s, s), ds).$$

(ii) Let $F(x, \lambda, t)$, $x \in D$, $\lambda \in \Lambda$ be a continuous $C^{m,\delta}(D\times\Lambda; R)$-semimartingale with the local characteristic belonging to the class $B^{m,\delta}$. Let f_t be a continuous predictable process with values in D. Set

$$K(\lambda, t) = \int_0^t F(f_s, \lambda, ds).$$

Then $K(\lambda, t)$ has a modification of a continuous $C^{m,\varepsilon}(\Lambda; R)$-semimartingale with the local characteristic of the class $B^{m,\delta}$, where $\varepsilon < \delta$. Furthermore, if g_t is a continuous predictable process with values in Λ, then

$$\int_0^t K(g_s, ds) = \int_0^t F(f_s, g_s, ds).$$

1.6. TIME CHANGE FORMULA OF STOCHASTIC INTEGRALS

Let T be a positive constant. Let B_t, $0 \le t \le T$ be a continuous $\{\mathcal{F}_t\}$-adapted strictly increasing process vanishing at zero. Define the inverse function τ_u of B_t by

$$\tau_u = \sup\{s \ge 0 : B_s \le u\} \qquad \text{on } u \ge 0.$$

It is an $\{\mathcal{F}_t\}$-stopping time for each u. Moreover, we have the following relations:

$$B_{\tau_u} = u \wedge B_T \quad \text{and} \quad \tau_{B_u} = u \quad \text{for all } u \ge 0.$$

Set

$$\tilde{\mathcal{F}}_u = \mathcal{F}_{\tau_u} \qquad \text{for } u \ge 0.$$

Then $\{\tilde{\mathcal{F}}_u\}_{u\ge0}$ is a filtration. It is easy to verify that for every t, B_t is an $\{\tilde{\mathcal{F}}_u\}$-stopping time. We now state a number of results on the time change of semimartingales by the random function τ_u.

Theorem 6.1 Let M_t, $0 \le t \le T$ be a continuous local martingale with respect to $\{\mathcal{F}_t\}$. Then $\tilde{M}_u := M_{\tau_u}$, $u \ge 0$ is a continuous local martingale with respect to $\{\tilde{\mathcal{F}}_u\}$.

Theorem 6.2 Let $F(x, t)$, $x \in D$, $0 \le t \le T$ be a continuous $C^{m,\delta}$-semimartingale with respect to $\{\mathcal{F}_t\}$ having the local characteristic $(a(x, y, t), b(x, t), A_t)$. Set

$$\tilde{F}(x, u) = F(x, \tau_u) \qquad \text{for } x \in D \text{ and } u \ge 0.$$

Then $\tilde{F}(x, u)$ is a continuous $C^{m,\delta}$-semimartingale with respect to $\{\tilde{\mathcal{F}}_u\}$ with the local characteristic $(\tilde{a}(x, y, u), \tilde{b}(x, u), \tilde{A}_u)$ given by

$$\tilde{a}(x, y, u) = a(x, y, \tau_u)\, \mathbb{1}_{\{u<B_T\}},$$
$$\tilde{b}(x, u) = b(x, \tau_u)\, \mathbb{1}_{\{u<B_T\}},$$
$$\tilde{A}_u = A_{\tau_u}.$$

Now let $F(x, t)$, $x \in D$, $0 \le t \le T$ be a continuous C-semimartingale with respect to $\{\mathcal{F}_t\}$ having the local characteristic of the class $B^{0,\delta}$. Let f_t, $0 \le t \le T$ be a continuous process adapted to $\{\mathcal{F}_t\}$. So the Itô integral

$$\int_0^t F(f_s, ds) \qquad \text{on } 0 \le t \le T$$

is well defined as a continuous $\{\mathcal{F}_t\}$-semimartingale. Set

$$\tilde{F}(x, u) = F(x, \tau_u) \quad \text{and} \quad \tilde{f}_u = f_{\tau_u}$$

for $u \ge 0$. Then the Itô integral

$$\int_0^u \tilde{F}(\tilde{f}_s, ds)$$

(with respect to the filtration $\{\tilde{\mathcal{F}}_u\}$) is also well defined as a continuous $\{\tilde{\mathcal{F}}_u\}$-semimartingale. Moreover we have the following formula.

Theorem 6.3 For all $u \ge 0$,

$$\int_0^u \tilde{F}(\tilde{f}_s, ds) = \int_0^{\tau_u} F(f_s, ds),$$

which is called the time change formula of the stochastic integral.

1.7. INEQUALITIES

This section introduces the theory of inequalities of Gronwall type and therefore forms a useful tool for the remaining chapters.

Since the discovery of the celebrated Gronwall inequality (cf. Bellman and Cooke (1963)), many mathematicians have devoted their time to further extensions and applications. We here mention Akinyele (1984), Bondge and Pachpatte (1979a, b), Dannan (1985), Pachpatte (1979), Thandapani and Agarwal (1982), Yeh (1980, 1982a, b, 1985a, b), Yeh and Shih (1982), Young (1973), Zahariev and Bainov (1981) among others. These results have been widely quoted in the theory of differential equations and integral equations for proving uniqueness, boundedness, comparison, continuous dependence, perturbation and stability results. The Lebesgue-Stieltjes integral inequalities of the Gronwall type were first obtained by the present author in order to deal with the stochastic stability of stochastic differential equations with respect to semimartingales in 1989 (cf. Mao (1989a, c, f)). In his book, Mao (1991f), there is one chapter devoted to these generalized inequalities. In this section, we shall introduce these generalized inequalities which will be used very frequently in the following chapters. All inequalities in this section are stated without proofs (for the proofs please see the author's book Mao (1991f)).

We first introduce some notations. Let T be a positive constant and $\mu(t)$ be a right continuous nondecreasing function defined on [0, T]. A function $f : [0, T] \to R$ is said to be *left limit* if there exists finite $\lim_{s\uparrow t} f(s)$ for all $0 < t \le T$. In this case, use $f(t-)$ to denote $\lim_{s\uparrow t} f(s)$ for $0 < t \le T$ and $a(0-) := a(0)$. Denote by $\mathcal{N}([0, T]; R_+)$ the family of all Borel measurable, bounded, nonnegative and left limit functions defined on [0, T]. Let us now start with the generalized Gronwall inequality.

Theorem 7.1 Let $u \in \mathcal{N}([0, T]; R_+)$ and c be a nonnegative constant. If

$$u(t) \le c + \int_0^t u(s-)\, d\mu(s) \qquad \text{for all } 0 \le t \le T,$$

then

$$u(t) \le c\, e^{\mu(t)} \qquad \text{for all } 0 \le t \le T.$$

In order to see this is a generalization of the well-known Gronwall inequality, we let u be a Borel measurable, bounded and nonnegative function defined on [0, T], v(t) be a nonnegative Lebesgue integrable function defined on [0, T], and let c be a nonnegative constant. Assume that

$$u(t) \le c + \int_0^t v(s)\,u(s)\,ds \qquad \text{for all } 0 \le t \le T.$$

Set

$$\mu(t) = \int_0^t v(s)\,ds \qquad \text{for } 0 \le t \le T,$$

which is a continuous nondecreasing function on [0, T]. Then the above inequality can be written as

$$u(t) \le c + \int_0^t u(s)\,d\mu(s) \qquad \text{for all } 0 \le t \le T.$$

Introduce

$$z(t) = c + \int_0^t u(s)\,d\mu(s) \qquad \text{for } 0 \le t \le T,$$

which is a continuous function. We see $u(t) \le z(t)$ and therefore

$$z(t) \le c + \int_0^t z(s)\,d\mu(s) = c + \int_0^t z(s-)\,d\mu(s) \qquad \text{for all } 0 \le t \le T.$$

It then follows, by Theorem 7.1, that

$$z(t) \le c\,e^{\mu(t)} \qquad \text{for all } 0 \le t \le T.$$

Hence

$$u(t) \le c \exp\Big(\int_0^t v(s)\,ds\Big) \qquad \text{for all } 0 \le t \le T,$$

which is the classical Gronwall inequality.

The following is the generalized Bihari inequality. For the classical Bihari inequality please see Bihari (1957).

Theorem 7.2 Let $u \in \mathcal{N}([0, T]; R_+)$ and $c > 0$. Let $K : R_+ \to R_+$ be a continuous nondecreasing function such that $K(t) > 0$ for all $t > 0$. Then the inequality

$$u(t) \le c + \int_0^t K(u(s-))\,d\mu(s) \qquad \text{for all } 0 \le t \le T$$

implies that

$$u(t) \le G^{-1}(\, G(c) + \mu(t)\,)$$

for all $0 \le t \le T$ such that

$$G(c) + \mu(t) \in \mathrm{Dom}(G^{-1}),$$

where

$$G(r) = \int_1^r (\, 1/\, K(v)\,)\, dv \qquad \text{on } r > 0$$

and G^{-1} is the inverse function of G.

The following is the generalized Wendroff inequality. For the classical Wendroff inequality please see page 154 of Beckenbach and Bellman (1961).

Theorem 7.3 Let $u \in \mathcal{N}([0, T]; R_+)$. Assume that $a : [0, T] \to (0, \infty)$ is right continuous and nondecreasing, and $K : R_+ \to R_+$ is continuous and nondecreasing such that $K(t) > 0$ for $t > 0$. If

$$u(t) \le a(t) + \int_0^t K(u(s-))\, d\mu(s)$$

holds for all $0 \le t \le T$. Then

$$u(t) \le G^{-1}\Big(\, G(a(0)) + \mu(t) + \int_0^t \frac{da(s)}{K(a(s-))}\,\Big)$$

for all $0 \le t \le T$ such that

$$G(a(0)) + \mu(t) + \int_0^t \frac{da(s)}{K(a(s-))} \in \mathrm{Dom}(G^{-1}),$$

where G and G^{-1} are the same as defined in Theorem 7.2.

More generally, we have the following inequality.

Theorem 7.4 Let $u \in \mathcal{N}([0, T]; R_+)$. Assume that $a : [0, T] \to (0, \infty)$ is right continuous and nondecreasing, and $K : R_+ \to R_+$ is continuous and nondecreasing such that $K(t) > 0$ for $t > 0$. Furthermore, assume that $f(t, s)$ is a bounded nonnegative Borel measurable function defined for $0 \le s \le t \le T$ and is nondecreasing in t. Then

$$u(t) \le a(t) + \int_0^t f(t, s)\, K(u(s-))\, d\mu(s) \qquad \text{for all } 0 \le t \le T$$

implies that

$$u(t) \le G^{-1}\Big(G(a(0)) + \int_0^t f(t, s)\, d\mu(s) + \int_0^t \frac{da(s)}{K(a(s-))} \Big)$$

for all $0 \le t \le T$ such that

$$G(a(0)) + \int_0^t f(t, s)\, d\mu(s) + \int_0^t \frac{da(s)}{K(a(s-))} \in \mathrm{Dom}(G^{-1}),$$

where G and G^{-1} are the same as in Theorem 7.2.

Using Theorem 7.4 one can easily get the following corollaries which it is more convenient to use sometimes.

Corollary 7.5 Let $u \in \mathcal{N}([0, T]; R_+)$, $c \ge 0$ and $0 < \alpha < 1$. Let f(t, s) be a bounded nonnegative Borel measurable function defined for $0 \le s \le t \le T$ which is non-decreasing in t. Assume that the inequality

$$u(t) \le c + \int_0^t f(t, s)\, (u(s-))^\alpha\, d\mu(s)$$

holds for all $0 \le t \le T$. Then

$$u(t) \le \Big(c^{1-\alpha} + (1-\alpha) \int_0^t f(t, s)\, d\mu(s) \Big)^{1/(1-\alpha)}$$

for all $0 \le t \le T$.

Corollary 7.6 Let $u \in \mathcal{N}([0, T]; R_+)$ and $c \ge 0$. Assume that

$$u(t) \le c + \int_0^t u(s-)\, (u(s-)+1)\, d\mu(s)$$

holds for all $0 \le t \le T$. Then

$$u(t) \le \frac{c\, e^{\mu(t)}}{1 - c(e^{\mu(t)}-1)}$$

holds for all $0 \le t \le T$ such that $c(e^{\mu(t)}-1) < 1$.

In order to state Lebesgue–Stieltjes integral inequalities with retardation which will be used in the analysis of stochastic differential delay equations, we need to introduce some new notations. Denote by $\mathfrak{D}$ the class of all Borel measurable functions σ defined on $[0, T]$ such that $0 \le \sigma(t) \le t$ for all $0 \le t \le T$. Let $\mathcal{H}$ denote the class of all functions $H : R_+ \to R_+$ satisfying

(i) $H(t)$ is nondecreasing and continuous for $t \ge 0$ and positive for $t > 0$;

(ii) there exists a continuous function Φ on R_+ such that $H(\alpha t) \le \Phi(\alpha)\, H(t)$ for all $\alpha > 0$ and $t \ge 0$.

In this case H is said to belong to the class $\mathcal{H}$ with the corresponding multiplier function Φ. It is obvious that H belongs to the class $\mathcal{H}$ with itself as the corresponding multiplier if H satisfies condition (i) and it is submultiplicative (i.e. $H(\alpha t) \le H(\alpha)\, H(t)$). We also denote by $\mathcal{Q}$ the family of all functions $Q : R_+ \to R_+$ satisfying

(i) $Q(t)$ is positive and nondecreasing for $t > 0$;

(ii) $Q(t)/\alpha \le Q(t/\alpha)$ for $\alpha \ge 1$ and $t \ge 0$.

Theorem 7.7 Let $u, h, q \in \mathcal{N}([0, T]; R_+)$ with h positive and nondecreasing, and $q(t) \ge 1$ for $0 \le t \le T$. Let $a : [0, T] \to (0, \infty)$ be a right continuous nondecreasing function. Let H belong to the class $\mathcal{H}$ with the corresponding multiplier function Φ. Let $f(t, s)$ be a bounded nonnegative Borel measurable function defined for $0 \le s \le t \le T$ which is nondecreasing in t. Let $\sigma \in \mathfrak{D}$. If

$$u(t) \le h(t)\, a(t) + q(t) \int_0^t f(t, s)\, H(u(\sigma(s)-))\, d\mu(s)$$

holds for all $0 \le t \le T$. Then

$$u(t) \le h(t)\, q(t)\, G^{-1}\Big(G(a(0)) + \int_0^t \frac{da(s)}{H(a(s-))}$$

$$+ \int_0^t \big[f(t, s)/h(s) \big]\, \Phi\big[h(\sigma(s)-)\, q(\sigma(s)-)\big]\, d\mu(s) \Big)$$

for all $0 \le t \le T$ such that

$$G(a(0)) + \int_0^t \frac{da(s)}{H(a(s-))}$$

$$+ \int_0^t \big[f(t, s)/h(s) \big]\, \Phi\big[h(\sigma(s)-)\, q(\sigma(s)-)\big]\, d\mu(s) \in \mathrm{Dom}(G^{-1}),$$

where

$$G(r) = \int_1^r (1/H(v))\,dv \qquad \text{on } r > 0$$

and G^{-1} is the inverse function of G.

Theorem 7.8 Let u, h, q ∈ $\mathcal{N}([0, T]; R_+)$ with both h and p nondecreasing, $h(t) \geq 1$ and $q(t) \geq 1$. Let $a : [0, T] \to (0, \infty)$ be a right continuous nondecreasing function. Let f(t, s) be a bounded nonnegative Borel measurable function defined for $0 \leq s \leq t \leq T$ which is nondecreasing in t. Let $Q \in \mathcal{Q}$ and $\sigma \in \mathcal{D}$. Assume that the inequality

$$u(t) \leq h(t)\,a(t) + q(t) \int_0^t f(t, s)\,Q(u(\sigma(s)-))\,d\mu(s)$$

holds for all $0 \leq t \leq T$. Then

$$u(t) \leq h(t)\,q(t)$$
$$\times\, G^{-1}\Big(G(a(0)) + \int_0^t \frac{da(s)}{Q(a(s-))} + \int_0^t f(t, s)\,q(\sigma(s)-)\,d\mu(s) \Big)$$

for all $0 \leq t \leq T$ such that

$$G(a(0)) + \int_0^t \frac{da(s)}{Q(a(s-))} + \int_0^t f(t, s)\,q(\sigma(s)-)\,d\mu(s) \;\in\; \mathrm{Dom}(G^{-1}),$$

where

$$G(r) = \int_1^r (1/Q(v))\,dv \qquad \text{on } r > 0$$

and G^{-1} is the inverse function of G.

In what follows we use the notation

$$\int_0^{s-} \ldots\, d\mu(r) = \lim_{z \uparrow s} \int_0^z \ldots\, d\mu(r).$$

Theorem 7.9 Let u ∈ $\mathcal{N}([0, T]; R_+)$. Assume that $a : [0, T] \to (0, \infty)$ is right continuous and nondecreasing, and $K : R_+ \to R_+$ is continuous and nondecreasing. Assume also that f(t, s) is a bounded nonnegative Borel measurable function defined for $0 \leq s \leq t \leq T$ which is nondecreasing in t. If the inequality

$$u(t) \le a(t) + \int_0^t f(t, s)\Big(u(s-) + \int_0^{s-} f(s, r)\, K(u(r-)\, d\mu(r) \Big) d\mu(s)$$

holds for all $0 \le t \le T$, then

$$u(t) \le G^{-1}\Big(G(a(0)) + \int_0^t \frac{da(s)}{a(s-) + K(a(s-))} + \int_0^t f(t, s)\, d\mu(s) \Big)$$

and also

$$u(t) \le a(t) + \int_0^t f(t, s)$$

$$\times\, G^{-1}\Big(G(a(0)) + \int_0^{s-} \frac{da(z)}{a(z-) + K(a(z-))} + \int_0^{s-} f(s, z)\, d\mu(z) \Big) d\mu(s)$$

for all $0 \le t \le T$ such that

$$G(a(0)) + \int_0^t \frac{da(s)}{a(s-) + K(a(s-))} + \int_0^t f(t, s)\, d\mu(s) \in \mathrm{Dom}(G^{-1}),$$

where

$$G(r) = \int_1^r \frac{dv}{v+K(v)}\, dv \qquad \text{on } r > 0$$

and G^{-1} is the inverse function of G.

Theorem 7.10 Let $u, p \in \mathcal{N}([0, T]; R_+)$ with $q(t) \ge 1$. Let $a : [0, T] \to (0, \infty)$ be a right continuous nondecreasing function. Let both f(t, s) and g(t, s) be bounded nonnegative Borel measurable functions defined for $0 \le s \le t \le T$ which are nondecreasing in t. Let $\sigma, \rho \in \mathfrak{D}$. Assume that the inequality

$$u(t) \le a(t) + q(t) \int_0^t f(t, s) \Big\{ u(\sigma(s)-) + \int_0^{s-} g(s, r)\, u(\rho(r)-)\, d\mu(r) \Big\} d\mu(s)$$

holds for all $0 \le t \le T$. Then for all $0 \le t \le T$,

$$u(t) \le q(t)\, a(0)$$

$$\times \exp\Big(\int_0^t \frac{da(s)}{a(s-)} + \int_0^t \big[f(t, s)\, q(\sigma(s)-) + g(t, s)\, q(\rho(s)-) \big] d\mu(s) \Big)$$

and also

$$u(t) \le q(t) \Big\{ a(t) + a(0) \int_0^t f(t, s)\, q(\sigma(s)-)$$

$$\times \exp\Big(\int_0^{s-} \frac{da(z)}{a(z-)} + \int_0^{s-} \Big[f(s, z)\, q(\sigma(z)-) + g(s, z)\, q(\rho(z)-) \Big] d\mu(z) \Big)\, d\mu(s) \Big\}.$$

We finally introduce two stochastic integral inequalities of the Gronwall type to close this section and hence this chapter. Before we state the inequalities, let us remind the reader that we always use the same probability space $(\Omega, \mathcal{F}, P, \{\mathcal{F}_t\}_{t\ge 0})$ introduced in Section 1.2 throughout this book.

Theorem 7.11 Let ρ be a finite stopping time, $K > 0$ and $c \ge 0$. Let $(A_t)_{0\le t\le \rho}$ be a nondecreasing continuous adapted process such that $A_0 = 0$ and $A_\rho \le K$ a.s. Let $(X_t)_{0\le t\le \rho}$ be a nondecreasing progressive process. If

$$EX_\tau \le c + E\int_0^\tau X_s\, dA_s$$

holds for any stopping time τ with $0 \le \tau \le \rho$ a.s., then

$$EX_\rho \le c\, e^K.$$

Theorem 7.12 Let ρ be a finite stopping time, $K > 0$ and $c > 0$. Let $(A_t)_{0\le t\le \rho}$ be a nondecreasing continuous adapted process such that $A_0 = 0$ and $A_\rho \le K$ a.s. Let $(X_t)_{0\le t\le \rho}$ be a nondecreasing nonnegative progressive process. Let g be a nondecreasing concave function defined on R_+ such that

$$g(r) > 0 \quad \text{for all } r > 0.$$

If the inequality

$$EX_\tau \le c + E\int_0^\tau g(X_s)\, dA_s$$

holds for any stopping time τ with $0 \le \tau \le \rho$ a.s., then

$$EX_\rho \le G^{-1}(G(c) + K),$$

where

$$G(r) = \int_1^r \frac{dv}{g(v)} \qquad \text{on } r > 0$$

and G^{-1} is the inverse function of G.

CHAPTER 2

Stochastic Differential Equations

2.1. INTRODUCTION

We repeat once again that throughout this book we let $(\Omega, \mathcal{F}, P, \{\mathcal{F}_t\}_{t\geq 0})$ be a complete probability space with a filtration $\{\mathcal{F}_t\}$ satisfying the usual conditions (i.e. increasing, right continuous and $\mathcal{F}_0$ contains all P-null sets). For the sake of convenience we sometimes use the notation $X = (X_t)_{t\geq 0}$ for a stochastic process like in the previous chapter but sometimes we use $X = (X(t))_{t\geq 0}$ like in this chapter.

The objective of this chapter is to discuss the solution of a stochastic differential equation driven by a nonlinear integrator of the form

$$x(t) = \xi + \int_{t_o}^{t} F(x(s), ds) \qquad \text{on } t \geq t_o, \tag{1.1}$$

where $t_o \geq 0$ and $F(x, t) = (F_1(x, t), \ldots, F_d(x, t))^T$ is a d-dimensional continuous C-semimartingale with parameter $x \in R^d$.

Let us first introduce assumptions for the continuous C-semimartingale F so that equation (1.1) is well defined. Let $F(x, t) = B(x, t) + M(x, t)$ be the decomposition such that $B(x, t) = (B_1(x, t), ..., B_d(x, t))^T$ is a d-dimensional continuous process of finite variation and $M(x, t) = (M_1(x, t), ..., M_d(x, t))^T$ is a d-dimensional continuous local martingale. Set $A_{ij}(x, y, t) = \langle M_i(x, t), M_j(y, t)\rangle$. Let A(t) be a continuous strictly increasing process vanishing at zero such that all $A_{ij}(x, y, t)$ and $B_i(x, t)$ are absolutely continuous with respect to A(t) a.s. for any $x, y \in R^d$. Then there exist predictable processes $a_{ij}(x, y, t)$ and $b_i(x, t)$ with parameters x, y such that

$$A_{ij}(x, y, t) = \int_0^t a_{ij}(x, y, s)\, dA(s)$$

and

$$B_i(x, t) = \int_0^t b_i(x, s)\, dA(s).$$

Set $a(x, y, t) = (a_{ij}(x, y, t))_{d\times d}$ and $b(x, t) = (b_1(x, t), ..., b_d(x, t))^T$. Then a(x, y, t) is a $d \times d$-matrix valued function with the following properties:

(i) *symmetric*: $a(x, y, t) = a(y, x, t)^T$ a.s. for any x, y;
(ii) *nonnegative definite*: $\sum_{i,j,k,l} a_{ij}(x_k, x_l, t)\, \xi_{ik}\xi_{jl} \geq 0$ a.s. for any $x_k \in R^d$, $\xi_{ik} \in R$, $1 \leq k, l \leq n$, $1 \leq i, j \leq d$.

As in the previous chapter the triple (a(x, y, t), b(x, t), A(t)) is called the *local characteristic* of F(x, t). The local characteristic is said to belong to the class $(B^{m,\delta}, B^{m',\delta'})$ if for every i = 1, ..., d the local characteristic of $F_i(x, t)$ belongs to the class $(B^{m,\delta}, B^{m',\delta'})$. When $m = m'$ and $\delta = \delta'$ the local characteristic is simply said to belong the class $B^{m,\delta}$. It is useful to see from the nonnegative definite property that, for example,

$$\begin{aligned} 0 &\leq a_{ii}(x, x, t)-2a_{ii}(x, y, t)+a_{ii}(y, y, t) \\ &\leq 2\Big(a_{ii}(x, x, t)-2a_{ii}(x, z, t)+a_{ii}(z, z, t) \Big) \\ &\quad + 2\Big(a_{ii}(z, z, t)-2a_{ii}(z, y, t)+a_{ii}(y, y, t) \Big) \qquad \text{a.s.} \end{aligned}$$

for any $x, y, z \in R^d$.

Now if F(x, t) is a continuous C-semimartingale with the local characteristic belonging to the class $B^{0,\delta}$, then Itô's stochastic integral

$$\int_0^t F(x(s), ds)$$

is well defined for any continuous R^d-valued predictable process x(t). To be precise, let us give the definition of the solution to equation (1.1).

Definition 1.1 A continuous R^d-valued adapted process x(t), $t \geq t_o$ is called a *solution of Itô's stochastic differential equation driven by a nonlinear integrator* F(x, t) *starting at* ξ *at time* t_o if it satisfies equation (1.1).

In this chapter we shall mainly study the existence and uniqueness of the solution of equation (1.1). In the literature (cf. Carmona & Nualart (1990), Kunita (1990), Mao (1990e)), the existence and uniqueness are usually proved under the well-known Lipschitz condition: there exists a positive constant K such that

$$|b(x, t) - b(y, t)|^2 \vee |a(x, x, t) - 2a(x, y, t) + a(y, y, t)| \leq K|x - y|^2 \tag{1.2}$$

for all x, y $\in R^d$. Moreover, the solution is constructed through the Picard iteration:

$$x_0(t) \equiv \xi \quad \text{and}$$

$$x_n(t) = \xi + \int_{t_o}^t F(x_{n-1}(s), ds) \qquad \text{for } t \geq t_o,\ n = 1, 2, \ldots . \tag{1.3}$$

We shall establish completely new results on the existence and uniqueness in the following sections. We shall not only prove the existence and uniqueness under weaker conditions e.g. the Hölder continuity, but also use a new approximation procedure rather than Picard's. As a matter of fact, it should be pointed out that it still remains open whether the Picard successive approximation is applicable under these weaker conditions.

We shall organize this chapter as follows: We first study the existence and uniqueness of the solution for a one dimensional stochastic differential equation driven by a nonlinear integrator in Section 2.2 and then extend the theory to a special multi-dimensional equation in Section 2.3. The theory for general multi-dimensional stochastic differential equations will be discussed in Section 2.4. The L^p-estimates for the solutions are obtained in Section 2.5, but the almost surely asymptotic bounds for the solutions will be discussed in Section 2.6. Finally, Section 2.7 is devoted to the discussion of the relation between stochastic flows and stochastic differential equations.

For the sake of simplification of notations, we shall let $t_o = 0$ in the following sections unless specified. Of course our theory is applicable for any $t_o \geq 0$ obviously.

2.2. ONE-DIMENSIONAL STOCHASTIC DIFFERENTIAL EQUATIONS

In order to make our theory more understandable, let us start with a one-dimensional stochastic differential equation of the form

$$x(t) = \xi + \int_0^t F(x(s), ds) \qquad \text{on } t \geq 0, \tag{2.1}$$

where $F(x, t)$, $x \in R$ is a one-dimensional continuous C-semimartingale with the local characteristic $(a(x, y, t), b(x, t), A(t))$ belonging to the class $B^{0,\delta}$ and the decomposition $F(x, t) = B(x, t) + M(x, t)$. Let us first introduce hypotheses:

(H2.1) $p \geq 2$ and ξ is an $\mathcal{F}_0$-measurable random variable such that $E|\xi|^p < \infty$.

(H2.2) Both $b(0, t)$ and $a(0, 0, t)$ are bounded.

(H2.3) For all $x, y \in R$ and $t \geq 0$,

$$|b(x, t)-b(y, t)| \leq \kappa(|x-y|) \qquad \text{a.s.}$$

where $\kappa : R_+ \to R_+$ is a continuous increasing concave function such that

$$\int_{0+} \frac{du}{\kappa(u)} = \infty.$$

(H2.4) For all $x, y \in R$ and $t \geq 0$

$$a(x, x, t) - 2a(x, y, t) + a(y, y, t) \leq \rho(|x - y|) \qquad \text{a.s.}$$

where $\rho : R_+ \to R_+$ is a continuous increasing concave function such that

$$\int_{0+} \frac{du}{\rho(u)} = \infty.$$

Example 2.1 There exist many examples of such functions κ and ρ used in the above hypotheses. For instance, let $\delta > 0$ be small enough and define

$$\kappa(u) = u,\ u\left(\log\frac{1}{u}\right),\ u\left(\log\frac{1}{u}\right)\left(\log\log\frac{1}{u}\right), \ldots, \quad \text{for } 0 \leq u \leq \delta,$$

$$\kappa(u) = \kappa(a) + \kappa'(a-)\,(u - a) \qquad \text{for } u > \delta.$$

Then they all satisfy the requirements of (H2.3).

We also introduce the inverse function $\tau(u)$ of $A(t)$, i.e.

$$\tau(u) = \inf\{t \geq 0: A(t) \geq u\}$$

for $u \geq 0$ (we use convention $\inf\emptyset = \infty$ throughout this book unless specified). For each u, $\tau(u)$ is a stopping time with respect to the filtration $\{\mathcal{F}_t\}$. Also

$$\tau(A(t)) = t \quad \text{and} \quad A(\tau(t)) = t \wedge A(\infty)$$

for all $t \geq 0$, where $A(\infty) = \lim_{t\to\infty} A(t)$.

Theorem 2.1 Let (H2.1)-(H2.4) hold. Then there exists a unique solution $x(t)$ to equation (2.1), and for any $T > 0$,

$$E\Big(\sup_{0\leq t<\tau(T)} |x(t)|^p\Big) < \infty. \tag{2.2}$$

Moreover

$$\lim_{n\to\infty} E\Big(\sup_{0\leq t<\tau(T)} |x(t) - x^n(t)|\Big) = 0, \tag{2.3}$$

where $x^n(t)$, $t \geq 0$ and $n \geq 1$, are defined as follows:

$$\begin{aligned} &x^n(0) = \xi \quad \text{and} \\ &x^n(t) = x^n(\tau(i/n)) + F(x^n(\tau(i/n)), t) - F(x^n(\tau(i/n)), \tau(i/n)) \\ &\qquad \text{if } \tau(i/n) < t \leq \tau((i+1)/n),\ i = 0, 1, \dots . \end{aligned} \tag{2.4}$$

In order to prove the theorem, we need to prepare several lemmas under an additional condition $A(t) \equiv t$. In this case (2.4) reduces to

$$\begin{aligned} &x^n(0) = \xi \quad \text{and} \\ &x^n(t) = x^n(i/n) + F(x^n(i/n), t) - F(x^n(i/n), i/n) \\ &\qquad \text{if } i/n < t \leq (i+1)/n,\ i = 0, 1, \dots . \end{aligned} \tag{2.5}$$

Introduce simple processes

$$\tilde{x}^n(t) = \sum_{i=0}^{\infty} x^n(i/n)\, \mathbb{1}_{[i/n,\,(i+1)/n)}(t). \tag{2.6}$$

Then (2.5) can be written as

$$x^n(t) = \xi + \int_0^t F(\tilde{x}^n(s), ds)$$

$$= \xi + \int_0^t b(\tilde{x}^n(s), s)\, ds + \int_0^t M(\tilde{x}^n(s), ds). \tag{2.7}$$

Lemma 2.2 Let (H2.1) hold and $A(t) \equiv t$. Assume that both $b(x, t)$ and $a(x, y, t)$ satisfy the linear growth condition, i.e. there exists a positive constant K such that

(H2.5) $|b(x, t)|^2 \vee a(x, x, t) \le K(1 + |x|)^2$ a.s. for all $x \in R$ and $t \ge 0$.

Then for any $T > 0$, there exists a $C = C(T) > 0$ such that

$$E|x^n(t) - \tilde{x}^n(t)|^p \le \frac{C}{n^{p/2}} \quad \text{for all } 0 \le t \le T \text{ and } n \ge 1. \tag{2.8}$$

Proof. It follows from (2.7) that

$$E\Big(\sup_{0\le s\le t} |x^n(s)|^p\Big) \le 3^{p-1} E|\xi|^p + 3^{p-1} E\Big(\int_0^t |b(\tilde{x}^n(u), u)|\, du\Big)^p$$

$$+ 3^{p-1} E\Big(\sup_{0\le s\le t} \Big|\int_0^s M(\tilde{x}^n(u), du)\Big|^p\Big).$$

But by Hölder's inequality and the linear growth condition,

$$3^{p-1} E\Big(\int_0^t |b(\tilde{x}^n(u), u)|\, du\Big)^p \le (3t)^{p-1} E\int_0^t |b(\tilde{x}^n(u), u)|^p\, du$$

$$\le (3t)^{p-1} E\int_0^t K^{p/2}(1 + |\tilde{x}^n(u)|)^p\, du \le (6t)^{p-1} K^{p/2} E\int_0^t (1 + |\tilde{x}^n(u)|^p)\, du$$

$$\le (6t)^{p-1} K^{p/2} \int_0^t \Big\{1 + E\Big(\sup_{0\le s\le u} |x^n(s)|^p\Big)\Big\} du.$$

Also by the Burkholder-Davis-Gundy inequality

$$3^{p-1} E\Big(\sup_{0\le s\le t} \Big|\int_0^s M(\tilde{x}^n(u), du)\Big|^p\Big)$$

$$\leq 3^{p-1}\, C_p\, E\Big(\int_0^t a(\tilde{x}^n(u), \tilde{x}^n(u), u)\, du \Big)^{p/2},$$

where C_p is a positive constant depending only on p (cf. Theorem 1-2.6). In view of Hölder's inequality and the linear growth condition we then see that

$$3^{p-1}\, E\Big(\sup_{0\leq s\leq t} \Big| \int_0^s M(\tilde{x}^n(u), du) \Big|^p \Big)$$

$$\leq 6^{p-1}\, C_p\, K^{p/2}\, t^{(p-2)/2} \int_0^t \Big\{ 1 + E\Big(\sup_{0\leq s\leq u} |x^n(s)|^p \Big) \Big\} du.$$

Therefore

$$E\Big(\sup_{0\leq s\leq t} |x^n(s)|^p \Big) \leq c_1 + c_2 \int_0^t E\Big(\sup_{0\leq s\leq u} |x^n(s)|^p \Big)\, du \quad \text{for } 0 \leq t \leq T,$$

where

$$c_2 = 6^{p-1}\, K^{p/2} \Big(T^{p-1} + C_p T^{(p-2)/2} \Big),$$

$$c_1 = 3^{p-1}\, E|\xi|^p + c_2 T.$$

By the well-known Gronwall inequality (cf. Theorem 1-7.1) we obtain

$$E\Big(\sup_{0\leq s\leq t} |x^n(s)|^p \Big) \leq c_1\, e^{c_2 t} \quad \text{for all } 0 \leq t \leq T. \tag{2.9}$$

We therefore have that, if $0 \leq s < t \leq T,\ t - s \leq 1$,

$$E|x^n(t) - x^n(s)|^p$$

$$\leq 2^{p-1}\, E\Big| \int_s^t b(\tilde{x}^n(u), u)\, du \Big|^p + 2^{p-1}\, E\Big| \int_s^t M(\tilde{x}^n(u), du) \Big|^p$$

$$\leq \Big[(4(t-s))^{p-1} K^{p/2} + 4^{p-1} C_p K^{p/2} (t-s)^{(p-2)/2} \Big] \int_s^t \Big(1 + E|\tilde{x}^n(u)|^2 \Big) du$$

$$\leq 4^{p-1} K^{p/2} (1 + C_p) \Big(1 + C_1\, e^{C_2 T} \Big) (t-s)^{p/2} := C(t-s)^{p/2}.$$

So the required inequality (2.8) follows by the definition of $\tilde{x}^n(t)$. The proof is complete.

Lemma 2.3 Assume that both (H2.1) and (H2.5) hold and $A(t) \equiv t$. Let $x(t)$ be a solution of equation (2.1). Then for any $T > 0$,

$$E\Big(\sup_{0\le s\le t} |x(s)|^p\Big) \le c_1\, e^{c_2 t} \quad \text{for all } 0 \le t \le T,$$

where c_1 and c_2 are defined as in the proof of Lemma 2.2.

This lemma can be proved in the same way as inequality (2.9). In order to state another lemma we introduce an auxiliary hypothesis:

(H2.6) For all $x, y \in R$ and $t \ge 0$,

$$a(x, x, t) - 2a(x, y, t) + a(y, y, t) \le \rho(|x - y|^{2\alpha}) \quad \text{a.s.}$$

where $\alpha \in [1/2, 1]$ and $\rho : R_+ \to R_+$ is a continuous increasing concave function such that

$$\int_{0+} \frac{du}{\rho(u^{2\alpha})} = \infty.$$

Note that (H2.6) is exactly the same as (H2.4) when $\alpha = 1/2$.

Lemma 2.4 Let (H2.1)–(H2.3) and (H2.6) hold and $A(t) \equiv t$. Then for any $T > 0$

$$\sup_{0\le t\le T} E|x^m(t) - x^n(t)| \to 0 \qquad \text{as } m, n \to \infty. \tag{2.10}$$

Proof. We first verify the linear growth condition (H2.5). Since both κ and ρ are concave functions with $\kappa(0)=\rho(0)=0$, we can choose two positive constants γ and λ such that

$$\kappa(u) \vee \rho(u) \le \gamma + \lambda u \qquad \text{for all } u \ge 0.$$

Let β be the bound of $|b(0, t)|^2$ and $a(0, 0, t)$. Then

$$|b(x, t)|^2 \le 2\beta + 2\kappa^2(|x|) \le 2\beta + 2(\gamma + \lambda|x|)^2.$$

Also note from the nonnegative definite property of $a(x, y, t)$ that

$$0 \le a(x, x, t) - 4a(x, 0, t) + 4a(0, 0, t).$$

Thus

$$a(x, x, t) \le 2a(0, 0, t) + 2\Big(a(x, x, t) - 2a(x, 0, t) + a(0, 0, t)\Big)$$

$$\le 2\beta + 2\rho(|x|^{2\alpha}) \le 2\beta + 2(\gamma + \lambda|x|^{2\alpha}) \le 2\beta + 2(\gamma + \lambda + \lambda|x|^2).$$

Hence (H2.5) holds if we set $K = 2\max\{\lambda^2, \beta + \gamma^2, \beta + \gamma + \lambda\}$.

We now let $a_0 = 1 > a_1 > a_2 > \cdots > a_k > \cdots$ be positive constants defined by

$$\int_{a_k}^{a_{k-1}} \frac{1}{\rho(u^{2\alpha})} du = k \quad \text{for } k = 1, 2, \dots .$$

Clearly $a_k \to 0$ as $k \to \infty$. For each $k \ge 1$, let $\psi_k(u)$ be a continuous function such that its support is contained in (a_k, a_{k-1}), and moreover,

$$0 \le \psi_k(u) \le \frac{2}{k\,\rho(u^{2\alpha})} \quad \text{and} \quad \int_{a_k}^{a_{k-1}} \psi_k(u)\, du = 1.$$

Such a function exists obviously. Set

$$\varphi_k(x) = \int_0^{|x|} dy \int_0^y \psi_k(u)\, du.$$

It is easy to see that $\varphi_k \in C^2(R; R)$, $\varphi_k(0) = 0$,

$$|\varphi_k'(x)| \le 1 \quad \text{for all } x \in R,$$

$$|\varphi_k''(x)| \le \frac{2}{k\,\rho(|x|^{2\alpha})} \quad \text{if } a_k < |x| < a_{k-1}, \text{ and } \varphi_k''(x) = 0 \text{ otherwise.}$$

In particular,

$$|x| - a_{k-1} \le \varphi_k(x) \le |x| \quad \text{for all } x \in R.$$

Let $m > n \ge 1$. Itô's formula implies that

$$\varphi_k(x^m(t) - x^n(t))$$

$$= \int_0^t \varphi_k'(x^m(s) - x^n(s)) \Big(b(\tilde{x}^m(s), s) - b(\tilde{x}^n(s), s) \Big) ds$$

$$+ \int_0^t \varphi_k'(x^m(s) - x^n(s)) \Big(M(\tilde{x}^m(s), ds) - M(\tilde{x}^n(s), ds) \Big)$$

$$+ \frac{1}{2}\int_0^t \varphi_k''(x^m(s) - x^n(s))$$

$$\times \Big(a(\tilde{x}^m(s), \tilde{x}^m(s), s) - 2a(\tilde{x}^m(s), \tilde{x}^n(s), s) + a(\tilde{x}^n(s), \tilde{x}^n(s), s)\Big)\, ds,$$

Hence

$$E\varphi_k(x^m(t) - x^n(t))$$

$$= E\int_0^t \varphi_k'(x^m(s) - x^n(s))\Big(b(\tilde{x}^m(s), s) - b(\tilde{x}^n(s), s)\Big)\, ds$$

$$+ \frac{1}{2} E\int_0^t \varphi_k''(x^m(s) - x^n(s))$$

$$\times \Big(a(\tilde{x}^m(s), \tilde{x}^m(s), s) - 2a(\tilde{x}^m(s), \tilde{x}^n(s), s) + a(\tilde{x}^n(s), \tilde{x}^n(s), s)\Big)\, ds$$

$$:= I_1(t) + I_2(t).$$

On the other hand, we estimate

$$I_1(t) \le E\int_0^t \Big(|b(\tilde{x}^m(s), s) - b(x^m(s), s)| + |b(x^m(s), s) - b(x^n(s), s)|$$

$$+ |b(x^n(s), s) - b(\tilde{x}^n(s), s)| \Big)\, ds$$

$$\le \int_0^t \kappa\Big(E|\tilde{x}^m(s) - x^m(s)|\Big)\, ds + \int_0^t \kappa\Big(E|x^m(s) - x^n(s)|\Big)\, ds$$

$$+ \int_0^t \kappa\Big(E|x^n(s) - \tilde{x}^n(s)|\Big)\, ds$$

$$\le \int_0^t \kappa\Big(\big[E|\tilde{x}^m(s) - x^m(s)|^p\big]^{1/p}\Big)\, ds + \int_0^t \kappa\Big(E|x^m(s) - x^n(s)|\Big)\, ds$$

$$+\int_0^t \kappa\left(\left[E|x^n(s) - \tilde{x}^n(s)|^p\right]^{1/p}\right) ds$$

$$\leq T\left\{ \kappa\left(C^{1/p}/m^{1/2}\right) + \kappa\left(C^{1/p}/n^{1/2}\right)\right\} + \int_0^t \kappa\left(E|x^m(s) - x^n(s)|\right) ds,$$

where Lemma 2.2 has been used. Also by the nonnegative definite property of $a(x, y, t)$ we can easily show that

$$a(\tilde{x}^m(s), \tilde{x}^m(s), s) - 2a(\tilde{x}^m(s), \tilde{x}^n(s), s) + a(\tilde{x}^n(s), \tilde{x}^n(s), s)$$

$$\leq 3\Big(a(\tilde{x}^m(s), \tilde{x}^m(s), s) - 2a(\tilde{x}^m(s), x^m(s), s) + a(x^m(s), x^m(s), s)\Big)$$

$$+ 3\Big(a(x^m(s), x^m(s), s) - 2a(x^m(s), x^n(s), s) + a(x^n(s), x^n(s), s)\Big)$$

$$+ 3\Big(a(x^n(s), x^n(s), s) - 2a(x^n(s), \tilde{x}^n(s), s) + a(\tilde{x}^n(s), \tilde{x}^n(s), s)\Big).$$

Thus

$$I_2(t) \leq \frac{3}{2} E \int_0^t |\varphi_k''(x^m(s) - x^n(s))|$$

$$\times \left\{\rho\left(|\tilde{x}^m(s) - x^m(s)|^{2\alpha}\right) + \rho\left(|x^n(s) - \tilde{x}^n(s)|^{2\alpha}\right) + \rho\left(|x^m(s) - x^n(s)|^{2\alpha}\right)\right\} ds$$

$$\leq \frac{3}{k\,\rho(a_k^{2\alpha})} \int_0^t \left\{ \rho\left(E|\tilde{x}^m(s) - x^m(s)|^{2\alpha}\right) + \rho\left(E|x^n(s) - \tilde{x}^n(s)|^{2\alpha}\right) \right\} ds$$

$$+ \frac{3}{2} E \int_0^t \frac{2}{k\,\rho\left(|x^m(s)-x^n(s)|^{2\alpha}\right)} \mathbb{1}_{\{a_k < |x^m(s)-x^n(s)| < a_{k-1}\}} \rho\left(|x^m(s) - x^n(s)|^{2\alpha}\right) ds$$

$$\leq \frac{3}{k\,\rho(a_k^{2\alpha})} \int_0^t \left\{\rho\left(\left[E|\tilde{x}^m(s)-x^m(s)|^p\right]^{2\alpha/p}\right) + \rho\left(\left[E|x^n(s)-\tilde{x}^n(s)|^p\right]^{2\alpha/p}\right)\right\} ds + \frac{3T}{k}$$

$$\leq \frac{3}{k\,\rho(a_k^{2\alpha})} T \left\{ \rho\left(C^{2\alpha/p}/m^{\alpha}\right) + \rho\left(C^{2\alpha/p}/n^{\alpha}\right) \right\} + \frac{3T}{k}.$$

Moreover

$$E\varphi_k(x^m(t) - x^n(t)) \ge E|x^m(t) - x^n(t)| - a_{k-1}.$$

Combining the above inequalities we obtain that

$$E|x^m(t) - x^n(t)| \le a_{k-1} + \frac{3T}{k} + \lambda_{m,n} + \frac{3}{k\,\rho(a_k^{2\alpha})}\,\gamma_{m,n}$$

$$+ \int_0^t \kappa\Big(E|x^m(s) - x^n(s)|\Big)\, ds \quad (2.11)$$

for $0 \le t \le T$, where

$$\lambda_{m,n} = T\Big\{\kappa\Big(C^{1/p}/m^{1/2}\Big) + \kappa\Big(C^{1/p}/n^{1/2}\Big)\Big\}$$

and

$$\gamma_{m,n} = T\,\Big\{\rho\Big(C^{2\alpha/p}/m^{\alpha}\Big) + \rho\Big(C^{2\alpha/p}/n^{\alpha}\Big)\Big\}.$$

By Bihari's inequality (cf. Theorem 1-7.2) we therefore obtain from (2.11) that

$$\sup_{0\le t\le T} E|x^m(t) - x^n(t)|$$

$$\le Q^{-1}\Big\{ Q\Big(a_{k-1} + \frac{3T}{k} + \lambda_{m,n} + \frac{3}{k\,\rho(a_k^{2\alpha})}\,\gamma_{m,n}\Big) + T \Big\} \tag{2.12}$$

provided

$$Q\Big(a_{k-1} + \frac{3T}{k} + \lambda_{m,n} + \frac{3}{k\,\rho(a_k^{2\alpha})}\,\gamma_{m,n}\Big) + T \in \mathrm{Dom}(Q^{-1}),$$

where

$$Q(u) = \int_1^u \frac{ds}{\kappa(s)} \qquad \text{for } u > 0,$$

and Q^{-1} is the inverse function of Q. By the condition on κ we see $(-\infty, 0] \subset \mathrm{Dom}(Q^{-1})$, $Q(u) \to -\infty$ as $u \to 0$ and $Q^{-1}(u) \to 0$ as $u \to -\infty$. Hence for any $\varepsilon > 0$, there exists a $\delta > 0$ such that

$$Q(u) + T < 0 \quad \text{and} \quad Q^{-1}\Big(Q(u) + T\Big) < \varepsilon \qquad \text{for all } 0 < u \le 2\delta.$$

Choose a k such that $a_{k-1} + 3T/k \le \delta$. For this fixed k, there exists an $N \ge 1$ such that

$$\lambda_{m,n} + \frac{3}{k\,\rho(a_k^{2\alpha})}\gamma_{m,n} \le \delta \quad \text{for all } m > n \ge N.$$

It therefore follows from (2.12) that

$$\sup_{0\le t\le T} E|x^m(t) - x^n(t)| < \varepsilon \quad \text{for all } m > n \ge N.$$

The proof is complete.

For the uniqueness of the solution we introduce one more auxiliary hypothesis:

(H2.7) For all $x, y \in R$ and $t \ge 0$

$$a(x, x, t) - 2a(x, y, t) + a(y, y, t) \le \theta(|x - y|) \quad \text{a.s.},$$

where $\theta : R_+ \to R_+$ is a continuous increasing function such that

$$\int_{0+} \frac{du}{\theta(u)} = \infty.$$

Please note that we do not require θ is concave, and therefore (H2.7) is weaker than (H2.4). The following lemma shows that the uniqueness holds under this weaker condition. Its version for classical one-dimensional Itô's equations is due to Yamada and Watanabe (1971).

Lemma 2.5 Assume (H2.1)-(H2.3) and (H2.7) hold and $A(t) \equiv t$. Then equation (2.1) has at most one solution.

Proof. Let $x(t)$ and $\hat{x}(t)$ be two solutions of equation (2.1). For each $k \ge 1$, define $\varphi_k(x)$ as we did in the proof of Lemma 2.4 except $\rho(u^{2\alpha})$ is replaced by $\theta(u)$. Applying Itô's formula to $\varphi_k(x(t)-\hat{x}(t))$ we derive that

$$E|x(t) - \hat{x}(t)| - a_{k-1} \le E\varphi_k(x(t) - \hat{x}(t))$$

$$\le E\int_0^t |b(x(s), s) - b(\hat{x}(s), s)|\, ds + \frac{1}{2}E\int_0^t \varphi_k''(x(s) - \hat{x}(s))\, \theta(|x(s) - \hat{x}(s)|)\, ds$$

$$\le \int_0^t \kappa\Big(E|x(s) - \hat{x}(s)|\Big)\, ds + t/k, \qquad 0 \le t < \infty.$$

Letting $k \to \infty$ gives

$$E|x(t) - \hat{x}(t)| \le \int_0^t \kappa\Big(E|x(s) - \hat{x}(s)|\Big)\, ds, \qquad 0 \le t < \infty$$

which implies by Bihari's inequality that

$$E|x(t) - \hat{x}(t)| = 0, \qquad 0 \le t < \infty.$$

The proof is complete.

Remark 2.1 The continuity condition for a(x, y, t) in Lemma 2.5 is the best possible in a sense. As a matter of fact, let us consider an Itô equation

$$dx(t) = g(x(t))\, dw(t) \quad \text{on } t \ge 0,$$

with initial data $x(0) = 0$, where w is a standard Brownian motion. Assume $g(0) = 0$ and $g(x) \ge 1$ for $|x| > \varepsilon$ (>0), and

$$\int_{-\varepsilon}^{\varepsilon} g^{-2}(x)\, dx < \infty.$$

Then the equation has infinitely many solutions (cf. Ikeda and Watanabe (1981)).

We now can start to prove Theorem 2.1.

Proof of Theorem 2.1. We first prove the theorem in the case $A(t) \equiv t$. By Burkholder-Davis-Gundy's inequality and Jensen's inequality, we derive that for any $T > 0$,

$$E\Big(\sup_{0\le t\le T} |x^m(t) - x^n(t)|\Big)$$

$$\le E\int_0^T \kappa\Big(|\tilde{x}^m(s) - \tilde{x}^n(s)|\Big)\, ds + 3\Big(\int_0^T \rho(E|\tilde{x}^m(s) - \tilde{x}^n(s)|)\, ds\Big)^{1/2}$$

$$\le \int_0^T \kappa\Big(E|\tilde{x}^m(s) - \tilde{x}^n(s)|\Big)\, ds + 3\Big(\int_0^T \rho(E|\tilde{x}^m(s) - \tilde{x}^n(s)|)\, ds\Big)^{1/2}.$$

But by Lemmas 2.2 and 2.4,

$$E|\tilde{x}^m(s) - \tilde{x}^n(s)|$$

$$\le E|\tilde{x}^m(s) - x^m(s)| + E|x^m(s) - x^n(s)| + E|x^n(s) - \tilde{x}^n(s)| \to 0$$

as m, n $\to \infty$. So

$$E\Big(\sup_{0\le t\le T} |x^m(t) - x^n(t)| \Big) \to 0 \qquad \text{as } m, n \to 0.$$

In other words, $\{x^n(t): 0 \le t \le T\}$ is a Cauchy sequence in $L(\Omega; C([0, T], R))$. Hence there exists a continuous process $x(t)$ on $t \ge 0$ such that for any $T \ge 0$, $\{x(t): 0 \le t \le T\} \in L(\Omega; C([0, T], R))$ and

$$\lim_{n\to\infty} E\Big(\sup_{0\le t\le T} |x(t) - x^n(t)| \Big) = 0. \tag{2.14}$$

It is also easy to show that $x(t)$ is $\{\mathcal{F}_t\}$-adapted. We now show $x(t)$ is a solution of equation (2.1). In fact,

$$E \int_0^t |b(x(s), s) - b(\tilde{x}^n(s), s)| \, ds$$

$$\le \int_0^t \kappa\Big(E|x(s) - x^n(s)|\Big) \, ds + \int_0^t \kappa\Big(E|x^n(s) - \tilde{x}^n(s)|\Big) \, ds \to 0$$

as $n \to \infty$. Also

$$E\Big| \int_0^t M(x(s), ds) - M(\tilde{x}^n(s), ds) \Big|$$

$$\le 3E\Big(\int_0^t 2\rho(|x(s) - x^n(s)|) + 2\rho(|x^n(s) - \tilde{x}^n(s)|) \, ds \Big)^{1/2}$$

$$\le 3 \Big(\int_0^t 2\rho(E|x(s) - x^n(s)|) + 2\rho(E|x^n(s) - \tilde{x}^n(s)|) \, ds \Big)^{1/2} \to 0.$$

Hence, we can let $n \to \infty$ in (2.7) to obtain that

$$x(t) = \xi + \int_0^t b(x(s), s) \, ds + \int_0^t M(x(s), ds) = \xi + \int_0^t F(x(s), ds).$$

In other words, $x(t)$ is a solution of equation (2.1). The uniqueness follows from Lemma 2.5.

In order to complete the proof we need to remove the additional condition $A(t) \equiv t$. To do so, we define $\mathcal{G}_u = \mathcal{F}_{\tau(u)}$ for $u \ge 0$, and also for each $k = 1, 2, \ldots$, $G^k(x, u) := F(x, \tau(u)\wedge k)$. In view of Theorem 1-6.2, $G^k(x, u)$ is a continuous C-semimartingale with respect to the filtration $\{\mathcal{G}_u\}$ and it has the local characteristic $(a(x, y, \tau(u))\, \mathbb{1}_{\{u<A(k)\}}, b(x, \tau(u))\, \mathbb{1}_{\{u<A(k)\}}, u)$. By the above proof we see that there exists a unique solution $y^k(u)$, which is continuous and $\{\mathcal{G}_u\}$-adapted, to the equation

$$y^k(u) = \xi + \int_0^u G^k(y^k(s), ds) \qquad \text{on } u \geq 0, \tag{2.15}$$

moreover,

$$\lim_{n\to\infty} E\Big(\sup_{0\leq u\leq T} |y^k(u) - y^{k,n}(u)| \Big) = 0 \quad \text{for any } T \geq 0, \tag{2.16}$$

and, particularly, the convergences are uniform in k, where

$$\begin{aligned} &y^{k,n}(0) = \xi \quad \text{and} \\ &y^{k,n}(u) = y^{k,n}(i/n) + G^k(y^{k,n}(i/n), u) - G^k(y^{k,n}(i/n), i/n) \\ &\qquad\qquad \text{if } i/n < u \leq (i+1)/n, \ i = 0, 1, \dots . \end{aligned}$$

It is easy to see that $y^k(u) = y^{k+1}(u)$ if $u \leq A(k)$ because $G^k(x, u) = G^{k+1}(x, u)$ whenever $u \leq A(k)$. Define $y(u)$ for $0 \leq u < A(\infty)$ as follows:

$$y(u) = y^k(u) \qquad \text{if } A(k-1) \leq u < A(k), \ k = 1, 2, \dots$$

Then $y(u)$ is a continuous $\{\mathfrak{G}_u\}$-adapted process, and moreover $y(u) = y^k(u)$ whenever $u \leq A(k)$. Set $x(t) = y(A(t))$ for $t \geq 0$. Then, for any $t \geq 0$, choose a $k \geq t$, and by the time change formula of stochastic integral (i.e. Theorem 1-6.3),

$$x(t) = y^k(A(t)) = \xi + \int_0^{A(t)} G^k(y^k(s), ds) = \xi + \int_0^t F(x(s), ds),$$

that is, $x(t)$ is a solution of equation (1.1). Also, for any $T > 0$,

$$E\Big(\sup_{0\leq t<\tau(T)\wedge k} |x(t)|^p \Big) = E\Big(\sup_{0\leq t<\tau(T)\wedge k} |y(A(t))|^p \Big)$$

$$= E\Big(\sup_{0\leq t<\tau(T)\wedge k} |y^k(A(t))|^p \Big) \leq E\Big(\sup_{0\leq u\leq T} |y^k(u)|^p \Big) \leq C,$$

where C is a positive constant independent of k by using Lemma 2.3. Letting $k \to \infty$, then the well-known Fatou convergence lemma gives the required (2.2). Furthermore, for each $n \geq 1$, we define $y^n(u)$ for $0 \leq u < A(\infty)$ as follows:

$$y^n(u) = y^{k,n}(u) \qquad \text{if } A(k-1) \leq u < A(k), \ k = 1, 2, \dots$$

It is easy to see that $y^n(u) = y^{k,n}(u)$ whenever $u \leq A(k)$. Set $x^n(t) = y^n(A(t))$ for $t \geq 0$. We now check $x^n(t)$ is the same as defined by (2.4). In fact, $x^n(0) = \xi$ obviously. If $\tau(i/n) < t \leq \tau((i+1)/n)$, then $i/n < A(t) \leq (i+1)/n$. Choose a $k > t$, then

$$x^n(t) = y^{k,n}(A(t)) = y^{k,n}(i/n) + G^k(y^{k,n}(i/n), A(t)) - G^k(y^{k,n}(i/n), i/n)$$
$$= x^n(\tau(i/n)) + F(x^n(\tau(i/n)), \tau(A(t))\wedge k) - F(x^n(\tau(i/n)), \tau(i/n)\wedge k)$$
$$= x^n(\tau(i/n)) + F(x^n(\tau(i/n)), t) - F(x^n(\tau(i/n)), \tau(i/n))$$

as required. Moreover, for any $T > 0$, by (2.16),

$$E\Big(\sup_{0\le t<\tau(T)\wedge k} |x(t) - x^n(t)|\Big) = E\Big(\sup_{0\le t<\tau(T)\wedge k} |y^k(A(t)) - y^{k,n}(A(t))|\Big)$$

$$\le E\Big(\sup_{0\le u\le T} |y^k(u) - y^{kn}(u)|\Big) \to 0 \qquad \text{as } n\to\infty \text{ uniformly for all } k > T.$$

Letting $k \to \infty$ we obtain the required (2.3). Finally, in order to show the uniqueness of the solution, let $x(t)$ and $\hat{x}(t)$ be two solutions of equation (2.1). For any $k \ge 1$, set $y^k(u) = x(\tau(u)\wedge k)$ and $\hat{y}^k(u) = \hat{x}(\tau(u)\wedge k)$. Then by Theorem 1-6.3 one can easily show that both $y^k(u)$ and $\hat{y}^k(u)$ are the solutions of equation (2.15), and so $y^k(u) = \hat{y}^k(u)$ a.s. which implies $x(t) = \hat{x}(t)$ a.s. since k is arbitrary. The proof is complete.

Now we introduce a new hypothesis:

(H2.8) For all $x, y \in R$ and $t \ge 0$

$$a(x, x, t) - 2a(x, y, t) + a(y, y, t) \le \rho(|x - y|^{2\alpha}) \qquad \text{a.s.}$$

where $1/2 \le \alpha \le 1$; $\rho : R_+ \to R_+$ is a continuous increasing concave function such that

$$\int_{0+} \frac{du}{\rho(u^{2\alpha})} = \infty,$$

and moreover, $u^{-1}\rho(u^{2\alpha}) \le \eta(u)$ and $\eta : R_+ \to R_+$ is another continuous increasing concave function with $\eta(0) = 0$.

Obviously (H2.8) is stronger than either (H2.6) or (H2.7). The examples for such ρ and η are as the follows.

Example 2.2 Let $\kappa(u)$ be any function defined in Example 2.1. Let $\rho(u) = \kappa(u)$ and $1/2 < \alpha \le 1$. Then

$$\int_{0+} \frac{du}{\rho(u^{2\alpha})} \ge \int_{0+} \frac{du}{\rho(u)} = \infty.$$

It is also easy to check $u^{-1}\rho(u^{2\alpha}) \le c\kappa(u^{2\alpha-1})$ for some $c > 0$. Set $\eta(u) = c\kappa(u^{2\alpha-1})$ which is a continuous increasing concave function with $\eta(0) = 0$. In other words, ρ and η satisfy the requirements of (H2.8).

Theorem 2.6 Let (H2.1)–(H2.3) and (H2.8) hold. Then the conclusions of Theorem 2.1 hold.

Proof. We only need to prove the theorem in the case $A(t) \equiv t$; the general case can be done in the same way as the proof of Theorem 2.1. By Burkholder–Davis–Gundy's inequality, Jensen's inequality and the hypotheses, we can derive that for any $T > 0$,

$$E\Big(\sup_{0\le t\le T} |x^m(t) - x^n(t)|\Big)$$

$$\le \int_0^T \kappa\Big(E|\tilde{x}^m(s) - \tilde{x}^n(s)|\Big)\, ds + 3E\Big\{\Big\langle \int_0^T \Big(M(\tilde{x}^m(s), ds) - M(\tilde{x}^n(s), ds)\Big) \Big\rangle\Big\}^{1/2},$$

where, and sometimes in the sequel of this book, we use the short notation $\langle N\rangle$ for $\langle N, N\rangle$. Note

$$3E\Big\{\Big\langle \int_0^T \Big(M(\tilde{x}^m(s), ds) - M(\tilde{x}^n(s), ds)\Big) \Big\rangle\Big\}^{1/2}$$

$$\le 6E\Big\{\Big\langle \int_0^T \Big(M(\tilde{x}^m(s), ds) - M(x^m(s), ds)\Big) \Big\rangle\Big\}^{1/2}$$

$$+ 6E\Big\{\Big\langle \int_0^T \Big(M(x^m(s), ds) - M(x^n(s), ds)\Big) \Big\rangle\Big\}^{1/2}$$

$$+ 6E\Big\{\Big\langle \int_0^T \Big(M(x^n(s), ds) - M(\tilde{x}^n(s), ds)\Big) \Big\rangle\Big\}^{1/2}$$

$$\le 6E\Big(\int_0^T \rho(|\tilde{x}^m(s) - x^m(s)|^{2\alpha})\, ds\Big)^{1/2} + 6E\Big(\int_0^T \rho(|x^n(s) - \tilde{x}^n(s)|^{2\alpha})\, ds\Big)^{1/2}$$

$$+ 6E\Big(\int_0^T \rho(|x^m(s) - x^n(s)|^{2\alpha})\, ds\Big)^{1/2}$$

$$\le 6\Big(\int_0^T E\rho(|\tilde{x}^m(s) - x^m(s)|^{2\alpha})\, ds\Big)^{1/2} + 6\Big(\int_0^T E\rho(|x^n(s) - \tilde{x}^n(s)|^{2\alpha})\, ds\Big)^{1/2}$$

$$+ 6E\Big(\int_0^T |x^m(s) - x^n(s)|\, \eta(|x^m(s) - x^n(s)|)\, ds\Big)^{1/2}$$

$$\leq 6\Big(\int_0^T \rho\big[E(|\tilde{x}^m(s) - x^m(s)|^{2\alpha})\big]\,ds\Big)^{1/2} + 6\Big(\int_0^T \rho\big[E(|x^n(s) - \tilde{x}^n(s)|^{2\alpha})\big]\,ds\Big)^{1/2}$$

$$+ 6E\Big\{\Big(\sup_{0\leq t\leq T}|x^m(t) - x^n(t)|\Big)\int_0^T \eta(|x^m(s) - x^n(s)|)\,ds\Big\}^{1/2}$$

$$\leq 6\Big(\int_0^T \rho\big[E(|\tilde{x}^m(s) - x^m(s)|^{2\alpha})\big]\,ds\Big)^{1/2} + 6\Big(\int_0^T \rho\big[E(|x^n(s) - \tilde{x}^n(s)|^{2\alpha})\big]\,ds\Big)^{1/2}$$

$$+ \frac{1}{2}E\Big(\sup_{0\leq t\leq T}|x^m(t) - x^n(t)|\Big) + 18\int_0^T \eta(E|x^m(s) - x^n(s)|)\,ds.$$

Consequently

$$E\Big(\sup_{0\leq t\leq T}|x^m(t) - x^n(t)|\Big)$$

$$\leq 2\int_0^T \kappa\Big(E|\tilde{x}^m(s) - \tilde{x}^n(s)|\Big)\,ds + 12\Big(\int_0^T \rho\big[E(|\tilde{x}^m(s) - x^m(s)|^{2\alpha})\big]\,ds\Big)^{1/2}$$

$$+ 12\Big(\int_0^T \rho\big[E(|x^n(s) - \tilde{x}^n(s)|^{2\alpha})\big]\,ds\Big)^{1/2} + 36\int_0^T \eta(E|x^m(s) - x^n(s)|)\,ds.$$

By Lemmas 2.2 and 2.3 we see that

$$E\Big(\sup_{0\leq t\leq T}|x^m(t) - x^n(t)|\Big) \to 0 \qquad \text{as } m, n \to \infty.$$

That is, $\{x^n(t): 0 \leq t \leq T\}$ is a Cauchy sequence in $L(\Omega; C([0, T], R))$. So there exists a continuous adapted process $x(t)$ on $t \geq 0$ such that $\{x(t): 0 \leq t \leq T\} \in L(\Omega; C([0, T], R))$,

$$\lim_{n\to\infty} E\Big(\sup_{0\leq t\leq T}|x(t) - x^n(t)|\Big) = 0 \quad \text{for any } T \geq 0.$$

Finally we show that $x(t)$ is a solution of equation (1.1). In fact, we see as before that

$$E\int_0^t |b(x(s), s) - b(\tilde{x}^n(s), s)|\,ds \to 0 \qquad \text{as } n \to \infty.$$

We also have

$$E\Big|\int_0^t M(x(s), ds) - M(\tilde{x}^n(s), ds)\Big|$$

$$\leq 3E\Big(\int_0^t 2\rho(|x(s) - x^n(s)|^{2\alpha}) + 2\rho(|x^n(s) - \tilde{x}^n(s)|^{2\alpha})\, ds \Big)^{1/2}$$

$$\leq 3E\Big(\int_0^t 2\rho(|x(s) - x^n(s)|^{2\alpha})\, ds \Big)^{1/2} + 3E\Big(\int_0^t 2\rho(|x^n(s) - \tilde{x}^n(s)|^{2\alpha})\, ds \Big)^{1/2}$$

$$\leq 6E\Big\{ \Big(\sup_{0\leq s\leq t} |x(s) - x^n(s)|\Big) \int_0^t \eta(|x(s) - x^n(s)|)ds\Big) \Big\}^{1/2}$$

$$+ 6\Big\{ \int_0^t \rho\Big(E\big[|x^n(s) - \tilde{x}^n(s)|^{2\alpha}\big]\Big)\, ds \Big\}^{1/2}$$

$$\leq 3E\Big(\sup_{0\leq s\leq t} |x(s) - x^n(s)|\Big) + 3\int_0^t \eta(E|x(s) - x^n(s)|)\, ds$$

$$+ 6\Big\{ \int_0^t \rho\Big(\big[E|x^n(s) - \tilde{x}^n(s)|^p\big]^{2\alpha/p}\Big)ds \Big\}^{1/2} \rightarrow 0 \quad \text{as } n \rightarrow 0.$$

Therefore, we can let $n \rightarrow \infty$ in (2.7) to obtain that

$$x(t) = \xi + \int_0^t b(x(s), s)\, ds + \int_0^t M(x(s), ds) \ = \xi + \int_0^t F(x(s), ds).$$

In other words, $x(t)$ is a solution of equation (2.1). The uniqueness follows from Lemma 2.5. The proof is complete.

Corollary 2.7 Let (H2.1) and (H2.2) hold. Assume that there exists a pair of constants $\alpha \in [1/2, 1]$ and $K > 0$ such that

$$|b(x, t) - b(y, t)| \leq K|x - y| \qquad \text{a.s.}$$

$$a(x, x, t) - 2a(x, y, t) + a(y, y, t) \leq K|x - y|^{2\alpha} \qquad \text{a.s.}$$

for all $x, y \in R$ and $t \geq 0$. Then the conclusions of Theorem 2.1 hold.

Obviously, if $\alpha = 1$, then the above condition reduces to the well-known Lipschitz condition. So our result is a generalization of the classical result. As an application of this corollary, we deduce an interesting result.

Corollary 2.8 Let (H2.1) hold and both b(x, t) and a(x, y, t) be bounded. Assume there exists a pair of constants $\alpha \geq 1/2$ and $K > 0$ such that

$$|b(x, t) - b(y, t)| \leq K|x - y| \quad \text{a.s.}$$

$$a(x, x, t) - 2a(x, y, t) + a(y, y, t) \leq K|x - y|^{2\alpha} \quad \text{a.s.}$$

for all $x, y \in R$ and $t \geq 0$. Then the conclusions of Theorem 2.1 hold.

Proof. We only need to prove the corollary in the case $\alpha > 1$, since it follows from Corollary 2.7 directly if $\alpha \in [1/2, 1]$. Let K_1 be the bound of a(x, y, t) and set $K_2 = K \vee 4K_1$. We claim that

$$a(x, x, t) - 2a(x, y, t) + a(y, y, t) \leq K_2|x - y|^2$$

for all $x, y \in R$ and $t \geq 0$. In fact, if $|x - y| \leq 1$, then

$$a(x, x, t) - 2a(x, y, t) + a(y, y, t) \leq K|x - y|^{2\alpha} \leq K_2|x - y|^2.$$

Otherwise, if $|x - y| > 1$, then

$$a(x, x, t) - 2a(x, y, t) + a(y, y, t) \leq 4K_1 \leq K_2|x - y|^2.$$

Therefore the conclusions follow from Corollary 2.7 again.

Example 2.3 Let us consider a classical Itô equation

$$dx(t) = g(x(t))\, dw(t) \quad \text{on } t \geq 0 \tag{2.17}$$

with initial condition $x(0) = x_o \in R$, where g is a real-valued function defined on R and w is a standard Brownian motion. Assume that g is not only bounded but also Hölder continuous of order $\alpha \geq 1/2$, i.e.

$$|g(x) - g(y)| \leq K|x - y|^{\alpha} \quad \text{for all } x, y \in R,$$

where K is a positive constant. Tanaka & Hasegawa (1964) and Skorohod (1965) independently proved the existence and uniqueness of the solution to equation (2.17) if $\alpha > 1/2$. Yamada & Watanabe (1971) strengthened the case and proved the same conclusion for $\alpha \geq 1/2$. However, it was treated by other methods rather than approximation procedure. It was only few years ago that Kaneko and Nakao (1988) showed that the Cauchy–Maruyama approximation converges to the solution. Now using our Corollary 2.8 we can easily obtain this result. In other words, if we define $x^n(t)$, $t \geq 0$ and $n \geq 1$ as follows:

$$x^n(0) = x_o \quad \text{and}$$
$$x^n(t) = x^n(i/n) + g(x^n(i/n))\,(w(t) - w(i/n))$$
$$\text{if } i/n < t \leq (i+1)/n, \ i = 0, 1, \ldots,$$

then $x^n(t)$ converges to the unique solution, say $x(t)$, of equation (2.17) in the sense that

$$\lim_{n\to\infty} E\Big(\sup_{0\leq t\leq T} |x(t) - x^n(t)| \Big) = 0 \quad \text{for any } T \geq 0.$$

As one more application of our results, let us consider an Itô equation

$$dx(t) = f(x(t), t)\,dt + \sum_{j=1}^{\infty} g_j(x(t), t)\,dw_j(t) \quad \text{on } t \geq 0 \tag{2.18}$$

with initial data $x(0) = \xi$, where $\{w_j(t): j \geq 1\}$ is a sequence of independent Brownian motions, and $f(x, t)$, $g_j(x, t)$ are all real-valued adapted processes with parameter $x \in R$. For this equation, we have the following result.

Corollary 2.9 Let (H2.1) hold. Assume there exists a pair of constants $\alpha \in [1/2, 1]$ and $K > 0$ such that

$$|f(0, t)|^2 + \sum_{j=1}^{\infty} |g_j(0, t)|^2 \leq K \quad \text{a.s.}$$

$$|f(x, t) - f(y, t)| \leq K|x - y| \quad \text{a.s}$$

$$\sum_{j=1}^{\infty} |g_j(x, t) - g_j(y, t)|^2 \leq K|x - y|^{2\alpha} \quad \text{a.s.}$$

for all $x, y \in R$ and $t \geq 0$. Then there exists a unique solution $x(t)$ to equation (2.18), and for any $T > 0$, $\{x(t): 0 \leq t \leq T\} \in L^p(\Omega; C([0, T], R))$. Moreover,

$$\lim_{n\to\infty} E\Big(\sup_{0\leq t\leq T} |x(t) - x^n(t)| \Big) = 0,$$

where $x^n(t)$, $t \geq 0$ and $n \geq 1$ are defined as follows: $x^n(0) = \xi$ and

$$x^n(t) = x^n(i/n) + \int_{i/n}^{t} f(x^n(i/n), s)\,ds + \sum_{j=1}^{\infty} \int_{i/n}^{t} g_j(x^n(i/n), s)\,dw_j(s)$$

if $i/n < t \leq (i+1)/n$, $i = 0, 1, \ldots$.

To close this section, let us make a comment. Suppose f and g_j are all time independent, i.e. $f(x, t) = f(x)$ and $g_j(x, t) = g_j(x)$. Then the above approximations reduce to

$$x^n(0) = \xi \quad \text{and}$$

$$x^n(t) = x^n(i/n) + f(x^n(i/n))\,(t - i/n) + \sum_{j=1}^{\infty} g_j(x^n(i/n))\,(w_j(t) - w_j(i/n))$$

$$\text{if } i/n < t \le (i+1)/n,\ i = 0, 1, \ldots,$$

which are the well-known Cauchy-Maruyama approximations (cf. Maruyama (1955) and McShane (1974)).

2.3. SPECIAL d-DIMENSIONAL STOCHASTIC DIFFERENTIAL EQUATIONS

The theory in Section 2.2 can be extended to a special multi-dimensional stochastic differential equation. Let $F(x, t)$, $x \in R^d$ be a d-dimensional continuous C-semimartingale with the local characteristic $(a(x, y, t), b(x, t), A(t))$ which belongs to the class $B^{0,\delta}$. Let $F(x, t) = B(x, t) + M(x, t)$ be the decomposition such that $B(x, t) = (B_1(x, t), \ldots, B_d(x, t))^T$ is a d-dimensional continuous process of finite variation and $M(x, t) = (M_1(x, t), \ldots, M_d(x, t))^T$ is a d-dimensional continuous local martingale. We assume in this section that

$$M_i(x, t) = M_i(x_i, t), \quad 1 \le i \le d,$$

where $x = (x_1, \ldots, x_d)^T$. So we have $a_{ij}(x, y, t) = a_{ij}(x_i, y_j, t)$ as well.

Consider a d-dimensional stochastic differential equation driven by $F(x, t)$ of the form

$$x(t) = \xi + \int_0^t F(x(s), ds) \qquad \text{on } t \ge 0 \tag{3.1}$$

where ξ is an $\mathcal{F}_0$-measurable R^d-valued random variable. We also introduce some hypotheses:

(H3.1) $p \ge 2$ and $E|\xi|^p < \infty$.

(H3.2) All $b_i(0, t)$ and $a_{ii}(0, 0, t)$ are bounded.

(H3.3) For all $x, y \in R^d$, $t \geq 0$ and $1 \leq i \leq d$,

$$|b_i(x, t) - b_i(y, t)| \leq \kappa(|x - y|) \qquad \text{a.s.}$$

where $\kappa : R_+ \to R_+$ is a continuous increasing concave function such that

$$\int_{0+} \frac{du}{\kappa(u)} = \infty.$$

(H3.4) For all $u, v \in R$, $t \geq 0$ and $1 \leq i \leq d$

$$a_{ii}(u, u, t) - 2a_{ii}(u, v, t) + a_{ii}(v, v, t) \leq \rho(|u - v|) \qquad \text{a.s.}$$

where $\rho : R_+ \to R_+$ is a continuous increasing concave function such that

$$\int_{0+} \frac{du}{\rho(u)} = \infty.$$

(H3.5) For all $u, v \in R$, $t \geq 0$ and $1 \leq i \leq d$

$$a_{ii}(u, u, t) - 2a_{ii}(u, v, t) + a_{ii}(v, v, t) \leq \rho(|u - v|^{2\alpha}) \qquad \text{a.s.}$$

where $\alpha \in (1/2, 1]$, and $\rho : R_+ \to R_+$ is a continuous increasing concave function such that

$$\int_{0+} \frac{du}{\rho(u^{2\alpha})} = \infty,$$

and moreover, $u^{-1}\rho(u^{2\alpha}) \leq \eta(u)$ and $\eta : R_+ \to R_+$ is another continuous increasing concave function with $\eta(0) = 0$.

The following theorems can be proved in the same way as Theorems 2.1 and 2.6 but we leave the details to the reader.

Theorem 3.1 Let (H3.1)-(H3.4) hold. Then there exists a unique solution $x(t)$ to equation (3.1), and for any $T > 0$,

$$E\left(\sup_{0 \leq t < \tau(T)} |x(t)|^p \right) < \infty.$$

Moreover

$$\lim_{n\to\infty} E\Big(\sup_{0\le t<\tau(T)} |x(t) - x^n(t)| \Big) = 0,$$

where $x^n(t)$, $t \ge 0$ and $n \ge 1$, are defined as follows:

$$x^n(0) = \xi \quad \text{and}$$

$$x^n(t) = x^n(\tau(i/n)) + F(x^n(\tau(i/n)), t) - F(x^n(\tau(i/n)), \tau(i/n))$$

$$\text{if } \tau(i/n) < t \le \tau((i+1)/n), \ \ i = 0, 1, \dots .$$

Theorem 3.2 The conclusions of Theorem 3.1 holds if (H3.4) is replaced by (H3.5).

As an application, let us consider a d-dimensional Itô equation

$$dx_i(t) = f_i(x(t), t)\, dt + \sum_{j=1}^{\infty} g_{ij}(x_i(t), t)\, dw_j(t) \quad \text{on } t \ge 0,\ \ 1 \le i \le d \tag{3.2}$$

with initial data $x(0) = \xi$, where $\{w_j(t): j \ge 1\}$ is a sequence of independent Brownian motions, $f_i(x, t)$ and $g_{ij}(u, t)$ are all real-valued adapted processes with parameters $x \in R^d$ and $u \in R$ respectively. For this equation, we have the following result.

Corollary 3.3 Let (H3.1) hold. Assume there exists a pair of constants $\alpha \in [1/2, 1]$ and $K > 0$ such that

$$|f_i(0, t)|^2 + \sum_{j=1}^{\infty} |g_{ij}(0, t)|^2 \le K \qquad \text{a.s.}$$

$$|f_i(x, t) - f_i(y, t)| \le K|x - y| \qquad \text{a.s.}$$

$$\sum_{j=1}^{\infty} |g_{ij}(u, t) - g_{ij}(v, t)|^2 \le K|u - v|^{2\alpha} \qquad \text{a.s.}$$

for all $x, y \in R^d$, $u, v \in R$, $t \ge 0$ and $1 \le i \le d$. Then there exists a unique solution $x(t)$ to equation (3.2), and for any $T > 0$, $\{x(t): 0 \le t \le T\} \in L^p(\Omega; C([0, T], R^d))$. Moreover,

$$\lim_{n\to\infty} E\Big(\sup_{0\le t\le T} |x(t) - x^n(t)| \Big) = 0,$$

where $x^n(t) = (x_1^n, \dots, x_d^n)^T$, $t \ge 0$ and $n \ge 1$ are defined as follows: $x^n(0) = \xi$ and

$$x_i^n(t) = x_i^n(k/n) + \int_{k/n}^{t} f_i(x^n(k/n), s)\, ds + \sum_{j=1}^{\infty} \int_{k/n}^{t} g_{ij}(x_i^n(k/n), s)\, dw_j(s)$$

if $k/n < t \le (k+1)/n$, $k = 0, 1, \dots,$ and $1 \le i \le d$.

2.4 GENERAL d-DIMENSIONAL STOCHASTIC DIFFERENTIAL EQUATIONS

In this section we shall discuss the existence and uniqueness of the solution for a general d-dimensional stochastic differential equation driven by a nonlinear integrator. Let $F(x, t)$, $x \in R^d$ be a d-dimensional continuous C-semimartingale with the local characteristic $(a(x, y, t), b(x, t), A(t))$ belonging to the class $B^{0,\delta}$. Consider a d-dimensional stochastic differential equation driven by $F(x, t)$ of the form

$$x(t) = \xi + \int_0^t F(x(s), ds) \qquad \text{on } t \geq 0, \tag{4.1}$$

where ξ is an $\mathcal{F}_0$-measurable R^d-valued random variable. We introduce hypotheses:

(H4.1) $p \geq 2$ and $E|\xi|^p < \infty$.

(H4.2) Both $b(0, t)$ and $a(0, 0, t)$ are bounded.

(H4.3) For all $x, y \in R^d$, $t \geq 0$,

$$|b(x, t) - b(y, t)|^2 \leq \kappa(|x - y|^2) \qquad \text{a.s.}$$

and

$$\sum_{i=1}^d \Big(a_{ii}(x, x, t) - 2a_{ii}(x, y, t) + a_{ii}(y, y, t)\Big) \leq \kappa(|x - y|^2) \qquad \text{a.s.}$$

where $\kappa : R_+ \to R_+$ is a continuous increasing concave function such that

$$\int_{0+} \frac{du}{\kappa(u)} = \infty.$$

Theorem 4.1 Let (H4.1)-(H4.3) hold. Then there exists a unique solution $x(t)$ to equation (4.1), and for any $T > 0$,

$$E\Big(\sup_{0 \leq t < \tau(T)} |x(t)|^p\Big) < \infty. \tag{4.2}$$

Moreover

$$\lim_{n \to \infty} E\Big(\sup_{0 \leq t < \tau(T)} |x(t) - x^n(t)|^2\Big) = 0, \tag{4.3}$$

where $x^n(t)$, $t \geq 0$ and $n \geq 1$, are defined as follows:

$$x^n(0) = \xi \quad \text{and}$$

$$x^n(t) = x^n(\tau(i/n)) + F(x^n(\tau(i/n)), t) - F(x^n(\tau(i/n)), \tau(i/n))$$

$$\text{if } \tau(i/n) < t \le \tau((i+1)/n), \ i = 0, 1, \dots . \tag{4.4}$$

Proof. As before we need only to prove the theorem in the case A(t) ≡ t. Again we define simple processes

$$\tilde{x}^n(t) = \sum_{i=0}^{\infty} x^n(i/n) \, \mathbb{1}_{[i/n,\,(i+1)/n)}(t) \quad \text{for } n = 1, 2, \dots .$$

Then

$$x^n(t) = \xi + \int_0^t F(\tilde{x}^n(s), ds) = \xi + \int_0^t b(\tilde{x}^n(s), s)\,ds + \int_0^t M(\tilde{x}^n(s), ds). \tag{4.5}$$

Let $T > 0$ be arbitrary. We first claim that there exists a $C = C(T) > 0$ such that

$$E|x^n(t) - \tilde{x}^n(t)|^p \le \frac{C}{n^{p/2}} \qquad \text{for all } 0 \le t \le T \text{ and } n \ge 1, \tag{4.6}$$

which can be proved in the same way as Lemma 2.2 since hypotheses (H4.2) and (H4.3) imply that b(x, t) and a(x, y, t) satisfy the linear growth condition. By Hölder's inequality and Doob's martingale inequality we can derive from (4.5) that if $m > n \ge 1$ and $0 \le t \le T$,

$$E\Big(\sup_{0\le s\le t} |x^m(s) - x^n(s)|^2\Big) \le 2TE\int_0^t |b(\tilde{x}^m(s), s) - b(\tilde{x}^n(s), s)|^2\,ds$$

$$+ 8E\sum_{i=1}^{d} < \int_0^t \Big(M_i(\tilde{x}^m(s), ds) - M_i(\tilde{x}^n(s), ds)\Big) >.$$

However

$$2TE\int_0^t |b(\tilde{x}^m(s), s) - b(\tilde{x}^n(s), s)|^2\,ds$$

$$\le 6TE\int_0^t \Big(|b(\tilde{x}^m(s), s) - b(x^m(s), s)|^2 + |b(x^m(s), s) - b(x^n(s), s)|^2$$

$$+ |b(x^n(s), s) - b(\tilde{x}^n(s), s)|^2\Big)\,ds$$

$$\le 6T\Big\{ \int_0^t \kappa\Big(E|\tilde{x}^m(s) - x^m(s)|^2\Big)\, ds + \int_0^t \kappa\Big(E|x^m(s) - x^n(s)|^2\Big)\, ds$$

$$+ \int_0^t \kappa\Big(E|x^n(s) - \tilde{x}^n(s)|^2\Big)\, ds \Big\}$$

$$\le 6T\Big\{ \int_0^t \kappa\Big(\big[E|\tilde{x}^m(s) - x^m(s)|^p\big]^{2/p}\Big)\, ds + \int_0^t \kappa\Big(E|x^m(s) - x^n(s)|^2\Big)\, ds$$

$$+ \int_0^t \kappa\Big(\big[E|x^n(s) - \tilde{x}^n(s)|^p\big]^{2/p}\Big)\, ds \Big\}$$

$$\le 6T^2 \Big\{ \kappa\Big(C^{2/p}/m\Big) + \kappa\Big(C^{2/p}/n\Big) \Big\} + 6T \int_0^t \kappa\Big(E|x^m(s) - x^n(s)|^2\Big)\, ds,$$

where (4.6) has been used. Also

$$8E \sum_{i=1}^d < \int_0^t \Big(M_i(\tilde{x}^m(s), ds) - M_i(\tilde{x}^n(s), ds)\Big) >$$

$$\le 24E \sum_{i=1}^d \int_0^t \Big(a_{ii}(\tilde{x}^m(s), \tilde{x}^m(s), s) - 2a_{ii}(\tilde{x}^m(s), x^m(s), s) + a_{ii}(x^m(s), x^m(s), s)$$

$$+ a_{ii}(x^m(s), x^m(s), s) - 2a_{ii}(x^m(s), x^n(s), s) + a_{ii}(x^n(s), x^n(s), s)$$

$$+ a_{ii}(x^n(s), x^n(s), s) - 2a_{ii}(x^n(s), \tilde{x}^n(s), s) + a_{ii}(\tilde{x}^n(s), \tilde{x}^n(s), s) \Big)\, ds$$

$$\le 24\Big\{ \int_0^t \kappa\Big(E|\tilde{x}^m(s) - x^m(s)|^2\Big)\, ds + \int_0^t \kappa\Big(E|x^m(s) - x^n(s)|^2\Big)\, ds$$

$$+ \int_0^t \kappa\Big(E|x^n(s) - \tilde{x}^n(s)|^2\Big)\, ds \Big\}$$

$$\le 24T \Big\{ \kappa\Big(C^{2/p}/m\Big) + \kappa\Big(C^{2/p}/n\Big) \Big\} + 24 \int_0^t \kappa\Big(E|x^m(s) - x^n(s)|^2\Big)\, ds.$$

Therefore we obtain that

$$E\Big(\sup_{0\le s\le t}|x^m(s)-x^n(s)|^2\Big)\le \delta_{m,n}+6(T+4)\int_0^t \kappa\Big(E|x^m(s)-x^n(s)|^2\Big)\,ds$$

$$\le \delta_{m,n}+6(T+4)\int_0^t \kappa\Big(E\Big[\sup_{0\le r\le s}|x^m(r)-x^n(r)|^2\Big]\Big)\,ds,$$

where

$$\delta_{m,n}=6T(T+4)\Big\{\kappa\Big(C^{2/p}/m\Big)+\kappa\Big(C^{2/p}/n\Big)\Big\}.$$

Using Bihari's inequality we then have

$$E\Big(\sup_{0\le t\le T}|x^m(t)-x^n(t)|^2\Big)\le Q^{-1}\Big(Q(\delta_{m,n})+6T(T+4)\Big) \tag{4.7}$$

provided $Q(\delta_{m,n})+6T(T+4)\in \mathrm{Dom}(Q^{-1})$, where

$$Q(u)=\int_1^u \frac{ds}{\kappa(s)} \qquad \text{for } u>0,$$

and Q^{-1} is the inverse function of Q. By the condition on κ we see $(-\infty, 0]\subset \mathrm{Dom}(Q^{-1})$, $Q(u)\to-\infty$ as $u\to 0$ and $Q^{-1}(u)\to 0$ as $u\to-\infty$. Also note $\delta_{m,n}\to 0$ as $m, n\to\infty$. So $Q(\delta_{m,n})\to-\infty$ as $m, n\to\infty$. Thus (4.7) gives

$$E\Big(\sup_{0\le t\le T}|x^m(t)-x^n(t)|^2\Big)\to 0 \qquad \text{as } m, n\to\infty.$$

In other words, $\{x^n(t): 0\le t\le T\}$ is a Cauchy sequence in $L^2(\Omega; C([0,T], R^d))$. Hence there exists a continuous process x(t) on $t\ge 0$ such that for any $T>0$, $\{x(t): 0\le t\le T\}\in L^2(\Omega; C([0,T], R^d))$ and

$$\lim_{n\to\infty} E\Big(\sup_{0\le t\le T}|x(t)-x^n(t)|^2\Big)=0.$$

It is also easy to see that x(t) is $\{\mathcal{F}_t\}$-adapted. We now show x(t) is a solution of equation (4.1). In fact,

$$E\Big|\int_0^t b(x(s),s)\,ds-\int_0^t b(\tilde{x}^n(s),s)\,ds\Big|^2$$

$$\le 2t\int_0^t \kappa\Big(E|x(s)-x^n(s)|^2\Big)\,ds+2t\int_0^t \kappa\Big(E|x^n(s)-\tilde{x}^n(s)|^2\Big)\,ds\to 0$$

as $n\to\infty$. Also

$$E\left|\int_0^t M(x(s), ds) - M(\tilde{x}^n(s), ds)\right|^2$$

$$\leq 2\int_0^t \kappa\left(E|x(s) - x^n(s)|^2\right) ds + 2\int_0^t \kappa\left(E|x^n(s) - \tilde{x}^n(s)|^2\right) ds \to 0.$$

Consequently, we can let $n \to \infty$ in (4.5) to obtain that

$$x(t) = \xi + \int_0^t b(x(s), s)\, ds + \int_0^t M(x(s), ds) = \xi + \int_0^t F(x(s), ds).$$

In other words, $x(t)$ is a solution of equation (4.1). It is then easy to show that

$$E\left(\sup_{0 \leq t \leq T} |x(t)|^p\right) < \infty \quad \text{for any } T > 0$$

since the hypotheses ensure the linear growth condition (cf. Lemma 2.3). Finally we need to show the uniqueness. Let $\hat{x}(t)$, $t \geq 0$ be another solution. Then we can derive that

$$E|x(t) - \hat{x}(t)|^2 \leq 2(t+1)\int_0^t \kappa\left(E|x(s) - \hat{x}(s)|^2\right) ds \quad \text{for all } t \geq 0,$$

which implies, by Bihari's inequality,

$$E|x(t) - \hat{x}(t)|^2 = 0 \quad \text{for all } t \geq 0.$$

So $x(t) = \hat{x}(t)$ a.s. for all $t \geq 0$. The proof is complete.

Remark 4.1 One can show that under the conditions of Theorem 4.1 the Picard approximations converge in L^2 to the solution of equation (4.1). In fact, Yamada (1981) proved this result for classical Itô equations. But, under the conditions of any other theorem in Sections 2.2–2.4, we do not know yet whether the Picard approximation procedure is applicable or not.

As a corollary we obtain the following classical result.

Corollary 4.2 Let (H4.1) and (H4.2) hold. Assume $b(x, t)$ and $a(x, y, t)$ satisfy the Lipschitz condition, i.e. there exists a $K > 0$ such that for all $x, y \in R^d$, $t \geq 0$,

$$|b(x, t) - b(y, t)| \leq K|x - y| \qquad \text{a.s.},$$

$$\sum_{i=1}^{d}\left(a_{ii}(x, x, t) - 2a_{ii}(x, y, t) + a_{ii}(y, y, t)\right) \le K|x - y|^2 \qquad \text{a.s.}$$

Then the conclusions of Theorem 4.1 hold.

Let us introduce a new hypothesis:

(H4.4) There exist three continuous increasing functions $\kappa(u)$, $\rho(u)$ and $\eta(u)$ on $u \ge 0$ such that $\kappa(0) = \rho(0) = \eta(0) = 0$, both $\kappa(u) + \rho(u)u^{-1}$ and $\eta(u)$ are concave,

$$\int_{0+}\left(\kappa(u) + \rho(u)u^{-1}\right)^{-1} du = \infty,$$

$$\rho(u) \le \eta(u^2) \quad \text{for } u \ge 0,$$

and moreover, for all $x, y \in R^d$ and $t \ge 0$,

$$|b(x, t) - b(y, t)| \le \kappa(|x - y|) \qquad \text{a.s.}$$

and

$$\sum_{i=1}^{d}\left(a_{ii}(x, x, t) - 2a_{ii}(x, y, t) + a_{ii}(y, y, t)\right) \le \rho(|x - y|) \qquad \text{a.s.}$$

Before we state a new result, let us give some examples for such functions κ, ρ and η. In fact, let $K > 0$ and define $\kappa(u) = Ku$, $\rho(u) = Ku^2$ and $\eta(u) = Ku$ for $u \ge 0$. Then they satisfy the requirements of (H4.4). Moreover, let $0 < \delta < e^{-1}$ and define for $u \ge 0$

$$\kappa(u) = u\left(\log\frac{1}{u}\right) \quad \text{if } 0 \le u \le \delta \text{ otherwise}$$

$$\kappa(u) = \delta\left(\log\frac{1}{\delta}\right) + \left(\log\frac{1}{\delta} - 1\right)(u - \delta),$$

$$\rho(u) = u\,\kappa(u),$$

$$\eta(u) = \left(\log\frac{1}{\delta}\right)\left(\log\frac{1}{\delta} - 1\right)^{-1}\kappa(u).$$

Then they also satisfy the requirements of (H4.4).

Theorem 4.3 Let (H4.1), (H4.2) and (H4.4) hold. Then there exists a unique solution $x(t)$ to equation (4.1), and for any $T > 0$,

$$E\Big(\sup_{0\le t<\tau(T)} |x(t)|^p\Big) < \infty. \tag{4.8}$$

Moreover

$$\lim_{n\to\infty} E\Big(\sup_{0\le t<\tau(T)} |x(t) - x^n(t)|\Big) = 0, \tag{4.9}$$

where $x^n(t)$ are defined in Theorem 4.1.

Proof. Again we need only to prove the theorem in the case $A(t) \equiv t$. Define $\tilde{x}^n(t)$ the same as in the proof of Theorem 4.1 so we have (4.5). Let $T > 0$ be arbitrary. We note that (4.6) still holds since hypotheses (H4.2) and (H4.4) ensure the linear growth condition.

Step 1. We first claim that

$$\sup_{0\le t\le T} E|x^m(t) - x^n(t)| \to 0 \quad \text{as } m, n \to \infty. \tag{4.10}$$

Let $a_0=1>a_1>a_2>\cdots>a_k>\cdots$ be positive constants defined by

$$\int_{a_k}^{a_{k-1}} \frac{u}{\rho(u)}\,du = 2, \qquad k = 1, 2, \dots .$$

Clearly $a_k \to 0$ as $k \to \infty$. For each $k \ge 1$, let $\psi_k(u)$ be a continuous function such that its support is contained in (a_k, a_{k-1}), and moreover,

$$0 \le \psi_k(u) \le \frac{u}{\rho(u)} \quad \text{and} \quad \int_{a_k}^{a_{k-1}} \psi_k(u)\,du = 1.$$

Such a function exists obviously. Set

$$\varphi_k(v) = \int_0^v dz \int_0^z \psi_k(u)\,du \qquad \text{for } v \ge 0.$$

It is easy to see that $\varphi_k \in C^2([0, \infty))$,

$$v - a_{k-1} \le \varphi_k(v) \le v \quad \text{for all } v \ge 0,$$

$$\varphi_k'(v) = \varphi_k''(v) = 0 \quad \text{if } v \notin (a_k, a_{k-1}),$$

$$0 \le \varphi_k'(v) \le 1 \quad \text{and} \quad 0 \le \varphi_k''(v) \le \frac{v}{\rho(v)} \qquad \text{if } a_k < v < a_{k-1}.$$

Let $\Phi_k(x) = \varphi_k(|x|)$ for $x \in R^d$ and it is continuously twice differentiable. Let $m > n \ge 1$.

Then Itô's formula implies that for $0 \le t \le T$,

$$E\Phi_k(x^m(t) - x^n(t))$$

$$= E\int_0^t \sum_{i=1}^d (\Phi_k)_{x_i}(x^m(s) - x^n(s)) \Big(b_i(\tilde{x}^m(s), s) - b_i(\tilde{x}^n(s), s) \Big) ds$$

$$+ \frac{1}{2} E\int_0^t \sum_{i,j=1}^d (\Phi_k)_{x_i x_j}(x^m(s) - x^n(s))\, d\langle N_i(s), N_j(s)\rangle$$

$$:= I_1(t) + I_2(t),$$

where

$$N_i(s) = \int_0^s \Big(M_i(\tilde{x}^m(r), dr) - M_i(\tilde{x}^n(r), dr)\Big).$$

Note

$$|(\Phi_k)_{x_i}(x)| \le 1,$$

$$(\Phi_k)_{x_i x_j}(x) = \varphi_k'(|x|)\,(\delta_{ij}|x|^2 - x_i x_j)\,|x|^{-3} + \varphi_k''(|x|)\, x_i x_j\, |x|^{-2},$$

and then

$$|(\Phi_k)_{x_i x_j}(x)| \le 2\Big(|x|^{-1} + \rho^{-1}(|x|)\,|x|\Big)\, \mathbb{1}_{\{a_k<|x|<a_{k-1}\}}.$$

Define

$$G(u) = d\Big(\kappa(u) + \rho(u)u^{-1}\Big) \quad \text{for } u \ge 0$$

which is concave and $G(0) = 0$ by the assumptions. So

$$I_1(t) \le E\int_0^t \sum_{i=1}^d |b_i(\tilde{x}^m(s), s) - b_i(\tilde{x}^n(s), s)|\, ds$$

$$\le dE\int_0^t \Big(\kappa(|\tilde{x}^m(s) - x^m(s)|) + \kappa(|x^m(s) - x^n(s)|) + \kappa(|x^n(s) - \tilde{x}^n(s)|) \Big) ds$$

$$\le \int_0^t G\Big(\big[E|\tilde{x}^m(s) - x^m(s)|^p\big]^{1/p}\Big)\,ds + dE\int_0^t \kappa\big(|x^m(s) - x^n(s)|\big)\,ds$$

$$+ \int_0^t G\Big(\big[E|x^n(s) - \tilde{x}^n(s)|^p\big]^{1/p}\Big)\,ds$$

$$\le T\Big\{ G\big(C^{1/p}/m^{1/2}\big) + G\big(C^{1/p}/n^{1/2}\big) \Big\} + dE\int_0^t \kappa\big(|x^m(s) - x^n(s)|\big)\,ds.$$

Furthermore,

$$I_2(t) \le \frac{1}{2} E \int_0^t \sum_{i,j=1}^d \Big(|x^m(s) - x^n(s)|^{-1} + \rho^{-1}(|x^m(s)-x^n(s)|)\,|x^m(s) - x^n(s)|\Big)$$

$$\times \mathbb{1}_{\{a_k<|x^m(s)-x^n(s)|<a_{k-1}\}} \Big(d\langle N_i(s), N_i(s)\rangle + d\langle N_j(s), N_j(s)\rangle \Big)$$

$$\le 3dE \int_0^t \sum_{i=1}^d \Big(|x^m(s) - x^n(s)|^{-1} + \rho^{-1}(|x^m(s) - x^n(s)|)\,|x^m(s) - x^n(s)|\Big)$$

$$\times \mathbb{1}_{\{a_k<|x^m(s)-x^n(s)|<a_{k-1}\}}$$

$$\times \Big(a_{ii}(\tilde{x}^m(s), \tilde{x}^m(s), s) - 2a_{ii}(\tilde{x}^m(s), x^m(s), s) + a_{ii}(x^m(s), x^m(s), s)$$

$$+ a_{ii}(x^m(s), x^m(s), s) - 2a_{ii}(x^m(s), x^n(s), s) + a_{ii}(x^n(s), x^n(s), s)$$

$$+ a_{ii}(x^n(s), x^n(s), s) - 2a_{ii}(x^n(s), \tilde{x}^n(s), s) + a_{ii}(\tilde{x}^n(s), \tilde{x}^n(s), s) \Big)\,ds$$

$$\le 3dE \int_0^t \Big(|x^m(s) - x^n(s)|^{-1} + \rho^{-1}(|x^m(s) - x^n(s)|)\,|x^m(s) - x^n(s)|\Big)$$

$$\times \mathbb{1}_{\{a_k<|x^m(s)-x^n(s)|<a_{k-1}\}}$$

$$\times \Big(\rho(|\tilde{x}^m(s) - x^m(s)|) + \rho(|x^m(s) - x^n(s)|) + \rho(|x^n(s) - \tilde{x}^n(s)|) \Big)\,ds$$

$$\le 3dE \int_0^t \Big(\rho(|x^m(s) - x^n(s)|)\,|x^m(s) - x^n(s)|^{-1} + a_{k-1}\Big)\,ds$$

$$+ 3dE \int_0^t \Big(\frac{1}{a_k} + \frac{a_{k-1}}{\rho(a_k)}\Big) \Big(\eta(|\tilde{x}^m(s) - x^m(s)|^2) + \eta(|x^n(s) - \tilde{x}^n(s)|^2) \Big)\,ds$$

$$\leq 3dE \int_0^t \rho(|x^m(s) - x^n(s)|)\, |x^m(s) - x^n(s)|^{-1}\, ds + 3dTa_{k-1}$$

$$+ 3dT\left(\frac{1}{a_k} + \frac{a_{k-1}}{\rho(a_k)}\right)\left\{\eta\left(C^{2/p}/m\right) + \eta\left(C^{2/p}/n\right)\right\}.$$

Combining the above inequalities we obtain that

$$E|x^m(t) - x^n(t)|$$

$$\leq a_{k-1} + T\left\{G\left(C^{1/p}/m^{1/2}\right) + G\left(C^{1/p}/n^{1/2}\right)\right\} + dE\int_0^t \kappa\left(|x^m(s) - x^n(s)|\right) ds$$

$$+ 3dE \int_0^t \rho(|x^m(s) - x^n(s)|)\, |x^m(s) - x^n(s)|^{-1}\, ds + 3dTa_{k-1}$$

$$+ 3dT\left(\frac{1}{a_k} + \frac{a_{k-1}}{\rho(a_k)}\right)\left\{\eta\left(C^{2/p}/m\right) + \eta\left(C^{2/p}/n\right)\right\}$$

$$\leq (3dT + 1)a_{k-1} + T\left\{G\left(C^{1/p}/m^{1/2}\right) + G\left(C^{1/p}/n^{1/2}\right)\right\}$$

$$+ 3dT\left(\frac{1}{a_k} + \frac{a_{k-1}}{\rho(a_k)}\right)\left\{\eta\left(C^{2/p}/m\right) + \eta\left(C^{2/p}/n\right)\right\}$$

$$+ 3\int_0^t G\left(E|x^m(s) - x^n(s)|\right) ds \tag{4.11}$$

for all $0 \leq t \leq T$. From this inequality we can deduce (4.10) in the same way as the proof of Lemma 2.4.

<u>Step 2</u>. We then show

$$E\left(\sup_{0\leq t\leq T} |x^m(t) - x^n(t)|\right) \to 0 \quad \text{as } m, n \to \infty. \tag{4.12}$$

In fact,

$$E\left(\sup_{0\leq t\leq T} |x^m(t) - x^n(t)|\right)$$

$$\leq E\int_0^T |b(\tilde{x}^m(s), s) - b(\tilde{x}^n(s), s)|\, ds$$

$$+ 3E\sum_{i=1}^{d} < \int_0^T \Big(M_i(\tilde{x}^m(s), ds) - M_i(\tilde{x}^n(s), ds)\Big) >^{1/2}$$

$$\leq E \int_0^T \Big(\kappa(|\tilde{x}^m(s) - x^m(s)|) + \kappa(|x^m(s) - x^n(s)|) + \kappa(|x^n(s) - \tilde{x}^n(s)|) \Big)\, ds$$

$$+ 3dE\Big\{ \int_0^T 3\Big(\rho(|\tilde{x}^m(s)-x^m(s)|) + \rho(|x^m(s)-x^n(s)|) + \rho(|x^n(s)-\tilde{x}^n(s)|) \Big)\, ds \Big\}^{1/2}$$

$$\leq T\Big\{ G\Big(C^{1/p}/m^{1/2}\Big) + G\Big(C^{1/p}/n^{1/2}\Big) \Big\} + E\int_0^T \kappa\Big(|x^m(s) - x^n(s)|\Big)\, ds$$

$$+ 6d\Big\{ \int_0^T \Big(\eta(E|\tilde{x}^m(s) - x^m(s)|^2) + \eta(E|x^n(s) - \tilde{x}^n(s)|^2) \Big)\, ds \Big\}^{1/2}$$

$$+ 6dE\Big\{ \Big(\sup_{0\leq t\leq T} |x^m(t) - x^n(t)|\Big) \int_0^T \rho(|x^m(s) - x^n(s)|)\, |x^m(s) - x^n(s)|^{-1}\, ds \Big\}^{1/2}$$

$$\leq T\Big\{ G\Big(C^{1/p}/m^{1/2}\Big) + G\Big(C^{1/p}/n^{1/2}\Big) \Big\} + E\int_0^T \kappa\Big(|x^m(s) - x^n(s)|\Big)\, ds$$

$$+ 6d\Big\{ T\eta\Big(C^{2/p}/m\Big) + T\eta\Big(C^{2/p}/n\Big) \Big\}^{1/2}$$

$$+ \frac{1}{2}\, E\Big(\sup_{0\leq t\leq T} |x^m(t) - x^n(t)|\Big) + 18d^2\, E\int_0^T \rho(|x^m(s)-x^n(s)|)\, |x^m(s)-x^n(s)|^{-1}\, ds$$

$$\leq T\Big\{G\Big(C^{1/p}/m^{1/2}\Big) + G\Big(C^{1/p}/n^{1/2}\Big)\Big\} + 6d\Big\{T\eta\Big(C^{2/p}/m\Big) + T\eta\Big(C^{2/p}/n\Big)\Big\}^{1/2}$$

$$+ \frac{1}{2}\, E\Big(\sup_{0\leq t\leq T} |x^m(t) - x^n(t)|\Big) + 18d\int_0^T G(E|x^m(s) - x^n(s)|)\, ds.$$

This, together with (4.11), implies (4.12).

Step 3. We now show that $x^n(t)$ converges to a solution of equation (4.1). As a matter of fact, step 2 shows that there exists a continuous $\{\mathcal{F}_t\}$-adapted process $x(t)$ on $t \geq 0$ such that, for any $T > 0$, $\{x(t): 0 \leq t \leq T\} \in L(\Omega; C([0, T], R^d))$ and

$$\lim_{n\to\infty} E\Big(\sup_{0\leq t\leq T} |x(t) - x^n(t)| \Big) = 0.$$

Note

$$E\left|\int_0^t b(x(s), s)\, ds - \int_0^t b(\tilde{x}^n(s), s)\, ds\right|$$

$$\le \int_0^t G\Big(E|x(s) - x^n(s)|^2\Big)\, ds + \int_0^t G\Big(E|x^n(s) - \tilde{x}^n(s)|^2\Big)\, ds \to 0$$

and

$$E\left|\int_0^t M(x(s), ds) - M(\tilde{x}^n(s), ds)\right|$$

$$\le 3dE\left\{\int_0^t 2\Big(\rho(|x(s) - x^n(s)|) + \rho(|x^n(s) - \tilde{x}^n(s)|)\Big)\, ds\right\}^{1/2}$$

$$\le 6dE\Big(\int_0^t \rho(|x(s) - x^n(s)|)\, ds\Big)^{1/2} + 6dE\Big(\int_0^t \eta(|x^n(s) - \tilde{x}^n(s)|^2)\, ds\Big)^{1/2}$$

$$\le 6dE\left\{\Big(\sup_{0\le s\le t} |x(s) - x^n(s)|\Big)\int_0^t G(|x(s) - x^n(s)|)ds\Big)\right\}^{1/2}$$

$$+ 6d\Big(\int_0^t \eta(E|x^n(s) - \tilde{x}^n(s)|^2)\, ds\Big)^{1/2}$$

$$\le 3dE\Big(\sup_{0\le s\le t} |x(s) - x^n(s)|\Big) + 3d\int_0^t G(E|x(s) - x^n(s)|)\, ds$$

$$+ 6d\Big(\int_0^t \eta(E|x^n(s) - \tilde{x}^n(s)|^2)\, ds\Big)^{1/2}$$

$$\to 0 \quad \text{as } n \to 0.$$

Consequently, we can let $n \to \infty$ in (4.5) to obtain that

$$x(t) = \xi + \int_0^t b(x(s), s)\, ds + \int_0^t M(x(s), ds) = \xi + \int_0^t F(x(s), ds).$$

In other words, $x(t)$ is a solution of equation (4.1). It is then easy to show that

$$E\Big(\sup_{0\le t\le T} |x(t)|^p\Big) < \infty \quad \text{for any } T > 0$$

since hypotheses (H4.2) and (H4.4) ensure the linear growth condition (cf. Lemma 2.3).

Step 4. Finally we need to show the uniqueness. Let $\hat{x}(t)$, $t \ge 0$ be another solution. Applying Itô's formula to $\Phi_k(x(t) - \hat{x}(t))$ we can derive in the same way as Step 1 that

$$E\Phi_k(x(t) - \hat{x}(t)) \le \int_0^t G\big(E|x(s) - \hat{x}(s)|\big)\, ds \qquad \text{for all } t \ge 0.$$

Letting $k \to \infty$ gives

$$E|x(t) - \hat{x}(t)| \le \int_0^t G\big(E|x(s) - \hat{x}(s)|\big)\, ds \qquad \text{for all } t \ge 0$$

which implies, by Bihari's inequality,

$$E|x(t) - \hat{x}(t)| = 0 \quad \text{for all } t \ge 0.$$

So $x(t) = \hat{x}(t)$ a.s. for all $t \ge 0$. The proof is complete.

As an application, let us consider a d-dimensional Itô equation

$$dx(t) = f(x(t), t)\, dt + \sum_{j=1}^{\infty} g_j(x(t), t)\, dw_j(t) \qquad \text{on } t \ge 0 \tag{4.13}$$

with initial data $x(0) = \xi$, where $\{w_j(t): j\ge1\}$ is a sequence of independent Brownian motions, and $f(x, t)$, $g_j(x, t)$ are all d-dimensional adapted processes with parameter $x \in R^d$. For this equation, we have the following corollaries which follow from Theorems 4.1 and 4.3 respectively.

Corollary 4.4 Let (H4.1) hold. Assume that for all $x, y \in R^d$ and $t \ge 0$,

$$|f(0, t)|^2 + \sum_{j=1}^{\infty} |g_j(0, t)|^2 \le K \quad \text{a.s.}$$

and

$$|f(x, t) - f(y, t)|^2 + \sum_{j=1}^{\infty} |g_j(x, t) - g_j(y, t)|^2 \le \kappa(|x - y|^2) \qquad \text{a.s.}$$

where K is a positive constant, and $\kappa : R_+ \to R_+$ is a continuous increasing concave function such that

$$\int_{0+} \frac{du}{\kappa(u)} = \infty.$$

Then there exists a unique solution x(t) to equation (4.13), and for any $T > 0$, $\{x(t): 0 \le t \le T\} \in L^p(\Omega; C([0, T], R^d))$. Moreover,

$$\lim_{n\to\infty} E\Big(\sup_{0\le t\le T} |x(t) - x^n(t)|^2 \Big) = 0,$$

where $x^n(t)$, $t \ge 0$ and $n \ge 1$ are defined as follows:

$$x^n(0) = \xi \quad \text{and}$$

$$x^n(t) = x^n(i/n) + \int_{i/n}^{t} f(x^n(i/n), s)\, ds + \sum_{j=1}^{\infty} \int_{i/n}^{t} g_j(x^n(i/n), s)\, dw_j(s)$$

$$\text{if } i/n < t \le (i+1)/n,\ i = 0, 1, \ldots .$$

Corollary 4.5 Let (H4.1) hold. Let $K > 0$ and

$$|f(0, t)|^2 + \sum_{j=1}^{\infty} |g_j(0, t)|^2 \le K \quad \text{a.s. for all } t \ge 0.$$

Assume furthermore that there exist three functions $\kappa(u)$, $\rho(u)$ and $\eta(u)$ on $u \ge 0$ such that $\kappa(0) = \rho(0) = \eta(0) = 0$, both $\kappa(u)+\rho(u)u^{-1}$ and $\eta(u)$ are concave,

$$\int_{0+} \Big(\kappa(u) + \rho(u)u^{-1}\Big)^{-1} du = \infty,$$

$$\rho(u) \le \eta(u^2) \quad \text{for } u \ge 0,$$

and moreover, for all $x, y \in R^d$ and $t \ge 0$,

$$|f(x, t) - f(y, t)| \le \kappa(|x - y|) \qquad \text{a.s.}$$

and

$$\sum_{j=1}^{\infty} |g_j(x, t) - g_j(y, t)|^2 \le \rho(|x - y|) \quad \text{a.s.}$$

Then there exists a unique solution x(t) to equation (4.13), and for any $T > 0$, $\{x(t): 0 \le t \le T\} \in L^p(\Omega, C([0, T], R^d))$. Moreover,

$$\lim_{n\to\infty} E\Big(\sup_{0\le t\le T} |x(t) - x^n(t)| \Big) = 0,$$

where $x^n(t)$ are defined in Corollary 4.4.

2.5. L^p-ESTIMATES OF SOLUTIONS

In the previous sections we established the theory of the existence and uniqueness of the solution for a stochastic differential equation driven by a nonlinear integrator. As a by-product we also obtained the L^p-estimate for the solution in fact. We shall now give a summary of this L^p-estimate and then establish some new L^p-estimates for the solution.

Let $F(x, t)$, $x \in R^d$ be a d-dimensional continuous C-semimartingale with the local characteristic $(a(x, y, t), b(x, t), A(t))$ belonging to the class $B^{0,\delta}$. Consider a d-dimensional stochastic differential equation driven by $F(x, t)$ of the form

$$x(t) = \xi + \int_0^t F(x(s), ds) \qquad \text{on } t \geq 0, \tag{5.1}$$

where ξ is an $\mathcal{F}_0$-measurable R^d-valued random variable. Throughout this section we assume that equation (5.1) has a unique solution $x(t)$ on $t \geq 0$. Also $\tau(u)$ still denote the inverse function of $A(t)$.

Theorem 5.1 Assume that $E|\xi|^p < \infty$ for some $p \geq 2$. Assume also that $a(x, x, t)$ and $b(x, t)$ satisfy the linear growth condition, i.e. there exists a constant $K > 0$ such that

$$|b(x, t)|^2 \vee \sum_{i=1}^{d} a_{ii}(x, x, t) \leq K(1 + |x|)^2$$

for all $x \in R^d$ and $t \geq 0$. Then for any $T > 0$,

$$E\Big(\sup_{0 \leq s < \tau(t)} |x(s)|^p\Big) \leq c_1 e^{c_2 t} \qquad \text{on } 0 \leq t \leq T, \tag{5.2}$$

where

$$c_1 = 3^{p-1} E|\xi|^p + c_2,$$

$$c_2 = 6^{p-1} K^{p/2} \Big(T^{p-1} + C_p (dT)^{(p/2-1)} \Big),$$

and C_p is the universal constant determined by Theorem 1-2.6.

This theorem has essentially been proved in the previous sections and therefore the details of the proof are left to the reader. An immediate application of this theorem gives the following useful estimate.

Corollary 5.2 In addition to the assumptions of Theorem 5.1, assume $A(t) \leq \mu(t)$

a.s. for all $t \geq 0$, where $\mu(t)$ is an increasing real-valued function defined on $t \geq 0$. Then for any $T > 0$,

$$E\Big(\sup_{0 \leq s \leq t} |x(s)|^p \Big) \leq c_1 \, e^{c_2 \mu(t)} \quad \text{on } 0 \leq t \leq T, \tag{5.3}$$

where

$$c_1 = 3^{p-1} \, E|\xi|^p + c_2,$$

$$c_2 = 6^{p-1} \, K^{p/2} \Big(\mu(T)^{p-1} + C_p \big[d\mu(T)\big]^{(p/2-1)} \Big),$$

and C_p is the universal constant determined by Theorem 1-2.6.

Proof. For $0 \leq t \leq T$,

$$E\Big(\sup_{0 \leq s \leq t} |x(s)|^p \Big) = E\Big(\sup_{0 \leq s \leq \tau(A(t))} |x(s)|^p \Big)$$

$$\leq E\Big(\sup_{0 \leq s < \tau(\mu(t))} |x(s)|^p \Big)$$

and then the required estimate (5.3) follows by applying Theorem 5.1. The proof is complete.

We now start to establish some more general L^p-estimates.

Theorem 5.3 Assume the following three conditions hold:

(i) $E|\xi|^p < \infty$ for some $p \geq 2$.

(ii) For all $x \in R^d$ and $t \geq 0$,

$$|b(x, t)|^p \vee \sum_{i=1}^{d} (a_{ii}(x, x, t))^{p/2} \leq \kappa(|x|^p),$$

where $\kappa : R_+ \to R_+$ is a continuous increasing concave function.

(iii) $A(t) \leq \mu(t)$ a.s. for all $t \geq 0$, where $\mu(t)$ is an increasing real-valued function defined on $t \geq 0$.

Then for any $T > 0$,

$$E\Big(\sup_{0 \leq s \leq T} |x(s)|^p \Big)$$

$$\leq G^{-1} \Big\{ G\Big(3^{p-1} \, E|\xi|^p\Big) + 3^{p-1} \, \mu(T) \Big(\mu(T)^{p-1} + C_p \big[d\mu(T)\big]^{(p/2-1)} \Big) \Big\} \tag{5.4}$$

where C_p is the universal constant determined by Theorem 1-2.6,

$$G(u) = \int_1^u \frac{dv}{\kappa(v)} \qquad \text{for } u > 0$$

and G^{-1} is the inverse function of G.

Proof. Using the hypotheses, Hölder's inequality and Burkholder–Davis–Gundy's, we can derive that for any stopping time τ with $0 \le \tau \le T$,

$$E\Big(\sup_{0\le s\le\tau} |x(s)|^p\Big) \le 3^{p-1} E|\xi|^p + 3^{p-1} E\Big(\int_0^\tau |b(x(s), s)|\, dA(s)\Big)^p$$

$$+ 3^{p-1} E\Big(\sup_{0\le s\le\tau} \Big|\int_0^s M(x(u), du)\Big|^p\Big)$$

$$\le 3^{p-1} E|\xi|^p + 3^{p-1} E\Big(A(\tau)^{p-1} \int_0^\tau |b(x(s), s)|^p\, dA(s)\Big)$$

$$+ 3^{p-1} d^{(p/2-1)} C_p \sum_{i=1}^d E\Big(\int_0^\tau a_{ii}(x(s), x(s), s)\, dA(s)\Big)^{p/2}$$

$$\le 3^{p-1} E|\xi|^p + 3^{p-1} \mu(T)^{p-1} E\int_0^t \kappa(|x(s)|^p)\, dA(s)$$

$$+ 3^{p-1} d^{(p/2-1)} C_p \sum_{i=1}^d E\Big(A(\tau)^{(p/2-1)} \int_0^\tau (a_{ii}(x(s), x(s), s))^{p/2}\, dA(s)\Big)$$

$$\le 3^{p-1} E|\xi|^p + 3^{p-1} \mu(T)^{p-1} E\int_0^\tau \kappa(|x(s)|^p)\, dA(s)$$

$$+ 3^{p-1} C_p \big[d\mu(T)\big]^{(p/2-1)} E\int_0^\tau \sum_{i=1}^d (a_{ii}(x(s), x(s), s))^{p/2}\, dA(s)$$

$$\le 3^{p-1} E|\xi|^p + 3^{p-1}\Big(\mu(T)^{p-1} + C_p\big[d\mu(T)\big]^{(p/2-1)}\Big)$$

$$\times E\int_0^\tau \kappa\Big(\sup_{0\le u\le s} |x(u)|^p\Big)\, dA(s)$$

An application of Theorem 1–7.12 gives the required inequality (5.4). The proof is complete.

The following estimate does not require condition (iii) of the previous theorem.

Theorem 5.4 Assume that only conditions (i) and (ii) of Theorem 5.3 hold. Then for any $T > 0$,

$$E\Big(\sup_{0\le s<\tau(T)} |x(s)|^p\Big) \le G^{-1}\Big\{G\Big(3^{p-1} E|\xi|^p\Big) + 3^{p-1}\, T\, \Big(T^{p-1} + C_p\, (dT)^{(p/2-1)}\Big)\Big\} \tag{5.5}$$

where C_p, G and G^{-1} are the same as in Theorem 5.3.

Proof. Set $\mathcal{G}_u = \mathcal{F}_{\tau(u)}$ for $u \ge 0$. For each $k = 1, 2, \ldots,$ define,

$$F^k(x, u) = F(x, \tau(u)\wedge k) \qquad \text{for } x \in R^d \text{ and } u \ge 0.$$

By Theorem 1–6.2, $F^k(x, u)$ is a continuous C–semimartingale with respect to $\{\mathcal{G}_u\}$ having the local characteristic $(a(x, y, \tau(u))\, \mathbb{1}_{\{u<A(k)\}}, b(x, \tau(u))\, \mathbb{1}_{\{u<A(k)\}}, u)$. Moreover, define

$$y^k(u) = x(\tau(u)\wedge k) \qquad \text{for } u \ge 0.$$

Then, by Theorem 1–6.3, one can derive that

$$y^k(u) = \xi + \int_0^{\tau(u)\wedge k} F(x(s), ds)$$

$$= \xi + \int_0^u F^k(y^k(s), ds).$$

Fix $T > 0$ arbitrarily. In the same way as the proof of Theorem 5.3 one can show that, for all $0 \le t \le T$,

$$E\Big(\sup_{0\le u\le t} |y^k(u)|^p\Big)$$

$$\le 3^{p-1} E|\xi|^p + 3^{p-1}\Big(T^{p-1} + C_p\, (dT)^{(p/2-1)}\Big)\, E\int_0^t \kappa\Big\{E\Big(\sup_{0\le u\le s} |y^k(u)|^p\Big)\Big\}\, ds.$$

This implies, by Theorem 1–7.2, that

$$E\Big(\sup_{0\le u\le T}|y^k(u)|^p\Big)$$
$$\le G^{-1}\Big\{G\Big(3^{p-1}E|\xi|^p\Big)+3^{p-1}T\Big(T^{p-1}+C_p(dT)^{(p/2-1)}\Big)\Big\}. \tag{5.6}$$

On the other hand, note that

$$\sup_{0\le s<\tau(T)}|x(s)|^p = \lim_{k\to\infty}\Big(\sup_{0\le s<\tau(T)\wedge k}|x(s)|^p\Big)$$
$$\le \limsup_{k\to\infty}\Big(\sup_{0\le u\le T}|x(\tau(u)\wedge k)|^p\Big)$$
$$= \limsup_{k\to\infty}\Big(\sup_{0\le u\le T}|y^k(u)|^p\Big).$$

By Fatou's lemma and inequality (5.6), we obtain that

$$E\Big(\sup_{0\le s<\tau(T)}|x(s)|^p\Big)\le \limsup_{k\to\infty}E\Big(\sup_{0\le u\le T}|y^k(u)|^p\Big)$$
$$\le G^{-1}\Big\{G\Big(3^{p-1}E|\xi|^p\Big)+3^{p-1}T\Big(T^{p-1}+C_p(dT)^{(p/2-1)}\Big)\Big\}.$$

as required. The proof is complete.

2.6. ALMOST SURELY ASYMPTOTIC BOUNDS OF SOLUTIONS

In this section we shall discuss the almost surely asymptotic bounds for the solution. We still let $F(x, t)$, $x \in R^d$ be a d-dimensional continuous C-semimartingale with the local characteristic $(a(x, y, t), b(x, t), A(t))$ belonging to the class $B^{0,\delta}$. Consider a stochastic differential equation

$$x(t) = \xi + \int_0^t F(x(t), dt) \quad \text{on } t \ge 0, \tag{6.1}$$

where ξ is an $\mathcal{F}_0$-measurable random vector. Assume the equation has a unique solution $x(t)$ on $t \ge 0$. For the convenience we introduce two operators L and H acting on C^2-functions:

$$LV(x) = \sum_{i=1}^d b_i(x, t)\frac{\partial}{\partial x_i}V(x) + \frac{1}{2}\sum_{i,j=1}^d a_{ij}(x, x, t)\frac{\partial^2}{\partial x_i \partial x_j}V(x),$$

$$HV(x) = \sum_{i,j=1}^{d} a_{ij}(x, x, t) \frac{\partial}{\partial x_i} V(x) \frac{\partial}{\partial x_j} V(x).$$

Theorem 6.1 Assume there exists a function $V \in C^2(R^d; R_+)$ and positive constants $c, \rho, K, \gamma, \eta, \mu_1, \mu_2$ such that

(i) $LV(x) \leq -cV(x) + \rho$,

(ii) $HV(x) \leq K\,V(x)$,

(iii) $\eta t^\gamma - \mu_1 \leq A(t) \leq \eta t^\gamma + \mu_2$.

Then

$$\limsup_{t\to\infty} \frac{V^*(x(t))}{\log t} \leq \frac{\gamma K}{2c} e^{c(\mu_1+\mu_2)+1} \quad \text{a.s.} \tag{6.2}$$

where $V^*(x(t)) = \sup_{0\leq s\leq t} V(x(s))$.

In order to prove this theorem we need to prepare two lemmas which in fact will be used very frequently in the sequel of this book.

Lemma 6.2 Let $N(t)$, $t \geq 0$ be a continuous (real-valued) local martingale vanishing at $t = 0$. Let $\{\tau_k\}$ and $\{\gamma_k\}$ be two sequences of positive numbers with $\tau_k \to \infty$. Let $g(t)$ be a positive increasing function on $t \geq 0$ such that

$$\sum_{k=1}^{\infty} (g(k))^{-\theta} < \infty \quad \text{for some } \theta > 1.$$

Then for almost all $\omega \in \Omega$ there exists a random integer $k_o(\omega)$ such that for all $k \geq k_o$

$$N(t) \leq \frac{1}{2} \gamma_k <N(t)> + \frac{\theta}{\gamma_k} \log(g(k)) \qquad \text{on } 0 \leq t \leq \tau_k.$$

Proof. By Theorem 1-2.4, we see that

$$P\Big(\omega : \sup_{0\leq t\leq \tau_k} \big(N(t) - \frac{1}{2} \gamma_k <N(t)> \big) > \frac{\theta}{\gamma_k} \log(g(k)) \Big) \leq (g(k))^{-\theta}.$$

for all $k \geq 1$. Hence the well-known Borel-Cantelli lemma yields that for almost all $\omega \in \Omega$ there exists a random integer $k_o(\omega)$ such that for all $k \geq k_o$

$$\sup_{0\leq t\leq \tau_k} \big(N(t) - \frac{1}{2} \gamma_k <N(t)> \big) \leq \frac{\theta}{\gamma_k} \log(g(k)),$$

that is,

$$N(t) \le \frac{1}{2}\,\gamma_k <N(t)> + \frac{\theta}{\gamma_k}\log(g(k)) \qquad \text{on } 0 \le t \le \tau_k$$

as required. The proof is complete.

Lemma 6.3 Let f(t), g(t) be two nonnegative real-valued functions defined on $t \ge 0$ with $g(t) \uparrow \infty$ as $t \to \infty$. Let

$$f^*(t) = \sup_{0 \le s \le t} f(s).$$

Then

$$\limsup_{t\to\infty}\left(f^*(t)/g(t)\right) = \limsup_{t\to\infty}\left(f(t)/g(t)\right).$$

Proof. If the right hand side equals ∞, the equality holds. Otherwise set

$$\sigma = \limsup_{t\to\infty}\left(f(t)/g(t)\right).$$

Assign $\varepsilon > 0$ arbitrarily. Then there exists a $T = T(\varepsilon) > 0$ such that

$$f(t)/g(t) \le \sigma + \varepsilon \quad \text{for all } t \ge T.$$

Therefore

$$f^*(t) \le f^*(T) + \sup_{T \le s \le t} f(s) \le f^*(T) + g(t)(\sigma + \varepsilon)$$

for all $t \ge T$. Hence

$$\limsup_{t\to\infty}\left(f^*(t)/g(t)\right) \le \sigma + \varepsilon.$$

Since $\varepsilon > 0$ is arbitrary,

$$\limsup_{t\to\infty}\left(f^*(t)/g(t)\right) \le \limsup_{t\to\infty}\left(f(t)/g(t)\right).$$

But $f^*(t) \ge f(t)$ and so the equality must hold, and the proof is complete.

Proof of Theorem 6.1. By Itô's formula and condition (i) we see that

$$e^{cA(t)}\,V(x(t))$$

$$= V(\xi) + \int_0^t e^{cA(s)}\left(cV(x(s)) + LV(x(s))\right) dA(s) + N(t)$$

$$\leq V(\xi) + (\rho/c)e^{cA(t)} + N(t) \qquad \text{for all } t \geq 0, \tag{6.3}$$

where

$$N(t) := \int_0^t e^{cA(s)} \sum_{i=1}^d \frac{\partial}{\partial x_i} V(x(s))\, M^i(x(s), ds),$$

which is a continuous local martingale. Assign $\alpha > 0$, $\beta > 0$ and $\theta > 1$ arbitrarily, and let $k = 1, 2, \ldots$. In view of Lemma 6.2 with $\tau_k = (k\alpha)^{1/\gamma}$, $\gamma_k = \beta e^{-c\eta k\alpha}$ and $g(t) \equiv t$ we see that for almost all $\omega \in \Omega$ there exists a random integer $k_o(\omega)$ such that for all $k \geq k_o$

$$N(t) \leq \frac{1}{2} \beta e^{-c\eta k\alpha} \langle N(t) \rangle + \theta\beta^{-1} e^{c\eta k\alpha} \log k$$

on $0 \leq t \leq (k\alpha)^{1/\gamma}$. Note from condition (ii) that

$$\langle N(t) \rangle = \int_0^t e^{2cA(s)}\, HV(x(s))\, dA(s)$$

$$\leq \int_0^t K e^{2cA(s)}\, V(x(s))\, dA(s).$$

We obtain that

$$N(t) \leq \frac{1}{2} \beta K e^{-c\eta k\alpha} \int_0^t K e^{2cA(s)}\, V(x(s))\, dA(s) + \theta\beta^{-1} e^{c\eta k\alpha} \log k$$

for all $0 \leq t \leq (k\alpha)^{1/\gamma}$, $k \geq k_o$ almost surely. Substituting this into (6.3) gives that

$$e^{cA(t)} V(x(t)) \leq V(\xi) + (\rho/c)e^{cA(t)} + \theta\beta^{-1} e^{c\eta k\alpha} \log k$$
$$+ \frac{1}{2} \beta K\, e^{-c\eta k\alpha} \int_0^t e^{2cA(s)}\, V(x(s))\, dA(s)$$

for all $0 \leq t \leq (k\alpha)^{1/\gamma}$, $k \geq k_o$ almost surely. By Theorem 1-7.1 we get

$$e^{cA(t)} V(x(t)) \leq \Big(V(\xi) + (\rho/c)e^{cA(t)} + \theta\beta^{-1} e^{c\eta k\alpha} \log k \Big)$$
$$\times \exp\Big(\frac{1}{2} (\beta K/c)\, e^{-c\eta k\alpha + cA(t)} \Big)$$

for all $0 \leq t \leq (k\alpha)^{1/\gamma}$, $k \geq k_o$ almost surely. Consequently, if $((k-1)\alpha)^{1/\gamma} \leq t \leq (k\alpha)^{1/\gamma}$ and $k \geq k_o$, we have

$$V(x(t)) \le \Big(V(\xi) + \rho/c + \theta\beta^{-1}e^{c\eta\alpha+c\mu_1} \log k \Big)$$
$$\times \exp\Big(\frac{1}{2} (\beta K/c)\, e^{c\mu_2} \Big) \qquad \text{a.s.}$$

where condition (iii) has been used. This implies

$$\limsup_{t\to\infty} \frac{V(x(t))}{\log t}$$
$$\le \limsup_{k\to\infty} \Big(\log[(k-1)\alpha)]^{1/\gamma} \Big)^{-1}$$
$$\times \Big(V(\xi) + \rho/c + \theta\beta^{-1}e^{c\eta\alpha+c\mu_1} \log k \Big) \exp\Big(\frac{1}{2} (\beta K/c)\, e^{c\mu_2} \Big)$$
$$\le \gamma\theta\beta^{-1} e^{c\eta\alpha+c\mu_1} \exp\Big(\frac{1}{2} (\beta K/c)\, e^{c\mu_2} \Big) \qquad \text{a.s.}$$

Letting $\theta \to 1$, $\alpha \to 0$ gives

$$\limsup_{t\to\infty} \frac{V(x(t))}{\log t} \le \gamma\beta^{-1} e^{c\mu_1} \exp\Big(\frac{1}{2} (\beta K/c)\, e^{c\mu_2} \Big) \qquad \text{a.s.}$$

Choosing $\beta^{-1} = (K/2c)\, e^{c\mu_2}$ we then obtain that

$$\limsup_{t\to\infty} \frac{V(x(t))}{\log t} \le \frac{\gamma K}{2c}\, e^{c(\mu_1+\mu_2)+1} \qquad \text{a.s.}$$

Finally, the required inequality (6.2) follows from Lemma 6.3.

Theorem 6.4 Assume there exists a function $V \in C^2(R^d; R_+)$ and positive constants ρ, K, γ, η such that

(i) $LV(x) \le \rho$,

(ii) $HV(x) \le K\, V(x)$,

(iii) $\limsup_{t\to\infty} \Big(A(t)/t^\gamma \Big) \le \eta$ a.s.

Then

$$\limsup_{t\to\infty} \frac{V^*(x(t))}{t^\gamma \log\log t} \le \frac{1}{2} K\eta e \qquad \text{a.s.} \tag{6.4}$$

Proof. By Itô's formula and condition (i) we see that

$$V(x(t)) \le V(\xi) + \rho A(t) + N(t) \qquad \text{for all } t \ge 0,$$

where

$$N(t) = \int_0^t \sum_{i=1}^d \frac{\partial}{\partial x_i} V(x(s))\, M^i(x(s), ds).$$

Assign $\beta > 0$ and $\theta > 1$ arbitrarily and let $k = 1, 2, \ldots$. Applying Lemma 6.2 with $\tau_k = \theta^{k/\gamma}$, $\gamma_k = \beta\theta^{-k}$ and $g(t) \equiv t$ we deduce that for almost all $\omega \in \Omega$ there exists a random integer $k_o(\omega)$ such that for all $0 \le t \le \theta^{k/\gamma}$ and $k \ge k_o$

$$N(t) \le \frac{1}{2}\beta\theta^{-k} \langle N(t) \rangle + \beta^{-1}\theta^{k+1} \log k$$

$$\le \frac{1}{2}\beta K\, \theta^{-k} \int_0^t V(x(s))\, dA(s) + \beta^{-1}\theta^{k+1} \log k,$$

where condition (ii) has been used. Therefore

$$V(x(t)) \le V(\xi) + \rho A(t) + \beta^{-1}\theta^{k+1} \log k + \frac{1}{2}\beta K\, \theta^{-k} \int_0^t V(x(s))\, dA(s)$$

for all $0 \le t \le \theta^{k/\gamma}$, $k \ge k_o$ almost surely. So

$$V(x(t)) \le \Big(V(\xi) + \rho A(t) + \beta^{-1}\theta^{k+1} \log k \Big) \times \exp\Big(\frac{1}{2}\beta K\, \theta^{-k} A(t) \Big) \tag{6.5}$$

for all $0 \le t \le \theta^{k/\gamma}$, $k \ge k_o$ almost surely. Note from condition (iii) that for any $\varepsilon > 0$ there exists a random $T = T(\varepsilon, \omega) > 0$, for almost all $\omega \in \Omega$, such that

$$A(t) \le (\eta + \varepsilon)t^\gamma \qquad \text{for all } t \ge T.$$

Hence, if $k \ge k_o$ is so large that $\theta^{(k-1)/\gamma} \ge T$ and $\theta^{(k-1)/\gamma} \le t \le \theta^{k/\gamma}$, then (6.5) implies

$$\frac{V(x(t))}{t^\gamma \log\log t} \le \Big(V(\xi) + \rho(\eta + \varepsilon)\theta^k + \beta^{-1}\theta^{k+1} \log k \Big) \exp\Big(\frac{1}{2}\beta K(\eta + \varepsilon) \Big) \times \Big(\theta^{k-1}(\log(k-1) - \log\gamma + \log\log\theta) \Big)^{-1} \quad \text{a.s.}$$

which implies that

$$\limsup_{t\to\infty} \frac{V(x(t))}{t^{\gamma}\log\log t} \le \beta^{-1}\theta^2 \exp\Big(\frac{1}{2}\beta K(\eta+\varepsilon)\Big) \quad \text{a.s.}$$

Letting $\theta \to 1$, $\varepsilon \to 0$ and choosing $\beta^{-1} = K\eta/2$ we obtain

$$\limsup_{t\to\infty} \frac{V(x(t))}{t^{\gamma}\log\log t} \le \frac{1}{2}K\eta e \quad \text{a.s.}$$

Then the required inequality (6.4) follows from Lemma 6.3. The proof is complete.

Theorem 6.5 Assume there exists a function $V \in C^2(R^d; R_+)$ and positive constants c, ρ, K such that

$$LV(x) \le cV(x) + \rho \quad \text{and} \quad HV(x) \le K\,V(x).$$

Then

$$\lim_{t\to\infty} \frac{V^*(x(t))}{e^{cA(t)}\log\log(g(t))} = 0 \quad \text{a.s.} \tag{6.6}$$

where g(t) is any positive increasing function defined on $t \ge 0$ which has the properties:

$$\sum_{k=1}^{\infty} (g(k))^{-\theta} < \infty \quad \text{for some } \theta > 1$$

and

$$\limsup_{k\to\infty} \frac{\log(g(k))}{\log(g(k-1))} < \infty.$$

In particular, we have

$$\lim_{t\to\infty} \frac{V^*(x(t))}{e^{cA(t)}\log\log t} = 0 \quad \text{a.s.} \tag{6.7}$$

if we choose $g(t) \equiv t$.

Proof. By Itô's formula and the hypothesis we see that

$$\begin{aligned} & e^{-cA(t)}V(x(t)) \\ &= V(\xi) + \int_0^t e^{-cA(s)}\Big(-cV(x(s)) + LV(x(s))\Big)\,dA(s) + N(t) \\ &\le V(\xi) + \rho/c + N(t) \end{aligned} \tag{6.8}$$

for all $t \geq 0$, where

$$N(t) := \int_0^t e^{-cA(s)} \sum_{i=1}^{d} \frac{\partial}{\partial x_i} V(x(s))\, M^i(x(s), ds),$$

Assign $p \geq 1$ arbitrarily and let $k = 1, 2, \ldots$. Using Lemma 6.2 with $\tau_k = 2^{g^p(k)}$ and $\gamma_k = 1$ we see that for almost all $\omega \in \Omega$ there exists a random integer $k_o(\omega)$ such that for all $k \geq k_o$

$$N(t) \leq \frac{1}{2} \langle N(t) \rangle + \theta \log(g(k))$$

on $0 \leq t \leq 2^{g^p(k)}$. Note from the hypothesis that

$$\langle N(t) \rangle = \int_0^t e^{-2cA(s)}\, HV(x(s))\, dA(s)$$

$$\leq \int_0^t K e^{-2cA(s)}\, V(x(s))\, dA(s).$$

It then follows

$$N(t) \leq \frac{1}{2} K \int_0^t e^{-2cA(s)}\, V(x(s))\, dA(s) + \theta \log(g(k))$$

for all $0 \leq t \leq 2^{g^p(k)}$, $k \geq k_o$ almost surely. Substituting this into (6.8) yields

$$e^{-cA(t)} V(x(t)) \leq V(\xi) + \rho/c + \theta \log(g(k)) + \frac{1}{2} K \int_0^t e^{-2cA(s)}\, V(x(s))\, dA(s).$$

for all $0 \leq t \leq 2^{g^p(k)}$, $k \geq k_o$ almost surely. Thus, by Theorem 1-7.1

$$e^{-cA(t)} V(x(t)) \leq \Big(V(\xi) + \rho/c + \theta \log(g(k)) \Big) \exp(K/2c)$$

for all $0 \leq t \leq 2^{g^p(k)}$, $k \geq k_o$ almost surely. Thus, if $2^{g^p(k-1)} \leq t \leq 2^{g^p(k)}$, $k \geq k_o$

$$\frac{V(x(t))}{e^{cA(t)} \log\log(g(t))} \leq \Big(V(\xi) + \rho/c + \theta \log(g(k)) \Big) \times \exp(K/2c) \Big(p \log(g(k-1)) + \log\log 2 \Big)^{-1},$$

which implies

$$\limsup_{t\to\infty} \frac{V(x(t))}{e^{cA(t)}\log\log(g(t))} \le \theta\sigma p^{-1} \exp(K/2c) \qquad \text{a.s.}$$

where

$$\sigma = \limsup_{k\to\infty} \frac{\log(g(k))}{\log(g(k-1))} .$$

Since $p > 1$ is arbitrary,

$$\lim_{t\to\infty} \frac{V(x(t))}{e^{cA(t)}\log\log(g(t))} = 0 \qquad \text{a.s.}$$

and then (6.6) follows from Lemma 6.3. Finally, let $g(t) \equiv t$ and then (6.6) reduces to (6.7). The proof is complete.

Let us now give several examples to illustrate the theory and to close this section as well.

Example 6.1 Consider a d-dimensional Itô equation

$$dx(t) = f(x(t), t)\, dt + \sum_{k=1}^{\infty} g_k(x(t), t)\, dw_k(t) \quad \text{on } t \ge 0, \tag{6.9}$$

with initial data $x(0) = \xi$, where $\{w_k: k \ge 1\}$ is a sequence of independent Brownian motions and $f = (f_1, ..., f_d)^T$, $g_k = (g_{k1}, ..., g_{kd})^T$ are all R^d-valued functions defined on $R^d \times R_+$. Suppose that the equation has a unique solution $x(t)$ on $t \ge 0$. Assume

$$\sum_{k=1}^{\infty} |g_k(x, t)|^2 \le K < \infty \quad \text{for all } (x, t) \in R^d \times R_+. \tag{6.10}$$

We also assume that

$$x^T f(x, t) \le -c|x|^2 + \rho \quad \text{for all } (x, t) \in R^d \times R_+, \tag{6.11}$$

where $c > 0$ and $\rho \ge 0$. Note

$$L|x|^2 = 2x^T f(x, t) + \sum_{k=1}^{\infty} |g_k(x, t)|^2 \le -2c|x|^2 + \rho + K,$$

and

$$H|x|^2 = \sum_{i,j=1}^{d} 4x_i x_j \sum_{k=1}^{\infty} g_{ki}(x, t)\, g_{kj}(x, t)$$

$$\le 2 \sum_{k=1}^{\infty} \sum_{i,j=1}^{d} \Big((x_i g_{kj}(x, t))^2 + (x_j g_{ki}(x, t))^2 \Big)$$

$$= 4 \sum_{k=1}^{\infty} |x|^2 \, |g_k(x, t)|^2 \le 4K|x|^2.$$

Hence, by Theorem 6.1, we have

$$\limsup_{t\to\infty} \frac{1}{\log t} \Big(\sup_{0\le s\le t} |x(s)|^2 \Big) \le Ke/c \qquad \text{a.s.} \tag{6.12}$$

If (6.11) is now replaced by a weaker condition

$$x^T f(x, t) \le \rho \quad \text{for all } (x, t) \in R^d \times R_+, \tag{6.13}$$

where $\rho \ge 0$, then in view of Theorem 6.4 we see

$$\limsup_{t\to\infty} \frac{1}{t \log\log t} \Big(\sup_{0\le s\le t} |x(s)|^2 \Big) \le 2Ke \qquad \text{a.s.} \tag{6.14}$$

Furthermore, if (6.11) is replaced by an even weaker condition

$$x^T f(x, t) \le c|x|^2 + \rho \quad \text{for all } (x, t) \in R^d \times R_+, \tag{6.15}$$

where $c > 0$ and $\rho \ge 0$, then Theorem 6.5 implies

$$\lim_{t\to\infty} \frac{1}{e^{2ct} \log\log t} \Big(\sup_{0\le s\le t} |x(s)|^2 \Big) = 0 \qquad \text{a.s.}$$

Example 6.2 Consider a d-dimensional nonlinear stochastic oscillator

$$\ddot{x} + f(x, \dot{x}) + \nabla G(x) = \sum_{k=1}^{\infty} g_k(x, \dot{x}, t)\, \dot{w}_k(t) + e(t) \quad \text{on } t \ge 0 \tag{6.16}$$

with suitable initial data, where $-\nabla G$ is a restoring force, $-f(x, \dot{x})$ is a dissipative force, $e(t)$ is an external driving force, and g_k represents the intensity of the stochastic disturbance based on the white noise $\dot{w}_k$. The corresponding Itô equation is

$$dx(t) = y(t)\, dt$$

$$dy(t) = \Big(-f(x(t), y(t)) - \nabla G(x(t)) + e(t)\Big)\, dt + \sum_{k=1}^{\infty} g_k(x(t), y(t), t)\, dw_k(t).$$

Define the energy of the system

$$U(t) = \frac{1}{2}|y(t)|^2 + G(x(t)).$$

We first assume that for all x, y $\in R^d$ and $t \ge 0$,

$$G(x) \ge 0, \quad y^T f(x, y, t) \ge -\rho, \quad e(t) \equiv 0$$

$$\text{and} \sum_{k=1}^{\infty} |g_k(x, y, t)|^2 \le K. \tag{6.17}$$

where ρ and K are positive constants. Introduce $V(x, y) = |y|^2/2 + G(x)$. Note

$$LV(x, y) = y^T\Big(-f(x, y) - \nabla G(x) + e(t)\Big) + y^T \nabla G(x)$$
$$+ \frac{1}{2}\sum_{k=1}^{\infty} |g_k(x, y, t)|^2 \le \rho + K/2,$$

and

$$HV(x, y) \le \sum_{k=1}^{\infty} |y|^2 |g_k(x, y, t)|^2 \le 2KV(x, y).$$

Then by Theorem 6.3 we obtain the following energy bound

$$\limsup_{t\to\infty} \frac{1}{t\log\log t}\Big(\sup_{0\le s\le t} U(s)\Big) \le Ke \quad \text{a.s.} \tag{6.18}$$

Now introduce weaker hypotheses: for all x, y $\in R^d$ and $t \ge 0$,

$$G(x) \ge 0, \quad y^T f(x, y, t) \ge -\rho - c_1 V(x, y), \quad |e(t)| \le c_2$$

$$\text{and} \sum_{k=1}^{\infty} |g_k(x, y, t)|^2 \le K. \tag{6.19}$$

where ρ, K, c_1, c_2 are all positive constants. Note under these conditions that

$$LV(x, y) \le \rho + c_1 V(x, y) + |e(t)|\,(1 + V(x, y)) + \frac{1}{2}\sum_{k=1}^{\infty} |g_k(x, y, t)|^2$$
$$= \rho + c_2 + K/2 + (c_1 + c_2)\, V(x, y)$$

and

$$HV(x, y) \le 2KV(x, y).$$

So by Theorem 6.4 we have

$$\lim_{t\to\infty} \frac{1}{e^{(c_1+c_2)t}\,\mathrm{loglog}t}\left(\sup_{0\le s\le t} U(s)\right) = 0 \qquad \text{a.s.} \tag{6.20}$$

Example 6.3 We now come back to equation (6.1). Assume the local characteristic $(a(x, y, t), b(x, t), A(t))$ of $F(x, t)$ satisfies the following two conditions

$$|a_{ij}(x, x, t)| \le K \qquad \text{and} \qquad x^T b(x, t) \le c|x|^2 \tag{6.21}$$

for all $x \in R^d$, $t \ge 0$ and $1 \le i, j \le d$, where $K > 0$ and $c > 0$. Note that

$$L|x|^2 = 2x^T b(x, t) + \sum_{i=1}^{d} a_{ii}(x, x, t) \le 2c|x|^2 + dK,$$

and

$$H|x|^2 = \sum_{i,j=1}^{d} 4x_i x_j\, a_{ij}(x, x, t) \le 4dK|x|^2.$$

Hence, by Theorem 6.5, the solution of equation (6.1) satisfies

$$\lim_{t\to\infty} \frac{1}{e^{2cA(t)}\,\mathrm{loglog}t}\left(\sup_{0\le s\le t} |x(s)|^2\right) = 0 \qquad \text{a.s.} \tag{6.22}$$

Although the lower bounds for the solutions are not presented in this section, it should be pointed out that there exist examples such that the limits in Theorems 6.1 and 6.4 are positive. As a matter of fact, the following is an example which satisfies the hypotheses of Theorem 6.1 and the corresponding limit is positive.

Example 6.4 Let us finally consider a simple one-dimensional equation

$$dx(t) = -\,cx(t)\,dt + K\,dw(t) \qquad \text{on } t \ge 0 \tag{6.23}$$

with initial data $x(0) = x_o \in R$, where $w(t)$ is a one-dimensional Brownian motion, $c > 0$ and $K > 0$. It is easy to show that

$$L|x|^2 = -\,2c|x|^2 + K^2,$$

and

$$H|x|^2 = 4K^2|x|^2.$$

Hence, by Theorem 6.1, we can have

$$\limsup_{t\to\infty} \frac{1}{\log t} \Big(\sup_{0\le s\le t} |x(s)|^2 \Big) \le K^2 e/c \qquad \text{a.s.}$$

On the other hand, note that

$$e^{ct} x(t) = x_o + N(t),$$

where

$$N(t) = K\int_0^t e^{cs}\, dw(s).$$

Note

$$\langle N(t)\rangle = K^2 \int_0^t e^{2cs}\, ds = \frac{K^2}{2c}(e^{2ct}-1) := \alpha(t) \to \infty \qquad \text{as } t\to\infty.$$

Use $\beta(u)$ to denote the inverse function of $\alpha(t)$, i.e. $\beta(u) = \sup\{t \ge 0\colon \alpha(t) \ge u\}$. Define $W(u) = N(\beta(u))$ and $\mathcal{G}_u = \mathcal{F}_{\beta(u)}$ on $u \ge 0$. In view of Theorem 1-6.2, one see that $W(u)$ is a continuous local martingale with respect to the filtration $\{\mathcal{G}_u\}$. Note that

$$\langle W(u)\rangle = \alpha(\beta(u)) = u \qquad \text{for all } u \ge 0.$$

So Theorem 1-2.5 implies that $W(u)$ is a Brownian motion with respect to the filtration $\{\mathcal{G}_u\}$. Therefore, by the well-known iterated logarithm law, we have

$$\limsup_{u\to\infty} \frac{|W(u)|}{\sqrt{2u\log\log u}} = 1 \qquad \text{a.s.}$$

Hence

$$\limsup_{t\to\infty} \frac{|N(t)|}{\sqrt{2\alpha(t)\log\log(\alpha(t))}} = 1 \qquad \text{a.s.}$$

Therefore

$$\limsup_{t\to\infty} \frac{1}{\log t} |x(t)|^2 = \limsup_{t\to\infty} \frac{1}{e^{2ct}\log t} e^{2ct}|x(t)|^2$$

$$= \limsup_{t\to\infty} \frac{1}{e^{2ct}\log t} |x_o + N(t)|^2 = K^2/c \qquad \text{a.s}$$

and then by Lemma 6.3

$$\limsup_{t\to\infty} \frac{1}{\log t}\left(\sup_{0\le s\le t} |x(s)|^2 \right) = K^2/c \qquad \text{a.s}$$

2.7. STOCHASTIC FLOWS AND STOCHASTIC DIFFERENTIAL EQUATIONS

In this section we shall use stochastic differential equations driven by nonlinear integrators to represent stochastic flows. Let us first give the definition of the stochastic flow of homomorphisms and the Brownian flow as well.

Definition 7.1 Let $\varphi_{s,t}(x)$, $s, t \in R_+$, $x \in R^d$ be a continuous R^d-valued random field defined on the probability space $(\Omega, \mathcal{F}, \{\mathcal{F}_t\}, P)$. It is called a *stochastic flow of homomorphisms* if it satisfies the following properties almost surely:

(i) $\varphi_{s,u} = \varphi_{t,u} \circ \varphi_{s,t}$ for any s, t ,u, where $\circ$ denotes the composition of maps;

(ii) $\varphi_{s,s}$ = identity map for any s;

(iii) for any s, t, $\varphi_{s,t}$ is a homomorphism of R^d onto itself.

It is called a *Brownian flow* if it still satisfies

(iv) for any $0 \le t_0 < t_1 < \dots < t_k$, $\varphi_{t_i, t_{i+1}}$, $i = 0, \dots, k-1$ are independent.

If $\varphi_{s,t}$ is only defined on $0 \le s \le t < \infty$, we call it a *forward stochastic flow of homomorphisms* or *forward Brownian flow* respectively.

One of the main problems in this section will be about the relation between stochastic flows and stochastic differential equations. It shall be shown that for a given forward stochastic flow $\varphi_{s,t}(x)$, $0 \le s \le t < \infty$, $x \in R^d$ with suitable regularity conditions, one can find a unique continuous C-semimartingale F(x, t) such that for every fixed s and x, $\varphi_t := \varphi_{s,t}(x)$ on $s \le t < \infty$ satisfies the Itô stochastic differential equation driven by F(x, t), i.e.

$$\varphi_t = x + \int_s^t F(\varphi_r, dr).$$

In particular, if $\varphi_{s,t}(x)$ is a forward Brownian flow, then the associate F(x, t) is a C-Brownian motion, the definition of which is as follows:

Definition 7.2 A random field F(x, t), $t \ge 0$ and $x \in R^d$ is called a C-Brownian motion if:

(i) For any $x_1, \dots, x_n$, the process $\{F(x_1, t), \dots, F(x_n, t): t \ge 0\}$ is a $d \times n$-dimensional Gaussian process with independent increments.

(ii) It has the decomposition $F(x, t) = M(x, t) + B(x, t)$, where $M(x, t)$ is a continuous martingale for any x, and there exists $\alpha : R^d \times R^d \times R_+ \to R^{d\times d}$ such that

$$<M(x, t), M(y, t)> = \int_0^t \alpha(x, y, s)\, ds;$$

and moreover, there exists $\beta : R^d \times R_+ \to R^d$ such that

$$B(x, t) = \int_0^t \beta(x, s)\, ds.$$

(iii) $F(x, 0) \equiv 0$.

It is easy to see that a C-Brownian motion is a continuous C-semimartingale which has the local characteristic $(\alpha(x, y, t), \beta(x, t), t)$ and a Gaussian distribution. We shall now propose the regularity conditions:

(H7.1) $\varphi_{s,t}(x)$ is square integrable for each $s, t \in R_+$, $x \in R^d$, and the infinitesimal mean $b(x, t)$ and the infinitesimal covariance $a(x, y, t)$ exist for all $x, y \in R^d$, $t \geq 0$, that is,

$$b(x, t) = (b_1, \ldots, b_d)^T = \lim_{h\to 0+} \frac{1}{h} E(\varphi_{t,t+h}(x) - x),$$

$$a(x, y, t) = (a_{ij})_{d\times d} = \lim_{h\to 0+} \frac{1}{h} E\{(\varphi_{t,t+h}(x) - x)(\varphi_{t,t+h}(y) - y)^T\}.$$

(H7.2) There exists a positive constant K such that

$$| E(\varphi_{s,t}(x) - x) | \leq K(1 + |x|)\, |t - s|,$$

$$| E\{(\varphi_{s,t}(x) - x)(\varphi_{s,t}(y) - y)^T\} | \leq K(1 + |x|)(1 + y|)\, |t - s|$$

for all $s, t \in R_+$, $x, y \in R^d$.

(H7.3) The $a(x, y, t)$ and $b(x, t)$ are δ-Hölder continuous ($\delta > 0$): For each compact subset D of R^d, there exists a $K > 0$ such that

$$| a(x, x, t) - 2a(x, y, t) + a(y, y, t) | \leq K\, |x - y|^{2\delta},$$

$$| b(x, t) - b(y, t) | \leq K\, |x - y|^{\delta}$$

hold for any x, y $\in$ D and $t \geq 0$.

The following theorem gives a representative of a forward Brownian flow. For the proof please see Kunita (1990).

Theorem 7.1 Let $\varphi_{s,t}(x)$, $0 \leq s \leq t < \infty$, $x \in R^d$ be a forward Brownian flow satisfying hypotheses (H7.1)–(H7.3). Then there exists a unique C–Brownian motion with the local characteristic $(\alpha(x, y, t), \beta(x, t), t)$ belonging to the class $B^{0,\gamma}$ $(\gamma < \delta)$ such that for every fixed s and x, $\varphi_t := \varphi_{s,t}(x)$ on $s \leq t < \infty$ satisfies the Itô stochastic differential equation driven by F(x, t), i.e.

$$\varphi_t = x + \int_s^t F(\varphi_r, dr).$$

Moreover $\alpha(x, y, t)$ and $\beta(x, t)$ coincide with a(x, y, t) and b(x, t) respectively, that is $\alpha(x, y, t) = a(x, y, t)$ and $\beta(x, t) = b(x, t)$.

Using this representative theorem, together with the theorems established in Section 2.6, one can obtain the almost surely asymptotic bounds for Brownian flows as described by the following theorem, where the operators L and H are introduced in the previous section.

Theorem 7.2 Let $\varphi_{s,t}(x)$, $0 \leq s \leq t < \infty$, $x \in R^d$ be a forward Brownian flow satisfying hypotheses (H7.1)–(H7.3). Assume there exists a function $V \in C^2(R^d; R_+)$ such that

$$HV(x) \leq K\,V(x)$$

holds for all $x \in R^d$ and $t \geq 0$, where $K > 0$.

(i) If for all $x \in R^d$ and $t \geq 0$,

$$LV(x) \leq -cV(x) + \rho,$$

where $c > 0$ and $\rho \geq 0$, then for every $x \in R^d$ and $s \geq 0$,

$$\limsup_{t\to\infty} \frac{1}{\log t} \Big(\sup_{s \leq u \leq t} V(\varphi_{s,u}(x)) \Big) \leq Ke/2c \quad \text{a.s.}$$

(ii) If for all $x \in R^d$ and $t \geq 0$,

$$LV(x) \leq \rho,$$

where $\rho \geq 0$, then for every $x \in R^d$ and $s \geq 0$,

$$\limsup_{t\to\infty} \frac{1}{t\log\log t}\left(\sup_{s\leq u\leq t} V(\varphi_{s,u}(x))\right) \leq Ke/2 \qquad \text{a.s.}$$

(iii) If for all $x \in R^d$ and $t \geq 0$,

$$LV(x) \leq cV(x) + \rho,$$

where $c > 0$ and $\rho \geq 0$, then for every $x \in R^d$ and $s \geq 0$,

$$\lim_{t\to\infty} \frac{1}{e^{ct}\log\log t}\left(\sup_{s\leq u\leq t} V(\varphi_{s,u}(x))\right) = 0 \qquad \text{a.s.}$$

Let us close this chapter by pointing out that with the help of Theorem 7.1 one can apply many results in this book to the forward Brownian flows as we did in the above theorem. It should also be mentioned that one can have a representative theorem for stochastic flows of homomorphisms, and we refer the interested reader to Kunita (1990) for the details.

CHAPTER 3

Stochastic Differential Delay Equations

3.1. INTRODUCTION

This chapter is devoted to the discussion of stochastic differential delay equations driven by nonlinear integrators. The local solutions are first studied under the local Lipschitz condition and the global solutions are obtained under the additional linear growth condition as usual. But, a more general concept of bounded stochastic integral contractors is then introduced to study the existence and uniqueness of the global solutions to the delay equations. In particular, the bounded stochastic integral contractors include the Lipschitz condition as a special case. As a by-product, a general result on the existence and uniqueness of the solutions to stochastic differential equations is also obtained by using bounded stochastic integral contractors. Approximate solutions to the delay equations are very useful in applications e.g. in obtaining numerical solutions, and they also represent an alternative to the procedure for establishing the existence and uniqueness of the solutions. Cauchy-Maruyama's approximation procedure as well as Carathéodory's will be discussed. Furthermore, the regularities of the solutions e.g.

L^p-estimates and almost surely asymptotic bounds will also be studied. As in the previous chapter, the initial time of all the equations discussed in this chapter is about to be set at zero but the theory holds for any positive initial time.

3.2. EXISTENCE AND UNIQUENESS OF SOLUTIONS

Let $F(x, y, t) = (F_1, ..., F_d)^T$ be a d-dimensional continuous C-semimartingale with spatial parameters $(x, y) \in R^d \times R^d$. It is assumed that F(x, y, t) has the local characteristic $(a(x, y, \bar{x}, \bar{y}, t), b(x, y, t), A(t))$ belonging to the class $B^{0,\delta}$ and has the decomposition $F(x, y, t) = M(x, y, t) + B(x, y, t)$ with $M = (M_1, ..., M_d)^T$ a d-dimensional continuous local martingale and $B = (B_1, ..., B_d)^T$ a d-dimensional continuous process of finite variation. Let $\delta(.)$ be a continuous nonnegative real-valued function defined on R_+ such that $t - \delta(t)$ is bounded below. Set $-\tau := \inf\{t - \delta(t) : t \geq 0\}$. Let $\xi(s)$, $-\tau \leq s \leq 0$ be a continuous R^d-valued stochastic process such that every $\xi(s)$ is $\mathcal{F}_0$-measurable and $E|\xi(s)|^2 < \infty$.

The objective of this section is to discuss the solution of a stochastic differential delay equation driven by the C-semimartingale F(x, y, t) of the form

$$x(t) = \xi(0) + \int_0^t F(x(s), x(s-\delta(s)), ds) \qquad \text{on } t \geq 0, \tag{2.1}$$

with initial data $x(s) = \xi(s)$ on $-\tau \leq s \leq 0$. We first state a classical result.

Theorem 2.1 Assume that $a(x, y, \bar{x}, \bar{y}, t)$ and $b(x, y, t)$ satisfy the Lipschitz condition, i.e. there exists a positive constant K such that

$$|b(x, y, t) - b(\bar{x}, \bar{y}, t)| \leq K(|x - \bar{x}| + |y - \bar{y}|) \qquad \text{a.s.},$$

$$|a(x, y, x, y, t) - 2a(x, y, \bar{x}, \bar{y}, t) + a(\bar{x}, \bar{y}, \bar{x}, \bar{y}, t)| \leq K(|x-\bar{x}|^2 + |y-\bar{y}|^2) \quad \text{a.s.}$$

for all $x, y, \bar{x}, \bar{y} \in R^d$ and $t \geq 0$. Then there exists a unique solution to equation (2.1).

Not only can the theorem be proved by the standard method of successive approximation but it is also a special case of our more general theory to be discussed in Section 3.3, so the proof is omitted here.

In the sequel of this section we will only assume the local Lipschitz condition for the local characteristic. Then as in the case of ordinary differential equation, the stochastic delay equation may not have a global solution, that is the explosion may occur at a finite time. Hence it is necessary to define a local solution for the delay equation as described by the following definition.

Definition 2.1 Let $x(t)$, $t \in [[-\tau, \sigma_\infty[[$ be a continuous local process with values in R^d adapted to $\{\mathcal{F}_t\}$ ($\mathcal{F}_s = \mathcal{F}_0$ for $-\tau \le s \le 0$). It is called a *local solution* of equation (2.1) if $x(t) = \xi(t)$ on $-\tau \le t \le 0$ and

$$x(t \wedge \sigma_k) = \xi(0) + \int_0^{t \wedge \sigma_k} F(x(s), x(s-\delta(s)), ds) \quad \text{on } t \ge 0 \tag{2.2}$$

is satisfied for any $k \ge 1$, where $\{\sigma_k\}$ is a nondecreasing sequence of finite stopping times such that $\sigma_k \uparrow \sigma_\infty$ a.s. Furthermore if $\limsup_{t \uparrow \sigma_\infty} |x(t)| = \infty$ is satisfied whenever $\sigma_\infty < \infty$, it is called a *maximal local solution* and σ_∞ is called the *explosion time*. A maximal local solution $x(t)$, $t \in [[-\tau, \sigma_\infty[[$ is said to be *unique* if for any other maximal local solution $\hat{x}(t)$, $t \in [[-\tau, \hat{\sigma}_\infty[[$ we have $x(t) = \hat{x}(t)$ and $\sigma_\infty = \hat{\sigma}_\infty$ a.s.

Theorem 2.2 Assume that there exists a sequence $\{C_k\}$ of positive numbers such that for each $k \ge 1$,

$$|b(x, y, t) - b(\bar{x}, \bar{y}, t)| \le C_k(|x - \bar{x}| + |y - \bar{y}|) \quad \text{a.s.},$$

$$|a(x, y, x, y, t) - 2a(x, y, \bar{x}, \bar{y}, t) + a(\bar{x}, \bar{y}, \bar{x}, \bar{y}, t)| \le C_k(|x-\bar{x}|^2 + |y-\bar{y}|^2) \quad \text{a.s.}$$

hold for all $x, y, \bar{x}, \bar{y} \in S_k := \{x \in R^d : |x| \le k\}$ and $t \ge 0$. Then there exists a unique maximal local solution to equation (2.1).

Proof. For each $k = 1, 2, \ldots$, let $\psi_k(x, y)$, $(x, y) \in R^d \times R^d$ be a smooth function such that $\psi_k(x, y) = 1$ if $|x| + |y| \le k/2$ and $0 \le \psi_k(x, y) \le 1$ if $k/2 < |x| + |y| \le k$ and $\psi_k(x, y) = 0$ if $|x| + |y| > k$. Define

$$F_k(x, y, t) = \psi_k(x, y)\, F(x, y, t)$$

which is a continuous C-semimartingale with the local characteristic

$$(a(x, y, \bar{x}, \bar{y}, t)\, \psi_k(x, y)\, \psi_k(\bar{x}, \bar{y}),\ b(x, y, t)\, \psi_k(x, y),\ A(t))$$

satisfying the Lipschitz condition. So Theorem 2.1 tells us that the stochastic differential delay equation

$$x_k(t) = \xi(0) + \int_0^t F_k(x_k(s), x_k(s-\delta(s)), ds) \quad \text{on } t \ge 0, \tag{2.3}$$

with initial data $x_k(s) = \xi(s)$ on $-\tau \le s \le 0$, has a unique global solution $x_k(t)$ on $t \ge -\tau$.

Now we introduce a notation: if $z(t)$, $t \geq -\tau$ is a predictable process such that $z(t) = \xi(t)$ on $-\tau \leq t \leq 0$, define the stopping time

$$\rho_k(z) := \inf\{ t \geq 0 : |z(t)| \geq k/4 \text{ or } |z(t-\delta(t))| \geq k/4\}.$$

We see that

$$x_{k+1}(t \wedge \rho_k(x_{k+1}))$$

$$= \xi(0) + \int_0^{t\wedge\rho_k(x_{k+1})} F_{k+1}\Big(x_{k+1}(s),\, x_{k+1}(s-\delta(s)),\, ds\Big)$$

$$= \xi(0) + \int_0^{t\wedge\rho_k(x_{k+1})} F_k\Big(x_{k+1}(s\wedge\rho_k(x_{k+1})),\, x_{k+1}(s\wedge\rho_k(x_{k+1}) - \delta(s)),\, ds\Big),$$

and

$$x_k(t \wedge \rho_k(x_{k+1}))$$

$$= \xi(0) + \int_0^{t\wedge\rho_k(x_{k+1})} F_k\Big(x_k(s\wedge\rho_k(x_{k+1})),\, x_k(s\wedge\rho_k(x_{k+1}) - \delta(s)),\, ds\Big).$$

Thanks to the uniqueness of the solution for an equation satisfying the Lipschitz condition, we obtain that

$$x_{k+1}(t\wedge\rho_k(x_{k+1})) = x_k(t\wedge\rho_k(x_{k+1})) \quad \text{for all } t \geq 0.$$

This implies immediately that

$$\rho_k(x_k) \leq \rho_k(x_{k+1})$$

and

$$x_k(t) = x_{k+1}(t) \quad \text{whenever } t \leq \rho_k(x_k). \tag{2.4}$$

Set $\sigma_k = \rho_k(x_k)$ and $\sigma_\infty = \lim_{k\to\infty}\sigma_k$. Define a local process $x(t)$, $t \in [[-\tau, \sigma_\infty[[$ as follows: $x(t) = \xi(t)$ on $-\tau \leq t \leq 0$ and

$$x(t) = x_k(t) \quad \text{if } \sigma_{k-1} < t \leq \sigma_k,\ k = 1, 2, \ldots .$$

It follows from (2.4) that

$$x(t) = x_k(t) \quad \text{whenever } t \le \sigma_k.$$

So for every $k \ge 1$,

$$x(t \wedge \sigma_k) = x_k(t \wedge \sigma_k)$$

$$= \xi(0) + \int_0^{t \wedge \sigma_k} F^k(x_k(s), x_k(s-\delta(s)), ds)$$

$$= \xi(0) + \int_0^{t \wedge \sigma_k} F(x(s), x(s-\delta(s)), ds) \quad \text{on } t \ge 0.$$

In other words, $x(t)$, $t \in [[-\tau, \sigma_\infty[[$ is a local solution of equation (2.1). In order to show it is a maximal local solution, we note that

$$\sigma_\infty = \lim_{k \to \infty} \rho_k(x_k) = \lim_{k \to \infty} \zeta_k,$$

where

$$\zeta_k := \sigma_\infty \wedge \inf\{\, t : |x(t)| \ge k/4 \text{ or } |x(t-\delta(t))| \ge k/4 \text{ for } 0 \le t < \sigma_\infty \}.$$

Note also that for almost every $\omega \in \Omega$, $\{|\xi(s)|: -\tau \le s \le 0\}$ is bounded. Therefore, for all k so large that $\{|\xi(s)|: -\tau \le s \le 0\} < k/4$ we see that

$$|x(\zeta_k)| = k/4 \qquad \text{whenever } \sigma_\infty < \infty.$$

So

$$\limsup_{t \to \sigma_\infty} |x(t)| \ge \lim_{k \to \infty} |x(\zeta_k)| = \infty \qquad \text{whenever } \sigma_\infty < \infty.$$

Hence $x(t)$, $t \in [[-\tau, \sigma_\infty[[$ is a maximal solution of equation (2.1).

We shall need to prove the uniqueness of the solution. To do so, we let $y(t)$, $t \in [[-\tau, \bar{\sigma}_\infty[[$ be any maximal solution of equation (2.1). Define

$$\pi_k := \bar{\sigma}_\infty \wedge \inf\{\, t : |y(t)| \ge k/4 \text{ or } |y(t-\delta(t))| \ge k/4 \text{ for } 0 \le t < \bar{\sigma}_\infty \}.$$

Then $\pi_k \uparrow \bar{\sigma}_\infty$ a.s. as $k \to \infty$. Set $\theta_k = \zeta_k \wedge \pi_k$, where ζ_k is defined above. Then it is easy to see that

$$x(t \wedge \theta_k) = \xi(0) + \int_0^{t \wedge \theta_k} F^k(x(s \wedge \theta_k), x(s \wedge \theta_k - \delta(s)), ds) \qquad \text{on } t \ge 0$$

and

$$y(t\wedge\theta_k) = \xi(0) + \int_0^{t\wedge\theta_k} F^k(y(s\wedge\theta_k),\ y(s\wedge\theta_k - \delta(s)),\ ds) \qquad \text{on } t \ge 0.$$

Thanks once again to the uniqueness of the solution for an equation satisfying the Lipschitz condition we have $x(t\wedge\theta_k) = y(t\wedge\theta_k)$ a.s. Letting $k \to \infty$, we get $x(t) = y(t)$ for $t < \sigma_\infty \wedge \bar\sigma_\infty$ a.s. We now claim that $\sigma_\infty = \bar\sigma_\infty$ a.s. In fact, for any $\omega \in \{\sigma_\infty < \bar\sigma_\infty\}$, we have

$$|y(\sigma_\infty)| = \lim_{k\to\infty} |y(\zeta_k)| = \lim_{k\to\infty} |x(\zeta_k)| = \infty$$

which is in contradiction with the fact that y(t) is continuous on $[-\tau, \bar\sigma_\infty)$. So we must have $\sigma_\infty \ge \bar\sigma_\infty$ a.s. Similarly, $\sigma_\infty \le \bar\sigma_\infty$ a.s., and therefore $\sigma_\infty = \bar\sigma_\infty$ a.s. The proof is complete.

A simple but sufficient condition for the existence of a global solution of equation (2.1) is that the local characteristic is of uniformly linear growth.

Theorem 2.3 In addition to the conditions of Theorem 2.2, assume that

$$|b(x, y, t)|^2 \le K(t)\,(1 + |x|^2 + |y|^2) \qquad \text{a.s.,}$$

$$\sum_{i=1}^{d} a_{ii}(x, y, x, y, t) \le K(t)\,(1 + |x|^2 + |y|^2) \qquad \text{a.s.}$$

hold for all $x, y \in R^d$ and $t \ge 0$, where $K(t)$, $t \ge 0$ is a positive continuous adapted process. Then equation (2.1) has a unique global solution $x(t)$, $t \ge -\tau$. Moreover, set $\xi(t) = \xi(0)$ for $t \ge 0$, and define stopping times

$$\zeta_n = \inf\{\, t \ge 0 : A(t) \ge n \text{ or } \int_0^t K(s)\, dA(s) \ge n\},$$

$$\theta_n = \inf\{\, t \ge 0 : |\xi(t-\delta(t))| \ge n\}$$

for $n = 1, 2, \ldots$. Then $\zeta_n \wedge \theta_n \uparrow \infty$ a.s. and

$$E\Big(\sup_{0\le t\le \zeta_n\wedge\theta_n} |x(t)|^2\Big) \le \Big[\, 3E|\xi(0)|^2 + 3n(4+n)(1+n^2)\Big]\, e^{6n(4+n)}. \tag{2.5}$$

Proof. Let $x(t)$, $t \in [[-\tau, \sigma_\infty[[$ be the maximal local solution of equation (2.1) and we need to show $\sigma_\infty = \infty$ a.s. Define stopping times

$$\sigma_k = \inf\{ t \geq 0 : |x(t)| \geq k \} \quad \text{for } k = 1, 2, \ldots .$$

Fix $n \geq 1$ arbitrarily and set $x_k(t) = x(t \wedge \sigma_k \wedge \zeta_n \wedge \theta_n)$ for $k = 1, 2, \ldots$. Then for any stopping time $\rho \leq \sigma_k \wedge \zeta_n \wedge \theta_n$ we have

$$x_k(\tau) = \xi(0) + \int_0^\rho F(x_k(s), x_k(s-\delta(s)), ds)$$

$$= \xi(0) + \int_0^\rho M(x_k(s), x_k(s-\delta(s)), ds) + \int_0^\rho b(x_k(s), x_k(s-\delta(s)), s)\, dA(s).$$

Therefore, by Doob's martingale inequality and Schwarz's inequality,

$$E\Big(\sup_{0\leq t\leq\rho} |x_k(t)|^2 \Big)$$

$$\leq 3E|\xi(0)|^2 + 12 \sum_{i=1}^d E \int_0^\rho a_{ii}(x_k(s), x_k(s-\delta(s)), s)\, dA(s)$$

$$+ 3n\, E \int_0^\rho | b(x_k(s), x_k(s-\delta(s)), s) |^2\, dA(s).$$

Thanks to the linear growth condition, we arrive at

$$E\Big(\sup_{0\leq t\leq\rho} |x_k(t)|^2 \Big)$$

$$\leq 3E|\xi(0)|^2 + 3(4+n)\, E \int_0^\rho K(s) \Big(1 + |x_k(s)|^2 + |x_k(s-\delta(s))|^2\Big)\, dA(s)$$

$$\leq 3E|\xi(0)|^2 + 3(4+n)\, E \int_0^\rho K(s) \Big\{ 1 + n^2 + 2\Big(\sup_{0\leq t\leq s} |x_k(t)|^2 \Big) \Big\}\, dA(s)$$

$$\leq 3E|\xi(0)|^2 + 3n(4+n)(1+n^2) + 6(4+n)\, E \int_0^\rho K(s) \Big(\sup_{0\leq t\leq s} |x_k(t)|^2 \Big)\, dA(s)$$

An application of Theorem 1-7.11 implies

$$E\Big(\sup_{0\leq t\leq\sigma_k\wedge\zeta_n\wedge\theta_n} |x(t)|^2 \Big) = E\Big(\sup_{0\leq t\leq\sigma_k\wedge\zeta_n\wedge\theta_n} |x_k(t)|^2 \Big)$$
$$\leq \Big[3E|\xi(0)|^2 + 3n(4+n)(1+n^2) + 6(4+n) \Big] e^{6n(4+n)}.$$

It follows by letting $k \to \infty$ that

$$E\Big(\sup_{0 \le t \le \sigma_\infty \wedge \zeta_n \wedge \theta_n} |x(t)|^2 \Big) \le \Big[3E|\xi(0)|^2 + 3n(4+n)(1+n^2) \Big] e^{6n(4+n)}. \quad (2.6)$$

Hence, $\sigma_\infty \ge \zeta_n \wedge \theta_n$ a.s. for every n, and so $\sigma_\infty = \infty$ a.s. Finally (2.5) follows from (2.6) and the proof is complete.

We now establish an L^p-estimate for the solution to close this section.

Theorem 2.4 In addition to the conditions of Theorem 2.2, assume:

(i) $E\Big[\sup_{-\tau \le s \le 0} |\xi(s)|^p \Big] < \infty$ for some $p \ge 2$.

(ii) For all $x, y \in R^d$, $t \ge 0$ and $1 \le i \le d$,

$$|b(x, y, t)|^2 \le K(1 + |x|^2 + |y|^2) \quad \text{a.s.},$$

$$a_{ii}(x, y, x, y, t) \le K(1 + |x|^2 + |y|^2) \quad \text{a.s.},$$

where K is a positive constant.

(iii) $A(t) \le \mu(t)$ a.s. for all $t \ge 0$, where $\mu(t)$ is an increasing real valued function defined on $t \ge 0$.

Then for any $T > 0$,

$$E\Big(\sup_{0 \le s \le T} |x(s)|^p \Big)$$

$$\le \Big\{ 3^{p-1} E|\xi(0)|^p + C\mu(T) \Big(1 + E\Big[\sup_{-\tau \le s \le 0} |\xi(s)|^p \Big] \Big) \Big\} e^{2C\mu(T)}, \quad (2.7)$$

where

$$C = 3^{3p/2-2} K^{p/2} \Big(\mu(T)^{p-1} + d^{p/2} C_p \, \mu(T)^{(p/2-1)} \Big)$$

and C_p is the universal constant determined by Theorem 1-2.6.

Proof. Using the hypotheses, Hölder's inequality and Burkholder-Davis-Gundy's inequality, we can derive that for any stopping time ρ such that $0 \le \rho \le T$,

$$E\Big(\sup_{0 \le t \le \rho} |x(s)|^p \Big)$$

$$\leq 3^{p-1} E|\xi(0)|^p + 3^{p-1} E\Big(\int_0^\rho |b(x(s), x(s-\delta(s)), s)|\, dA(s)\Big)^p$$

$$+ 3^{p-1} E\Big(\sup_{0\leq t\leq\rho} |\int_0^s M(x(u), x(u-\delta(u)), du)|^p\Big)$$

$$\leq 3^{p-1} E|\xi(0)|^p + 3^{p-1} E\Big(A(\tau)^{p-1} \int_0^\rho |b(x(s), x(s-\delta(s)), s)|^p\, dA(s)\Big)$$

$$+ 3^{p-1} d^{(p/2-1)} C_p \sum_{i=1}^d E\Big(\int_0^\rho a_{ii}(x(s), x(s-\delta(s)), x(s), x(s-\delta(s)), s) dA(s)\Big)^{p/2}$$

$$\leq 3^{p-1} E|\xi(0)|^p + 3^{p-1} \mu(T)^{p-1} E\int_0^\rho \Big(K(1 + |x(s)|^2 + |x(s-\delta(s))|^2)\Big)^{p/2} dA(s)$$

$$+ 3^{p-1} d^{p/2} C_p E\Big(A(\tau)^{(p/2-1)} \int_0^\rho \Big(K(1 + |x(s)|^2 + |x(s-\delta(s))|^2)\Big)^{p/2} dA(s)\Big)$$

$$\leq 3^{p-1} E|\xi(0)|^p + C\, E\int_0^\rho \Big(1 + |x(s)|^p + |x(s-\delta(s))|^p\Big)\, dA(s)$$

$$\leq 3^{p-1} E|\xi(0)|^p + C\mu(T)\Big(1 + E\Big[\sup_{-\tau\leq s\leq 0} |\xi(s)|^p\Big]\Big)$$

$$+ 2C\, E\int_0^\rho \Big(\sup_{0\leq u\leq s} |x(u)|^p\Big)\, dA(s)$$

An application of Theorem 1-7.11 gives the required inequality (2.7). The proof is complete.

3.3. BOUNDED STOCHASTIC INTEGRAL CONTRACTORS AND SOLUTIONS

The concept of contractors was introduced by Altman (1973 & 1977) as a functional analytic tool for solving deterministic operator equations in Banach spaces. Many authors e.g. Lee and Padgett (1977a & b), Padgett and Rao (1979), Rao and Padgett (1982), Rao and Tsokos (1975) extended the contractor concept to the random contractors which were applied to study the solutions of random nonlinear operator equations and random integral equations. Another important extension is the concept of bounded stochastic integral contractors in studying the existence and uniqueness of the solutions of Itô's stochastic differential equations (see Kuo (1979), Padgett (1983)) and stochastic differential-functional equations (see Mao (1992e)).

Let $F(x, t) = (F_1(x, t), ..., F_d(x, t))^T$, $x \in R^m$ be a d-dimensional continuous C-semimartingale with the local characteristic $(a(x, y, t), b(x, t), A(t))$ belonging to the class $B^{0,\delta}$, and let $F(x, t) = M(x, t) + B(x, t)$ be the decomposition such that $M(x, t)$ is a continuous local martingale and $B(x, t)$ is a continuous process of bounded variation. Let $\tau > 0$ and denote by $C([-\tau, 0]; R^d)$ the space of all continuous functions from $[-\tau, 0]$ to R^d with norm $\|u\|_\tau = \sup\{|u(s)| : -\tau \le s \le 0\}$. Set $\mathcal{F}_s = \mathcal{F}_0$ on $-\tau \le s \le 0$. If a process $z(.)$ is defined at least on $[t-\tau, t]$ we set $\hat{z}_t = \{z(t-s) : -\tau \le s \le 0\}$. Let $T > 0$. Let D be an operator from $C([-\tau, 0]; R^d)\times[0, T]$ to R^m which is Lipschitz continuous, that is there exists a positive constant γ such that

$$|D(u, t) - D(v, t)| \le \gamma \|u - v\|_\tau \quad \text{for all } u, v \in C([-\tau, 0]; R^d) \text{ and } 0 \le t \le T.$$

In this section we shall use the bounded stochastic integral contractors to investigate the existence and uniqueness of the solution of a general stochastic differential-functional equation

$$dx(t) = F(D(\hat{x}_t, t), dt) \quad \text{on } 0 \le t \le T \tag{3.1}$$

with initial data $\hat{x}_0 = \xi$, where $\xi = \{\xi(s) : -\tau \le s \le 0\}$ is a $C([-\tau, 0]; R^d)$-valued $\mathcal{F}_0$-measurable random variable. Let $C([-\tau, T] \times \Omega; R^d)$ denote the family of all continuous $\{\mathcal{F}_t\}$-adapted R^d-valued stochastic processes defined on $[-\tau, T]$. A stochastic process $x(.) \in C([-\tau, T] \times \Omega; R^d)$ is said to be a solution of the equation if it satisfies the initial condition $\hat{x}_0 = \xi$ and almost surely

$$x(t) = \xi(0) + \int_0^t F(D(\hat{x}_s, s), ds) \quad \text{on } 0 \le t \le T. \tag{3.1)'$$

In order to see the generalization of equation (3.1), let us look at a few special but important cases of the equation. First of all, let $\delta : R_+ \to [0, \tau]$ and $m = 2d$. Define the operator D by

$$D(u, t) = (u(0), u(-\delta(t))) \quad \text{for } u \in C([-\tau, 0]; R^d).$$

Then equation (3.1) becomes to a stochastic differential equation with time variable delay

$$dx(t) = F(x(t), x(t-\rho(t)), dt). \tag{3.2}$$

This sort of stochastic delay equations has been studied in the previous section. Next we let $0 < \tau_1 < ... < \tau_k = \tau$ and $m = (k+1)d$. Define

$$D(u, t) = (u(0), u(-\tau_1), ..., u(-\tau_k)).$$

Then equation (3.1) reduces to a stochastic differential equation with constants delay

$$dx(t) = F(x(t), x(t-\tau_1), ..., x(t-\tau_k), dt). \tag{3.3}$$

Moreover, if $\tau = 0$, $m = d$ and $D =$ the identity map, we arrive at a stochastic differential equation (without delay)

$$dx(t) = F(x(t), dt) \tag{3.4}$$

which has been studied in Chapter 2.

We now introduce the concept of bounded stochastic integral contractors associated with equation (3.1).

Definition 3.1 Let

$$\Phi(x, t) = (\Phi_{ij})_{d\times d}, \quad \Gamma^k(x, t) = (\Gamma^k_{ij})_{d\times d}, \quad 1 \le k \le d, \ (x, t) \in R^m \times [0, T].$$

Suppose

$$|\Phi(x, t)|^2 + \sum_{k=1}^{d} |\Gamma^k(x, t)|^2 a_{kk}(x, x, t)$$

is bounded on $R^m \times [0, T]$. Suppose also that there exists a positive constant K such that for any $\varphi, \psi \in C([-\tau, T] \times \Omega; R^d)$ the following inequalities hold for all $0 \le t \le T$:

$$|b(D(\hat{\varphi}_t, t), t) - b(D(\hat{\pi}_t(\varphi, \psi), t), t) - \Phi(D(\hat{\varphi}_t, t), t)\,\psi(t)|^2$$

$$\le K \sup_{-\tau\le s\le t} |\psi(s)|^2 \qquad \text{a.s.} \tag{3.5}$$

and

$$\sum_{i=1}^{d} \Big(a_{ii}(D(\hat{\varphi}_t, t), D(\hat{\varphi}_t, t), t) - 2a_{ii}(D(\hat{\varphi}_t, t), D(\hat{\pi}_t(\varphi, \psi), t), t)$$

$$+ a_{ii}(D(\hat{\pi}_t(\varphi, \psi), t), D(\hat{\pi}_t(\varphi, \psi), t), t)$$

$$- 2\sum_{k=1}^{d} \Gamma_i^k(D(\hat{\varphi}_t, t), t)\, \psi(t) \Big[a_{ik}(D(\hat{\varphi}_t, t), D(\hat{\varphi}_t, t), t)$$

$$- a_{ik}(D(\hat{\varphi}_t, t), D(\hat{\pi}_t(\varphi, \psi), t), t) \Big]$$

$$+ \sum_{k,j=1}^{d} \Gamma_i^k(D(\hat{\varphi}_t, t), t)\, \psi(t)\, \Gamma_i^j(D(\hat{\varphi}_t, t), t)\, \psi(t)\, a_{kj}(D(\hat{\varphi}_t, t), D(\hat{\varphi}_t, t), t) \Big)$$

$$\leq K \sup_{-\tau \leq s \leq t} |\psi(s)|^2 \qquad \text{a.s.,} \tag{3.6}$$

where $\Gamma_i^k = (\Gamma_{i\,1}^k, \ldots, \Gamma_{i\,d}^k)$ and

$$\pi(t, \varphi, \psi) = \varphi(t) + \psi(t) + \int_0^{t \vee 0} \Phi(D(\hat{\varphi}_s, s), s)\, \psi(s)\, dA(s)$$

$$+ \sum_{k=1}^{d} \int_0^{t \vee 0} \Gamma^k(D(\hat{\varphi}_s, s), s)\, \psi(s)\, M^k(D(\hat{\varphi}_s, s), ds), \quad -\tau \leq t \leq T.$$

Then we shall say that the C-semimartingale F has a *bounded stochastic integral contractor* $\{I + \int \Phi dA + \sum\int \Gamma^k M^k(ds)\}$.

Definition 3.2 F is said to be *stochastically closed* if for any predictable processes z_n and z such that

$$z^n \to z \quad \text{in } L^2([0, T] \times \Omega; R^m)$$

and

$$\int_0^{\cdot} F(z_n(s), ds) \to \psi \quad \text{in } L^2([0, T] \times \Omega; R^d),$$

then

$$\psi(t) = \int_0^t F(z(s), ds), \qquad 0 \leq t \leq T.$$

Definition 3.3 A bounded stochastic integral contractor $\{I + \int \Phi dA + \sum\int \Gamma^k M^k(ds)\}$ is said to be *regular* if the linear equation, on $0 \leq t \leq T$,

$$\psi(t) + \int_0^t \Phi(D(\hat{\varphi}_s, s), s)\psi(s)dA(s)$$

$$+ \sum_{k=1}^{d} \int_0^t \Gamma^k(D(\hat{\varphi}_s, s), s)\psi(s)M^k(D(\hat{\varphi}_s, s), ds) = z(t), \tag{3.7}$$

has a solution ψ in $C([0, T] \times \Omega; R^d)$ for any $\varphi \in C([-\tau, T] \times \Omega; R^d)$, $z \in C([0, T] \times \Omega; R^d)$.

Theorem 3.1 Assume that F is stochastically closed and has a regular bounded stochastic integral contractor $\{I + \int \Phi dA + \sum\int \Gamma^k M^k(ds)\}$. Let $\xi = \{\xi(s) : -\tau \le s \le 0\}$ be a $C([-\tau, 0]; R^d)$-valued $\mathcal{F}_0$-measurable random variable such that

$$E \int_0^T \left(| b(D(\hat{x}_s^0), s) |^2 + \sum_{i=1}^{d} a_{ii}(D(\hat{x}_s^0), D(\hat{x}_s^0), s) \right) dA(s) < \infty,$$

where $x^0(t) = \xi(t)$ for $-\tau \le t \le 0$ and $x^0(t) = \xi(0)$ for $0 \le t \le T$. Then equation (3.1) has a unique solution in $C([-\tau, T] \times \Omega; R^d)$.

Before the proof of the theorem, it is interesting to observe that if b(x, t) and a(x, y, t) are Lipschitzian, that is

$$| b(x, t) - b(y, t) |^2 \le \delta |x - y|^2,$$

$$\mathrm{trace}\Big(a(x, x, t) - 2a(x, y, t) + a(y, y, t) \Big) \le \delta |x - y|^2$$

for some $\delta > 0$, then F is stochastically closed and has a regular bounded stochastic integral contractor $\{I + \int \Phi dA + \sum\int \Gamma^k M^k(ds)\}$ with $\Phi = \Gamma^k = 0$. Hence we have the following corollary.

Corollary 3.2 Assume that b(x, t) and a(x, y, t) are Lipschitzian and ξ is the same as in Theorem 3.1. Then equation (3.1) has a unique solution in $C([-\tau, T] \times \Omega; R^d)$.

Proof of Theorem 3.1. For convenience we write D(u, t) =D(u) in the proof. We first prove the theorem under an additional condition:

(H3.1) $A(t) \equiv t$ for all $0 \le t \le T$.

Proof of Existence. Define $x^n(t)$, $y^n(t)$, $-\tau \le t \le T$, $n \ge 0$ as follows (x^0 has been defined above):

$$x^n(t) = \xi(t) \quad \text{and} \quad y^n(t) = 0 \qquad \text{for } -\tau \le t \le 0,$$

$$x^{n+1}(t) = x^n(t) - y^n(t) - \int_0^t \Phi(D(\hat{x}_s^n), s)\, y^n(s)\, ds$$

$$- \sum_{k=1}^d \int_0^t \Gamma^k(D(\hat{x}_s^n), s)\, y^n(s)\, M^k(D(\hat{x}_s^n), ds), \tag{3.8}$$

$$y^n(t) = x^n(t) - \xi(0) - \int_0^t F(D(\hat{x}_s^n), ds) \qquad \text{for } 0 \le t \le T. \tag{3.9}$$

<u>Step 1</u>. We claim

$$E\Big(\sup_{0 \le s \le t} |y^n(s)|^2 \Big) \le \alpha\, (n!)^{-1}\, (2K(T+1))^n\, t^n, \quad 0 \le t \le T \tag{3.10}$$

where α is a positive constant independent of n. In fact, it follows from the definitions above that

$$y^{n+1}(t) = x^{n+1}(t) - \xi(0) - \int_0^t F(D(\hat{x}_s^{n+1}), ds)$$

$$= x^n(t) - y^n(t) - \int_0^t \Phi(D(\hat{x}_s^n), s)\, y^n(s)\, ds$$

$$- \sum_{k=1}^d \int_0^t \Gamma^k(D(\hat{x}_s^n), s)\, y^n(s)\, M^k(D(\hat{x}_s^n), ds)$$

$$- \xi(0) - \int_0^t b(D(\hat{x}_s^{n+1}, s), s)\, ds - \int_0^t M(D(\hat{x}_s^{n+1}, s), ds)$$

$$= \int_0^t \Big(b(D(\hat{x}_s^n), s) - b(D(\hat{x}_s^{n+1}), s) - \Phi(D(\hat{x}_s^n), s)\, y^n(s) \Big)\, ds$$

$$+ \int_0^t \Big(M(D(\hat{x}_s^n), ds) - M(D(\hat{x}_s^{n+1}), ds) - \sum_{k=1}^d \Gamma^k(D(\hat{x}_s^n, s))\, y^n(s) M^k(D(\hat{x}_s^n), ds) \Big).$$

By Hölder's inequality and Doob's martingale inequality we have

$$E\Big(\sup_{0 \le s \le t} |y^{n+1}(s)|^2 \Big)$$

$$\leq 2T\, E \int_0^t |\, b(D(\hat{x}_s^n), s) - b(D(\hat{x}_s^{n+1}), s) - \Phi(D(\hat{x}_s^n), s)\, y^n(s)\, |^2\, ds$$

$$+ 8 \sum_{i=1}^{d} \int_0^t \Big(a_{ii}(D(\hat{x}_s^n), D(\hat{x}_s^n), s) - 2a_{ii}(D(\hat{x}_s^n), D(\hat{x}_s^{n+1}), s)$$

$$+ a_{ii}(D(\hat{x}_s^{n+1}), D(\hat{x}_s^{n+1}), s)$$

$$- 2\sum_{k=1}^{d} \Gamma_i^k(D(\hat{x}_s^n), s)\, \psi^n(s) \Big[a_{ik}(D(\hat{x}_s^n), D(\varphi_s^n), s) - a_{ik}(D(\hat{x}_s^n), D(\hat{x}_s^{n+1}), s) \Big]$$

$$+ \sum_{k,j=1}^{d} \Gamma_i^k(D(\hat{x}_s^n), s)\, y^n(s)\, \Gamma_i^j(D(\hat{x}_s^n), s)\, y^n(s)\, a_{kj}(D(\hat{x}_s^n), D(\hat{x}_s^n), s) \Big)\, ds$$

By conditions (3.5) and (3.6) with $\varphi = x^n$ and $\psi = -y^n$ we therefore have

$$E\Big(\sup_{0 \leq s \leq t} |y^{n+1}(s)|^2 \Big) \leq 2K(T + 4) \int_0^t E\Big(\sup_{0 \leq r \leq s} |y^n(r)|^2 \Big)\, ds.$$

Repeating integration gives

$$E\Big(\sup_{0 \leq s \leq t} |y^n(s)|^2 \Big)$$

$$\leq \Big((n-1)! \Big)^{-1} (2K(T + 4))^n \int_0^t (t-s)^{n-1}\, E\Big(\sup_{0 \leq r \leq s} |y^0(r)|^2 \Big)\, ds.$$

But we can see from (3.9) with n = 0 that

$$E\Big(\sup_{0 \leq r \leq s} |y^0(r)|^2 \Big)$$

$$\leq 2T \int_0^T E|\, b(D(\hat{x}_r^0), r)\, |^2\, dr + 2\sum_{k=1}^{d} \int_0^T E a_{kk}(D(\hat{x}_r^0), D(\hat{x}_r^0), r)\, dr$$

$$:= \alpha < \infty$$

for all $0 \leq s \leq T$. Therefore

$$E\Big(\sup_{0 \leq s \leq t} |y^n(s)|^2 \Big) \leq \alpha\, (n!)^{-1}\, (2K(T + 4))^n\, t^n$$

as required.

<u>Step 2</u>. Let β be a positive constant such that

$$| \Phi(x, t) |^2 + \sum_{k=1}^{d} | \Gamma^k(x, t) |^2 a_{kk}(x, x, t) \le \beta \tag{3.11}$$

for $(x, t) \in R^m \times [0, T]$. Then we see from (3.8) and (3.10) that

$$E\Big(\sup_{0\le t\le T} | x^{n+1}(t) - x^n(t) |^2 \Big)$$

$$\le (d+2)\, E\Big(\sup_{0\le t\le T} |y^n(t)| \Big) + (d+2)T\, E\int_0^T | \Phi(D(\hat{x}_s^n), s) |^2\, |y^n(s)|^2\, ds$$

$$+\ 4(d+2) \sum_{k=1}^{d} E\int_0^T | \Gamma^k(D(\hat{x}_s^n), s) |^2\, |y^n(s)|^2\, a_{kk}(D(\hat{x}_s^n), D(\hat{x}_s^n), s)\, ds$$

$$\le (d+2)\, E\Big(\sup_{0\le t\le T} |y^n(t)| \Big) + \beta(d+2)(T+4)\int_0^T E|y^n(s)|^2\, ds$$

$$\le (d+2)\big[1 + \beta T(T+4) \big]\, E\Big(\sup_{0\le t\le T} |y^n(t)| \Big)$$

$$\le (d+2)\big[1 + \beta T(T+4) \big]\, \alpha\, (n!)^{-1}\, (2KT(T+4))^n. \tag{3.12}$$

Hence

$$P\Big(\sup_{0\le t\le T} | x^{n+1}(t) - x^n(t) | > 2^{-n} \Big)$$

$$\le (d+2)\big[1 + \beta T(T+4) \big]\, \alpha\, (n!)^{-1}\, (8KT(T+4))^n.$$

By the Borel-Cantelli lemma, we have, with probability 1,

$$\sup_{0\le t\le T} | x^{n+1}(t) - x^n(t) | \le 2^{-n} \qquad \text{for all large } n.$$

Therefore, $x^n(t)$ converges uniformly in $t \in [0, T]$ almost surely. Denote the limit by $x(t)$, $t \in [0, T]$. Then $x(t)$ is a continuous adapted process. Extend $x(t)$ to $[-\tau, 0]$ by defining $x(t) = \xi(t)$ and we have $x \in C([-\tau, T] \times \Omega; R^d)$. It is easy to see from (3.12) that

$$x^n \to x \qquad \text{in } L^2([-\tau, T] \times \Omega; R^d).$$

Then

$$D(\hat{x}^n_\cdot) \to D(\hat{x}_\cdot) \qquad \text{in } L^2([0, T] \times \Omega; R^m)$$

since D is Lipschitz continuous.

Step 3. Set

$$\theta^n(t) = \int_0^t F(D(\hat{x}^n_s), ds) \qquad \text{and} \qquad \theta(t) = \int_0^t F(D(\hat{x}_s), ds)$$

on $0 \le t \le T$. We shall show that $\theta^n \to \theta$ in $L^2([0, T] \times \Omega; R^d)$. To do so we note from (3.5) with $\varphi = x^n$ and $\psi = -y^n$ that

$$| b(D(\hat{x}^n_s), s) - b(D(\hat{x}^{n+1}_s), s) |^2$$

$$\le | \Phi(D(\hat{x}^n_s), s)\, y^n(s) |^2 + K \sup_{0 \le r \le s} |y^n(r)|^2$$

$$\le (\beta + K) \sup_{0 \le r \le s} |y^n(r)|^2 .$$

Hence, by (3.3), we have

$$E \int_0^T \left| \int_0^t \Big(b(D(\hat{x}^n_s), s) - b(D(\hat{x}^{n+1}_s), s) \Big)\, ds \right|^2 dt$$

$$\le E \int_0^T t \int_0^t | b(D(\hat{x}^n_s), s) - b(D(\hat{x}^{n+1}_s), s) |^2\, ds\, dt$$

$$\le T^2 E \int_0^T | b(D(\hat{x}^n_s), s) - b(D(\hat{x}^{n+1}_s), s) |^2\, ds$$

$$\le (\beta + K)\, \alpha\, ((n + 1)!)^{-1} (2K(T + 1))^n\, T^{n+3}. \tag{3.13}$$

Moreover we have

$$E \int_0^T \left| \int_0^t M(D(\hat{x}^n_s), ds) - \int_0^t M(D(\hat{x}^{n+1}_s), ds) \right|^2 dt$$

$$= \sum_{i=1}^d \int_0^T E \Big< \int_0^t M_i(D(\hat{x}^n_s), ds) - \int_0^t M_i(D(\hat{x}^{n+1}_s), ds) \Big> dt$$

$$\leq T \sum_{i=1}^{d} E < \int_0^T \Big(M_i(D(\hat{x}_s^n), ds) - M_i(D(\hat{x}_s^{n+1}), ds) \Big) >,$$

where we use the notation $<U> = <U, U>$ as before. Note that the quadratic variations of martingales have the property

$$<U> \leq 2<U-V> + 2<V>$$

whenever U, V are continuous martingales. We have

$$< \int_0^T \Big(M_i(D(\hat{x}_s^n), ds) - M_i(D(\hat{x}_s^{n+1}), ds) \Big) >$$

$$\leq 2< \int_0^T \Big(M_i(D(\hat{x}_s^n), ds) - M_i(D(\hat{x}_s^{n+1}), ds)$$

$$- \sum_{k=1}^{d} \Gamma_i^k(D(\hat{x}_s^n), s)\, y^n(s)\, M_k(D(\hat{x}_s^n), ds) \Big) >$$

$$+ 2d \sum_{k=1}^{d} < \int_0^T \Gamma_i^k(D(\hat{x}_s^n), s)\, y^n(s)\, M_k(D(\hat{x}_s^n), ds) >$$

But by condition (3.6) (with $\varphi = x^n$ and $\psi = -y^n$),

$$\sum_{i=1}^{d} < \int_0^T \Big(M_i(D(\hat{x}_s^n), ds) - M_i(D(\hat{x}_s^{n+1}), ds)$$

$$- \sum_{k=1}^{d} \Gamma_i^k(D(\hat{x}_s^n), s)\, y^n(s)\, M^k(D(\hat{x}_s^n), ds) \Big) >$$

$$\leq K \int_0^T \sup_{0 \leq r \leq s} |y^n(r)|^2 \, ds.$$

Also from (3.11)

$$\sum_{k=1}^{d} < \int_0^T \Gamma_i^k(D(\hat{x}_s^n), s)\, y^n(s)\, M_k(D(\hat{x}_s^n), ds) >$$

$$\leq \sum_{k=1}^{d} \int_0^T |\Gamma^k(D(\hat{x}_s^n), s)|^2 \, |y^n(s)|^2 \, a_{kk}(D(\hat{x}_s^n), D(\hat{x}_s^n), s)\, ds$$

$$\leq \beta \int_0^T |y^n(s)|^2 \, ds.$$

We therefore find that

$$E \int_0^T \Big| \int_0^t M(D(\hat{x}_s^n), ds) - \int_0^t M(D(\hat{x}_s^{n+1}), ds) \Big|^2 dt$$

$$\leq 2T(K+\beta d^2) \, E \int_0^T \sup_{0 \leq r \leq s} |y^n(r)|^2 \, ds$$

$$\leq 2T(K+\beta d^2) \, \alpha \, ((n+1)!)^{-1} \, (2K(T+1))^n \, T^{n+1}. \tag{3.14}$$

Note that

$$E \int_0^T | \theta^n(t) - \theta^{n+1}(t) |^2 \, dt$$

$$\leq 2 \, E \int_0^T \Big| \int_0^t \Big(b(D(\hat{x}_s^n), s) - b(D(\hat{x}_s^{n+1}), s) \Big) ds \Big|^2 dt$$

$$+ \, 2 \, E \int_0^T \Big| \int_0^t M(D(\hat{x}_s^n), ds) - \int_0^t M(D(\hat{x}_s^{n+1}), ds) \Big|^2 dt \, .$$

This, together with (3.13) and (3.14), implies that θ^n converges in $L^2([0, T] \times \Omega, R^d)$. Since F is stochastically closed and

$$D(\hat{x}_{\cdot}^n) \to D(\hat{x}_{\cdot}) \qquad \text{in } L^2([0, T] \times \Omega; R^m),$$

we see that

$$\theta^n \to \theta \qquad \text{in } L^2([0, T] \times \Omega; R^d).$$

Step 4. We can now show that x is a solution of equation (3.1). In fact, taking the limits of both sides of (3.9) in $L^2([0, T] \times \Omega; R^d)$ we see that

$$x(t) = \xi(0) + \int_0^t F(D(\hat{x}_s), ds) \tag{3.15}$$

holds almost surely for any $t \in [0, T]$. But both sides of (3.15) have continuous sample paths, hence (3.15) holds for all $t \in [0, T]$ almost surely.

Proof of Uniqueness. Let x and w be two solutions of equation (3.1). By the regularity condition (3.7) with $\varphi = x$ and $z = x - w$, there exists $y \in C([0, T] \times \Omega; R^d)$ such that

$$y(t) + \int_0^t \Phi(D(\hat{x}_s), s)\, y(s)\, ds + \sum_{k=1}^{d} \int_0^t \Gamma^k(D(\hat{x}_s), s)\, y(s)\, M^k(D(\hat{x}_s), ds)$$

$$= x(t) - w(t). \tag{3.16}$$

Note $y(0) = 0$ and we can extend y to $[-\tau, 0]$ by defining $y(t) = 0$ for $-\tau \le t \le 0$ in order to have that $y \in C([-\tau, T] \times \Omega; R^d)$. For any $n \ge 1$ introduce a stopping time

$$\sigma_n = T \wedge \inf\{t: |y(t)|^2 \ge n, 0 \le t \le T\}.$$

We then have from (3.16) and (3.1) that

$$y(t \wedge \sigma_n)$$

$$= \int_0^{t\wedge\sigma_n} \Big(b(D(\hat{x}_s), s) - b(D(\hat{w}_s), s) - \Phi(D(\hat{x}_s), s)\, y(s) \Big)\, ds$$

$$+ \int_0^{t\wedge\sigma_n} \Big(M(D(\hat{x}_s), ds) - M(D(\hat{w}_s), ds)$$

$$- \sum_{k=1}^{d} \Gamma^k(D(\hat{x}_s), s)\, y(s)\, M^k(D(\hat{x}_s), ds) \Big).$$

In the same way as Step 1 we can derive

$$E\Big(\sup_{0\le s\le t} |y(s\wedge\sigma_n)|^2 \Big)$$

$$\le 2K(T + 4)\, E \int_0^{t\wedge\sigma_n} \sup_{0\le r\le s} |y(r)|^2\, ds$$

$$\le 2K(T + 4) \int_0^t E\Big(\sup_{0\le r\le s} |y(r\wedge\sigma_n)|^2 \Big)\, ds.$$

The well-known Gronwall inequality therefore gives

$$E\Big(\sup_{0 \le s \le T} |y(s \wedge \sigma_n)|^2 \Big) = 0$$

which implies that $y(t) = 0$ for $0 \le t \le \sigma_n$ almost surely. Note $\sigma_n \to T$ as $n \to \infty$. We have $y(t)=0$ on $0 \le t \le T$ almost surely. Therefore, by (3.16), $x = w$ and the solution is unique.

To complete the proof we shall remove the additional condition (H3.1) by changing the scale of time. For each $r \ge 0$ define the inverse function of $A(t)$:

$$\sigma_r = T \wedge \inf\{s: A(s) \ge r, 0 \le s \le T\}$$

which is an $\{\mathcal{F}_t\}$-stopping time. Let $\tilde{\mathcal{F}}_r = \mathcal{F}_{\sigma_r}$ and $\tilde{F}(x, r) = F(x, \sigma_r)$. Then, by Theorem 1-6.2, $\tilde{F}(x, r)$ is a continuous C-semimartingale adapted to $\{\tilde{\mathcal{F}}_r\}$ with the local characteristic of $\tilde{F}$ is $(a(x, y, \sigma_r)\mathbb{1}_{\{r<A(T)\}}, b(x, \sigma_r)\mathbb{1}_{\{r<A(T)\}}, r)$. Define an operator $\tilde{D} : C([-\tau, 0]; R^d) \times [0, \infty) \to R^m$ by

$$\tilde{D}(u, r) = D(u, \sigma_r),$$

which is Lipschitz continuous obviously. It is not difficult to check that $\tilde{F}$, restricted on any finite time interval $[0, \tilde{T}]$, is stochastically closed and has a bounded stochastic integral contractor. Therefore the stochastic differential-functional equation

$$dy(r) = \tilde{F}(\tilde{D}(\hat{y}_r), r), dr) \qquad \text{on } r \ge 0,$$

with initial data $\hat{y}_0 = \xi$, has a unique solution $y(r)$. Define $x(t) = \xi(t)$ on $-\tau \le t \le 0$ and $x(t) = y(A(t))$ on $0 < t \le T$. Then, by Theorem 1-6.3, we see

$$x(t) = \xi(0) + \int_0^{A(t)} \tilde{F}(\tilde{D}(\hat{y}_s, s), ds) = \xi(0) + \int_0^t F(D(\hat{x}_s, s), ds)$$

for $0 \le t \le T$. Therefore x is a solution of equation (3.1). In order to show the uniqueness of the solution without the additional condition (H3.1), let φ be an arbitrary solution of equation (3.1). Set $\psi(r) = \xi(r)$ on $-\tau \le r \le 0$ and $\psi(r) = \varphi(\sigma_r)$ on $r \ge 0$. Then, by Theorem 1-6.3 again, we have

$$\psi(r) = \xi(0) + \int_0^{\sigma_r} F(D(\hat{\varphi}_s, s), ds) = \xi(0) + \int_0^r \tilde{F}(\tilde{D}(\hat{\psi}_s, s), ds)$$

for all $r \ge 0$. Hence $\psi = y$, proving $\varphi = x$. The proof is complete.

As an important application, we shall now use the bounded stochastic integral contractor to deduce a new result on the existence and uniqueness of the solution for a stochastic differential equation (without delay). Let $F(x, t) = (F_1(x, t), ..., F_d(x, t))^T$, $(x, t) \in R^d \times [0, T]$, be a continuous C-semimartingale with the local characteristic $(a(x, y, t), b(x, t), A(t))$. Consider a stochastic differential equation (without delay)

$$dx(t) = F(x(t), dt) \quad \text{on } 0 \le t \le T, \tag{3.17}$$

with initial data $x(0) = \zeta$, where ζ is an $\mathcal{F}_0$-measurable random vector. We have pointed out that equation (3.17) is a special case of equation (3.1) and has been discussed in the previous chapter. As an application of Theorem 3.1 we shall obtain a generalized result for equation (3.17). To be precise let us give analogous definitions of bounded stochastic integral contractor and regularity associated with the equation.

Definition 3.4 Let

$$\Phi(x, t) = (\Phi_{ij})_{d\times d}, \quad \Gamma^k(x, t) = (\Gamma^k_{ij})_{d\times d}, \quad 1 \le k \le d, \ (x, t) \in R^d \times [0, T].$$

Suppose

$$|\Phi(x, t)|^2 + \sum_{k=1}^{d} |\Gamma^k(x, t)|^2 a_{kk}(x, x, t)$$

is bounded on $R^d \times [0, T]$. Suppose also that there exists a positive constant K such that for any $\varphi, \psi \in C([0, T] \times \Omega, R^d)$ the following inequalities hold for all $0 \le t \le T$:

$$|b(\varphi(t), t) - b(\pi(t, \varphi, \psi), t) - \Phi(\varphi(t), t)\psi(t)|^2 \le K \sup_{0\le s\le t} |\psi(s)|^2 \quad \text{a.s.}$$

and

$$\sum_{i=1}^{d} \Big(a_{ii}(\varphi(t), \varphi(t), t) - 2a_{ii}(\varphi(t), \pi(t, \varphi, \psi), t) + a_{ii}(\pi(t, \varphi, \psi), \pi(t, \varphi, \psi), t)$$

$$- 2\sum_{k=1}^{d} \Gamma^k_i(\varphi(t), t)\psi(t)\big[a_{ik}(\varphi(t), \varphi(t), t) - a_{ik}(\varphi(t), \pi(t, \varphi, \psi), t) \big]$$

$$+ \sum_{k,j=1}^{d} \Gamma^k_i(x(t), t)\, y(t)\, \Gamma^j_i(x(t), t)\, y(t)\, a^{kj}(x(t), x(t), t) \Big)$$

$$\le K \sup_{0\le s\le t} |\psi(s)|^2 \quad \text{a.s.,}$$

where $\Gamma^k_i = (\Gamma^k_{i1}, ..., \Gamma^k_{id})$ and

$$\pi(t, \varphi, \psi) = \varphi(t) + \psi(t) + \int_0^t \Phi(\varphi(s), s)\, \psi(s)\, dA(s)$$

$$+ \sum_{k=1}^{d} \int_0^t \Gamma^k(\varphi(s), s)\, \psi(s)\, M^k(\varphi(s), ds), \qquad 0 \le t \le T.$$

Then we say that the C-semimartingale F has a *bounded stochastic integral contractor* $\{I + \int \Phi dA + \sum \int \Gamma^k M^k(ds)\}$.

Definition 3.5 A bounded stochastic integral contractor $\{I + \int \Phi dA + \sum \int \Gamma^k M^k(ds)\}$ is said to be *regular* if the linear equation, on $0 \le t \le T$,

$$\psi(t) + \int_0^t \Phi(\varphi(s), s)\, \psi(s)\, ds$$

$$+ \sum_{k=1}^{d} \int_0^t \Gamma^k(\varphi(s), s)\, \psi(s)\, M^k(\varphi(s), ds) = z(t),$$

has a solution ψ in $C([0, T] \times \Omega, R^d)$ for any φ, $z \in C([0, T] \times \Omega; R^d)$.

The following theorem follows from Theorem 3.1.

Theorem 3.3 Assume F is stochastically closed and has a regular bounded stochastic integral contractor. Assume ζ is an $\mathcal{F}_0$-measurable random vector such that

$$E \int_0^T \left(| b(\zeta, s) |^2 + \text{trace } a(\zeta, \zeta, s) \right) dA(s) < \infty.$$

Then equation (3.17) has a unique solution in $C([0, T] \times \Omega, R^d)$.

Let us now give a simple example to illustrate our result and also to close this section.

Example 3.1 Let $F(x, t) = (F^1, ..., F^d)^T$, $(x, t) \in R^d \times [0, T]$, be a C-Brownian motion, which has the characteristic $(a(x, y, t), b(x, t), t)$. Consider a stochastic differential delay equation

$$dx(t) = F(x(t-\tau), dt) \quad \text{on } 0 \le t \le T, \tag{3.18}$$

with initial data $\hat{x}_0 = \xi$, where $\xi = \{\xi(s) : -\tau \le s \le 0\}$ is a $C([-\tau, 0], R^d)$-valued $\mathcal{F}_0$-

measurable random variable. Suppose that there exists a positive constant K_1 such that for any $\varphi, \psi \in C([-\tau, T] \times \Omega, R^d)$ the following inequality holds for all $0 \le t \le T$:

$$| b(\varphi(t-\tau), t) - b(\pi(t-\tau, \varphi, \psi), t) - \psi(t) |^2$$

$$\le K_1 \sup_{-\tau \le s \le t} |\psi(s)|^2 \quad \text{a.s.} \tag{3.19}$$

where

$$\pi(t, \varphi, \psi) = \varphi(t) + \psi(t) + \int_0^{t \vee 0} \psi(s)\, ds \quad \text{on } -\tau \le t \le T. \tag{3.20}$$

We also assume that there exists a positive constant K_2 such that for all $u, v \in R^d$ and $0 \le t \le T$,

$$\sum_{i=1}^{d} \Big(a_{ii}(u, u, t) - 2a_{ii}(u, v, t) + a_{ii}(v, v, t) \Big) \le K_2 |u - v|^2 \quad \text{a.s.} \tag{3.21}$$

Therefore we see that for any $\varphi, \psi \in C([-\tau, T] \times \Omega, R^d)$ the following inequality holds for all $0 \le t \le T$:

$$\sum_{i=1}^{d} \Big(a_{ii}(\varphi(t-\tau), \varphi(t-\tau), t) - 2a_{ii}(\varphi(t-\tau), \pi(t-\tau, \varphi, \psi), t)$$

$$+ a_{ii}(\pi(t-\tau, \varphi, \psi), \pi(t-\tau, \varphi, \psi), t) \Big)$$

$$\le K_2 \Big| \psi(t-\tau) + \int_0^{(t-\tau) \vee 0} \psi(s)\, ds \Big|^2$$

$$\le 2K_2 \Big(|\psi(t-\tau)|^2 + t \int_0^t |\psi(s)|^2\, ds \Big)$$

$$\le 2K_2 \Big(|\psi(t-\tau)|^2 + t^2 \sup_{0 \le s \le t} |\psi(s)|^2 \Big)$$

$$\le 2K_2(T^2 + 1) \sup_{-\tau \le s \le t} |\psi(s)|^2 \quad \text{a.s.} \tag{3.22}$$

That is, F has a bounded stochastic integral contractor $\{I + \int \Phi dA + \sum \int \Gamma^k M^k(ds)\}$ with $\Phi = I$ (= the identity matrix) and $\Gamma^k = 0$. We further assume that $b(x, t)$ has a property: if $z^n \to z$ in $L^2([0, T] \times \Omega; R^m)$ then

$$\int_0^{\cdot} b(z^n(s), ds) \to \int_0^{\cdot} b(z(s), ds) \quad \text{in } L^2([0, T] \times \Omega; R^d).$$

On the other hand, we have by condition (3.21) that for such z^n and z

$$E\int_0^T \left| \int_0^t M(z^n(s), ds) - \int_0^t M(z(s), ds) \right|^2 dt$$

$$= \sum_{i=1}^d \int_0^T E < \int_0^t M_i(z^n(s), ds) - \int_0^t M_i(z(s), ds) > dt$$

$$\le T \sum_{i=1}^d E < \int_0^T \left(M_i(z^n(s), ds) - M_i(z(s), ds) \right) >$$

$$= T \sum_{i=1}^d E \int_0^T \left(a_{ii}(z^n(s), z^n(s), s) - 2a_{ii}(z^n(s), z(s), s) + a_{ii}(z(s), z(s), s) \right) ds$$

$$\le TK_2\, E \int_0^T |z^n(s) - z(s)|^2\, ds \to 0 \quad \text{as } n \to \infty.$$

Hence, if $z^n \to x$ in $L^2([0, T] \times \Omega, R^m)$ then

$$\int_0^{\cdot} F(z^n(s), ds) \to \int_0^{\cdot} F(z^n(s), ds) \quad \text{in } L^2([0, T] \times \Omega; R^d),$$

that is F is stochastically closed. Note that for any $z \in C([0, T] \times \Omega; R^d)$, the equation

$$\psi(t) + \int_0^t \psi(s)\, ds = z(t) \tag{3.23}$$

has an explicit solution

$$\psi(t) = z(t) - \int_0^t z(s)\, e^{-(t-s)}\, ds.$$

In other words the bounded stochastic integral contractor $\{I + \int \Phi dA + \sum \int \Gamma^k M^k(ds)\}$ is regular. Therefore, by Theorem 3.1, the delay equation (3.18) has a unique solution in $C([-\tau, T] \times \Omega; R^d)$.

3.4. CAUCHY-MARUYAMA'S APPROXIMATE SOLUTIONS

There are several well-known procedures, for example, Cauchy-Maruyama's and Carathéodory's, for approximating the solution of a stochastic differential delay equation. In this section, we shall study the Cauchy-Maruyama approximation procedure, which in fact also represents an alternative to the procedure for establishing the existence and uniqueness of the solutions.

Let $F(x, y, t)$, $x, y \in R^d$ be a d-dimensional continuous C-semimartingale such that it has the decomposition $F(x, y, t) = M(x, y, t) + B(x, y, t)$ and the local characteristic $(a(x, y, \bar{x}, \bar{y}, t), b(x, y, t), A(t))$ belonging to the class $B^{0,\delta}$. Let $\tau > 0$ be a constant. Consider a d-dimensional stochastic differential delay equation driven by $F(x, y, t)$ of the form

$$x(t) = \xi(0) + \int_0^t F(x(s), x(s-\tau), ds) \qquad \text{on } t \geq 0 \tag{4.1}$$

with initial data $x(s) = \xi(s)$ on $-\tau \leq s \leq 0$, where $\xi(.)$ is a $C([-\tau, 0]; R^d)$-valued $\mathcal{F}_0$-measurable random variable. We introduce the standing hypotheses:

(H4.1) $E\|\xi\|_\tau^2 < \infty$ and $E|\xi(t) - \xi(s)|^2 \leq \alpha(t-s)$ for all $-\tau \leq s < t \leq 0$, where $\alpha > 0$.

(H4.2) Both $b(0, 0, t)$ and $a(0, 0, 0, 0, t)$ are bounded.

(H4.3) For all $x, y, \bar{x}, \bar{y} \in R^d$, $t \geq 0$,

$$|b(x, y, t) - b(\bar{x}, \bar{y}, t)|^2 \leq \kappa(|x - \bar{x}|^2 + |y - \bar{y}|^2) \qquad \text{a.s.},$$

$$\sum_{i=1}^d \Big(a_{ii}(x, y, x, y, t) - 2a_{ii}(x, y, \bar{x}, \bar{y}, t) + a_{ii}(\bar{x}, \bar{y}, \bar{x}, \bar{y}, t)\Big)$$

$$\leq \kappa(|x - \bar{x}|^2 + |y - \bar{y}|^2) \qquad \text{a.s.}$$

where $\kappa : R_+ \to R_+$ is a continuous increasing concave function such that

$$\int_{0+} \frac{du}{\kappa(u)} = \infty.$$

(H4.4) $A(t)$ is deterministic and $A(t) - A(s) \leq \beta(t - s)$ for all $0 \leq s < t < \infty$, where $\beta > 0$.

Theorem 4.1 Let (H4.1)-(H4.4) hold. Then there exists a unique solution $x(t)$ on $[-\tau, \infty)$ to equation (4.1), and for any $T > 0$,

$$E\Big(\sup_{-\tau \leq t \leq T} |x(t)|^2\Big) < \infty. \tag{4.2}$$

Moreover

$$\lim_{n\to\infty} E\Big(\sup_{-\tau \leq t \leq T} |x(t) - x^n(t)|^2\Big) = 0, \tag{4.3}$$

where $x^n(t)$, $t \geq -\tau$ and $n \geq 1$, are defined as follows:

$$x^n(t) = \xi(t) \quad \text{if } -\tau \leq t \leq 0 \quad \text{and}$$

$$x^n(t) = x^n(i/n) + F(x^n(i/n), x^n(i/n-\tau), t) - F(x^n(i/n), x^n(i/n-\tau), i/n)$$

$$\text{if } i/n < t \leq (i+1)/n, \ i = 0, 1, 2, \dots . \tag{4.4}$$

Proof. We divide the whole proof into five steps.

Step 1. For $n = 1, 2, \dots$, introduce simple processes

$$\hat{x}^n(t) = \sum_{i=0}^{\infty} x^n(i/n)\, \mathbb{1}_{[i/n,\,(i+1)/n)}(t) \qquad \text{on } t \geq 0,$$

$$\tilde{x}^n(t) = \sum_{i=0}^{\infty} x^n(i/n-\tau)\, \mathbb{1}_{[i/n,\,(i+1)/n)}(t) \qquad \text{on } t \geq 0.$$

Then (4.4) can be written as

$$x^n(t) = \xi(0) + \int_0^t F(\hat{x}^n(s), \tilde{x}^n(s), ds)$$

$$= \xi(0) + \int_0^t b(\hat{x}^n(s), \tilde{x}^n(s), s)\, dA(s) + \int_0^t M(\hat{x}^n(s), \tilde{x}^n(s), ds). \tag{4.5}$$

We first claim that for any $T > 0$, there exists a $C_1 = C_1(T) > 0$ such that

$$E\Big(\sup_{-\tau \leq t \leq T} |x^n(t)|^2 \Big) \leq C_1 \quad \text{for all } n = 1, 2, 3, \dots . \tag{4.6}$$

In fact, since $\kappa(.)$ is a concave function and $\kappa(0) = 0$, one can find a positive constant θ such that

$$\kappa(u) \leq \theta(1 + u) \quad \text{for all } u \geq 0. \tag{4.7}$$

By this inequality and hypotheses (H4.2) and (H4.3) it is easy to show that there exists a constant $K > 0$ such that

$$|b(x, y, t)|^2 \vee \sum_{i=1}^{d} a_{ii}(x, y, x, y, t) \leq K(1 + |x|^2 + |y|^2). \tag{4.8}$$

Using Hölder's inequality, Doob's inequality (i.e. Theorem 1-2.3) etc. we then derive from (4.5) that for any $0 \leq t \leq T$,

$$E\Big(\sup_{0 \leq s \leq t} |x^n(s)|^2 \Big) \leq 3E|\xi(0)|^2 + 3E\Big(A(t) \int_0^t |b(\hat{x}^n(s), \tilde{x}^n(s), s)|^2\, dA(s) \Big)$$

$$+ 12E \int_0^t \sum_{i=1}^d a_{ii}(\hat{x}^n(s), \tilde{x}^n(s), \hat{x}^n(s), \tilde{x}^n(s), s)\, dA(s)$$

$$\le 3E|\xi(0)|^2 + 3K(\beta T + 4)\, E\int_0^t \Big[1 + |\hat{x}^n(s)|^2 + |\tilde{x}^n(s)|^2 \Big]\, dA(s)$$

$$\le 3E|\xi(0)|^2 + 3K(\beta T + 4)\Big(\beta T + \beta\tau E\|\xi\|_\tau^2 + 2\int_0^t E\Big[\sup_{0\le r\le s} |x^n(r)|^2 \Big]\, dA(s) \Big).$$

Applying Theorem 1-7.11 we see that

$$E\Big(\sup_{0\le t\le T} |x^n(t)|^2 \Big) \le \Big(3E|\xi(0)|^2 + 3K(\beta T + 4)\big[\beta T + \beta\tau E\|\xi\|_\tau^2 \big] \Big)\, e^{6\beta TK(\beta T+4)}.$$

So (4.7) follows since

$$E\Big(\sup_{-\tau\le t\le T} |x^n(t)|^2 \Big) \le E\|\xi\|_\tau^2 + E\Big(\sup_{0\le t\le T} |x^n(t)|^2 \Big).$$

<u>Step 2</u>. For any $n \ge 1$, $0 \le s < t \le T$, we compute directly

$$E\Big(\sup_{s\le r\le t} |x^n(r) - x^n(s)|^2 \Big)$$

$$\le 2K\big[\beta(t - s) + 4\big]\, E\int_s^t \Big[1 + |\hat{x}^n(r)|^2 + |\tilde{x}^n(r)|^2 \Big]\, dA(r)$$

$$\le 2K\big[\beta T + 4\big]\, E\int_s^t \Big[1 + 2 \sup_{-\tau\le r\le T} |x^n(r)|^2 \Big]\, dA(r)$$

$$\le 2K\big[\beta T + 4\big]\, E\Big(\Big[1 + 2 \sup_{-\tau\le r\le T} |x^n(r)|^2 \Big] (A(t) - A(s)) \Big)$$

$$\le 2K\big[\beta T + 4\big]\, \beta(t - s)\, E\Big[1 + 2 \sup_{-\tau\le r\le T} |x^n(r)|^2 \Big]$$

$$\le 2\beta K(\beta T + 4)(1 + 2C_1)(t - s) = C_2(t - s), \tag{4.8}$$

where $C_2 = 2\beta K(\beta T + 4)(1 + 2C_1)$. Hence we see from the definition of $\hat{x}^n(t)$ that

$$E|\hat{x}^n(t) - x^n(t)|^2 \le C_2/n \qquad \text{if } 0 \le t \le T. \tag{4.9}$$

Similarly,

$$E|\tilde{x}^n(t) - x^n(t-\tau)|^2 \le (\alpha \vee C_2)/n \qquad \text{if } 0 \le t \le T, \tag{4.10}$$

where hypothesis (H4.1) has also been used.

Step 3. We now show that for any $T > 0$, $\{x^n(t): -\tau \le t \le T\}_{n\ge 1}$ is a Cauchy sequence in $L^2(\Omega; C([-\tau, T], R^d))$. Let $m > n \ge 1$ be arbitrary. By Hölder's inequality, Doob's martingale inequality and hypotheses (H4.3) and (H4.4) we can derive from (4.5) that for any $0 \le t \le T$,

$$E\Big(\sup_{0\le s\le t} |x^m(s) - x^n(s)|^2\Big)$$

$$\le 2\beta T E \int_0^t |b(\hat{x}^m(s), \tilde{x}^m(s), s) - b(\hat{x}^n(s), \tilde{x}^n(s), s)|^2 \, dA(s)$$

$$+ 8E \sum_{i=1}^d \int_0^t \Big[a_{ii}(\hat{x}^m(s), \tilde{x}^m(s), \hat{x}^m(s), \tilde{x}^m(s), s)$$

$$- 2a_{ii}(\hat{x}^m(s), \tilde{x}^m(s), \hat{x}^n(s), \tilde{x}^n(s), s) + a_{ii}(\hat{x}^n(s), \tilde{x}^n(s), \hat{x}^n(s), \tilde{x}^n(s), s) \Big] ds.$$

$$\le 2(\beta T + 4) \int_0^t \kappa\Big(E|\hat{x}^m(s) - \hat{x}^n(s)|^2 + E|\tilde{x}^m(s) - \tilde{x}^n(s)|^2 \Big) dA(s). \tag{4.11}$$

On the other hand, by (4.9) we compute

$$E|\hat{x}^m(s) - \hat{x}^n(s)|^2$$

$$\le 3E|\hat{x}^m(s) - x^m(s)|^2 + 3E|x^m(s) - x^n(s)|^2 + 3E|x^n(s) - \hat{x}^n(s)|^2$$

$$\le 3C_2\Big[\frac{1}{m} + \frac{1}{n}\Big] + 3E|x^m(s) - x^n(s)|^2.$$

Similarly, by (4.10) and (H4.1),

$$E|\tilde{x}^m(s) - \tilde{x}^n(s)|^2 \le 3C_3\Big[\frac{1}{m} + \frac{1}{n}\Big] + 3E|x^m(s-\tau) - x^n(s-\tau)|^2,$$

where $C_3 = \alpha \vee C_2$. So

$$E|\hat{x}^m(s) - \hat{x}^n(s)|^2 + E|\tilde{x}^m(s) - \tilde{x}^n(s)|^2$$

$$\le 6C_3\Big[\frac{1}{m} + \frac{1}{n}\Big] + 6E\Big[\sup_{0\le r\le s} |x^m(r) - x^n(r)|^2\Big].$$

Note also that $\kappa(u + v) \le \kappa(u) + \kappa(v)$ for all $u, v \ge 0$ since κ is concave and $\kappa(0) = 0$. We therefore obtain from (4.11) that

$$E\Big(\sup_{0\le s\le t}|x^m(s)-x^n(s)|^2\Big)$$

$$\le \delta_{m,n} + 12(\beta T+4)\int_0^t \kappa\Big(E\Big[\sup_{0\le r\le s}|x^m(r)-x^n(r)|^2\Big]\Big)\,dA(s),$$

where

$$\delta_{m,n} = 2\beta T(\beta T+4)\,\kappa\Big(6C_3\Big[\frac{1}{m}+\frac{1}{n}\Big]\Big).$$

Using Theorem 1–7.2 we get

$$E\Big(\sup_{0\le t\le T}|x^m(t)-x^n(t)|^2\Big) \le Q^{-1}\Big(Q(\delta_{m,n}) + 12\beta T(\beta T+4)\Big) \tag{4.12}$$

provided

$$Q(\delta_{m,n}) + 12\beta T(\beta T+4) \in \mathrm{Dom}(Q^{-1}),$$

where

$$Q(u) = \int_1^u \frac{ds}{\kappa(s)} \qquad \text{for } u > 0,$$

and Q^{-1} is the inverse function of Q. By the condition on κ we see $(-\infty, 0] \subset \mathrm{Dom}(Q^{-1})$, $Q(u) \to -\infty$ as $u \to 0$ and $Q^{-1}(u) \to 0$ as $u \to -\infty$. Also note $\delta_{m,n} \to 0$ as $m, n \to \infty$. So $Q(\delta_{m,n}) \to -\infty$ as $m, n \to \infty$. Thus (4.12) gives

$$E\Big(\sup_{0\le t\le T}|x^m(t)-x^n(t)|^2\Big) \to 0 \qquad \text{as } m, n \to \infty.$$

Hence, $\{x^n(t): -\tau \le t \le T\}_{n\ge1}$ is a Cauchy sequence in $L^2(\Omega; C([-\tau, T], R^d))$ for any $T > 0$. Therefore, there exists a continuous process $x(t)$ on $t \ge -\tau$ such that for any $T > 0$, $\{x(t): -\tau \le t \le T\} \in L^2(\Omega; C([-\tau, T], R^d))$ and

$$\lim_{n\to\infty} E\Big(\sup_{-\tau\le t\le T}|x(t)-x^n(t)|^2\Big) = 0. \tag{4.13}$$

Letting $n \to \infty$ in (4.7) gives

$$E\Big(\sup_{-\tau\le t\le T}|x(t)|^2\Big) \le C_1. \tag{4.14}$$

Step 4. Now we show $x(t)$ is a solution of equation (4.1). In fact, it is obvious that $x(t)$ is $\{\mathcal{F}_t\}$-adapted and $x(t) = \xi(t)$ on $-\tau \le t \le 0$. Moreover

$$E\Big|\int_0^t b(x(s), x(s-\tau), s)\,dA(s) - \int_0^t b(\hat{x}^n(s), \tilde{x}^n(s), s)\,dA(s)\Big|^2$$

$$+ E\left| \int_0^t M(x(s), x(s-\tau), ds) - \int_0^t M(\hat{x}^n(s), \tilde{x}^n(s), ds) \right|^2$$

$$\le (\beta t + 1) \int_0^t \kappa\left(E|x(s) - \hat{x}^n(s)|^2 + E|x(s-\tau) - \tilde{x}^n(s)|^2 \right) dA(s)$$

$$\le (\beta t + 1) \int_0^t \kappa\Big(2E|x(s) - x^n(s)|^2 + 2E|x^n(s) - \hat{x}^n(s)|^2$$
$$+ 2E|x(s-\tau) - x^n(s-\tau)|^2 + 2E|x^n(s-\tau) - \tilde{x}^n(s)|^2 \Big) dA(s)$$

$$\to 0 \qquad \text{as } n \to \infty.$$

Consequently, we can let $n \to \infty$ in (4.5) to obtain that

$$x(t) = \xi(0) + \int_0^t F(x(s), x(s-\tau), ds).$$

Step 5. In order to show the uniqueness, let $\bar{x}(t)$, $t \ge -\tau$ be another solution. Then we can derive that

$$E\Big(\sup_{0 \le s \le t} |x(s) - \bar{x}(s)|^2 \Big) \le 4(\beta t + 4) \int_0^t \kappa\Big(E\Big[\sup_{0 \le r \le s} |x(r) - \bar{x}(r)|^2 \Big] \Big) dA(s)$$

for all $t \ge 0$, which implies, by Theorem 1-7.2, that

$$E\Big(\sup_{0 \le s \le t} |x(s) - \bar{x}(s)|^2 \Big) = 0 \quad \text{for all } t \ge 0.$$

So $x(t) = \bar{x}(t)$ a.s. for all $t \ge 0$. The proof is complete.

Hypothesis (H4.1) requires the mean square continuity of the initial data $\xi(t)$ on $-\tau \le t \le 0$. This condition can be removed by a slight modification of the approximate solutions, which is described in the following theorem.

Theorem 4.2 Let (H4.2)-(H4.4) hold and $E(\|\xi\|_\tau)^2 < \infty$. Then there exists a unique solution $x(t)$ on $[-\tau, \infty)$ to equation (4.1), and for any $T > 0$,

$$E\Big(\sup_{-\tau \le t \le T} |x(t)|^2 \Big) < \infty.$$

Moreover

$$\lim_{n\to\infty} E\Big(\sup_{-\tau\le t\le T} |x(t)-x^n(t)|^2 \Big) = 0,$$

where $x^n(t)$ on $t \ge -\tau$, for $n > 1/\tau$, are defined by:

$$x^n(t) = \xi(t) \quad \text{if } -\tau \le t \le 0 \quad \text{and}$$

$$x^n(t) = x^n(i/n) + \int_{i/n}^{t} F(x^n(i/n), x^n(s-\tau), ds\,)$$

$$\text{if } i/n < t \le (i+1)/n, \ \ i = 0, 1, 2, \dots .$$

The theorem can be proved in the same way as Theorem 4.1 and the details are left to the reader.

3.5. CARATHÉODORY'S APPROXIMATE SOLUTIONS

Carathéodory developed an interesting approximation scheme via small delays for an ordinary differential equation. The scheme is described as follows (cf. Coddington and Levison (1955), Chapter 2): For an given ordinary differential equation

$$\dot{x}(t) = g(x(t)) \qquad \text{on } t \ge 0$$

with initial data $x(0) = x_o$, define the approximate solution $x^n(t)$ for each $n \ge 1$ via the delay equation

$$\dot{x}^n(t) = g(x(t - 1/n)) \quad \text{on } t \ge 0$$

with initial data $x^n(t) = x_o$ on $-1/n \le t \le 0$. Carathéodory showed that $x^n(t)$ converges to $x(t)$ uniformly in any finite time interval as $n \to \infty$. The advantage of this approximation procedure is that the delay equation can be solved explicitly by successive integrations over intervals of length $1/n$. The Carathéodory scheme was extended to the Itô stochastic differential equations by Bell and Mohammed (1989) and to the Itô stochastic differential delay equations by Mao (1991g).

In this section, we shall discuss the Carathéodory approximation procedure for a d-dimensional stochastic differential delay equation driven by $F(x, y, t)$ of the form

$$x(t) = \xi(0) + \int_0^t F(x(s), x(s-\delta(s)), ds) \qquad \text{on } t \ge 0 \tag{5.1}$$

with initial data $x(s) = \xi(s)$ on $-\tau \le s \le 0$, where $F(x, y, t)$ and ξ are the same as defined in Section 3.4, $\delta(.)$ is a continuous nonnegative function defined on R_+ such that $t-\delta(t)$ is bounded below and $-\tau := \inf\{t-\delta(t): t \ge 0\}$. For each integer $n \ge 2/\tau$, define $x^n(t)$, $t \ge -\tau$ as follows:

$$x^n(t) = \xi(t) \quad \text{on } -\tau \le t \le 0 \quad \text{and}$$

$$x^n(t) = \xi(0) + \int_0^t 1\!\!1_{D_n^c}(s)\, F(x^n(s-n^{-1}),\ x^n(s-\delta(s)),\ ds)$$

$$+ \int_0^t 1\!\!1_{D_n}(s)\, F(x^n(s-n^{-1}),\ x^n(s-\delta(s)-n^{-1}),\ ds) \qquad \text{on } t \ge 0, \tag{5.2}$$

where

$$D_n = \{t \ge 0 : \tau(t) < n^{-1}\} \quad \text{and} \quad D_n^c = \{t \ge 0 : \tau(t) \ge n^{-1}\}.$$

Note that each x^n can be determined by step-wise iterated integrals over the intervals $[0, 1/n]$, $(1/n, 2/n]$, ..., etc. The aim of this section is to show that x^n converges to the unique solution $x(t)$ of equation (5.1) as described by the following theorem.

Theorem 5.1 Let (H4.2)-(H4.4) hold and $E(\|\xi\|_\tau)^2 < \infty$. Then there exists a unique solution $x(t)$ on $[-\tau, \infty)$ to equation (5.1), and for any $T > 0$,

$$E\Big(\sup_{-\tau\le t\le T} |x(t)|^2\Big) < \infty. \tag{5.3}$$

Moreover

$$\lim_{n\to\infty} E\Big(\sup_{-\tau\le t\le T} |x(t) - x^n(t)|^2\Big) = 0, \tag{5.4}$$

where $x^n(t)$, $t \ge -\tau$ and $n \ge 2/\tau$, are defined by (5.2).

Proof. We divide the whole proof into five steps.

Step 1. We first claim that for any $T > 0$, there exists a $C_1 = C_1(T) > 0$ such that

$$E\Big(\sup_{-\tau\le t\le T} |x^n(t)|^2\Big) \le C_1 \quad \text{for all } n \ge 2/\tau. \tag{5.5}$$

In fact, we know as before that there exists a constant $K > 0$ such that

$$|b(x, y, t)|^2 \vee \sum_{i=1}^d a_{ii}(x, y, x, y, t) \le K(1 + |x|^2 + |y|^2).$$

We then apply Hölder's inequality, Doob's martingale inequality etc. to derive from (5.2) that for any $0 \le t \le T$,

$$E\Big(\sup_{0\le s\le t}|x^n(s)|^2\Big) \le 5E|\xi(0)|^2$$

$$+ 5E\Big(A(t)\int_0^t 1\!\!1_{D_n^c}(s)\,|b(x^n(s-n^{-1}),\ x^n(s-\delta),\ s)|^2\,dA(s)\Big)$$

$$+ 20E\int_0^t 1\!\!1_{D_n^c}(s)\sum_{i=1}^d a_{ii}(x^n(s-n^{-1}),\ x^n(s-\delta),\ x^n(s-n^{-1}),\ x^n(s-\delta),\ s)\,dA(s)$$

$$+ 5E\Big(A(t)\int_0^t 1\!\!1_{D_n}(s)\,|b^n(x(s-n^{-1}),\ x^n(s-\delta-n^{-1}),\ s)|^2\,dA(s)\Big)$$

$$+ 20E\int_0^t 1\!\!1_{D_n}(s)\sum_{i=1}^d a_{ii}(x^n(s-n^{-1}),\ x^n(s-\delta-n^{-1}),\ x^n(s-n^{-1}),\ x^n(s-\delta-n^{-1}),\ s)\,dA(s)$$

$$\le 5E|\xi(0)|^2 + 5K(\beta T+4)\,E\int_0^t 1\!\!1_{D_n^c}(s)\Big[1 + |x^n(s-n^{-1})|^2 + |x^n(s-\delta)|^2\Big]\,dA(s)$$

$$+ 5K(\beta T+4)\,E\int_0^t 1\!\!1_{D_n}(s)\Big[1 + |x^n(s-n^{-1})|^2 + |x^n(s-\delta-n^{-1})|^2\Big]\,dA(s)$$

$$\le 5E|\xi(0)|^2 + 5K(\beta T+4)\,E\int_0^t\Big[1 + 2\|\xi\|_\tau^2 + 2\sup_{0\le r\le s}|x^n(r)|^2\Big]\,dA(s)$$

$$\le 5E|\xi(0)|^2 + 5K(\beta T+4)\Big(\beta T(1 + 2E\|\xi\|_\tau^2) + 2\int_0^t E\Big[\sup_{0\le r\le s}|x^n(r)|^2\Big]\,dA(s)\Big).$$

Applying Theorem 1–7.11 we see that

$$E\Big(\sup_{0\le t\le T}|x^n(t)|^2\Big) \le \Big(5E|\xi(0)|^2 + 5\beta TK(\beta T+4)(1 + 2E\|\xi\|_\tau^2)\Big)\,e^{10\beta TK(\beta T+4)}.$$

So (5.5) follows since

$$E\Big(\sup_{-\tau\le t\le T}|x^n(t)|^2\Big) \le E\|\xi\|_\tau^2 + E\Big(\sup_{0\le t\le T}|x^n(t)|^2\Big).$$

Step 2. For any $n \ge 2/\tau$, $0 \le s < t \le T$, we compute directly

$$E\Big(\sup_{s\le r\le t}|x^n(r) - x^n(s)|^2\Big)$$

$$\leq 4K\Big[\beta(t-s)+4\Big] E \int_s^t 1\!\!1_{D_n^c}(r)\Big[\,1 + |x^n(r-n^{-1})|^2 + |x^n(r-\delta(r))|^2\,\Big]\, dA(r)$$

$$+\,4K\Big[\beta(t-s)+4\Big] E \int_s^t 1\!\!1_{D_n}(r)\Big[\,1 + |x^n(r-n^{-1})|^2 + |x^n(r-\delta(r)-n^{-1})|^2\,\Big]\, dA(r)$$

$$\leq 4K\Big[\beta T+4\Big] E \int_s^t \Big[\,1 + 2\sup_{-\tau\leq r\leq T} |x^n(r)|^2\,\Big]\, dA(r)$$

$$\leq 4K\Big[\beta T+4\Big] E\bigg(\Big[\,1 + 2\sup_{-\tau\leq r\leq T} |x^n(r)|^2\,\Big](A(t)-A(s))\bigg)$$

$$\leq 4K\Big[\beta T+4\Big]\,\beta(t-s)\, E\Big[\,1 + 2\sup_{-\tau\leq r\leq T} |x^n(r)|^2\,\Big]$$

$$\leq 4\beta K(\beta T+4)(1+2C_1)(t-s) = C_2(t-s), \tag{5.6}$$

where $C_2 = 4\beta K(\beta T+4)(1+2C_1)$.

Step 3. We now show that for any $T > 0$, $\{x^n(t): -\tau \leq t \leq T\}_{n\geq 1}$ is a Cauchy sequence in $L^2(\Omega; C([-\tau, T], R^d))$. Let $m > n \geq \tau/2$ be arbitrary. Write

$$x^m(t) - x^n(t) = J_1(t) + J_2(t), \qquad 0 \leq t \leq T, \tag{5.7}$$

where

$$J_1(t) = \int_0^t \Big(1\!\!1_{D_m^c}(s)\, F(x^m(s-m^{-1}),\, x^m(s-\delta(s)),\, ds) - 1\!\!1_{D_n^c}(s)\, F(x^n(s-n^{-1}),\, x^n(s-\delta(s)),\, ds) \Big),$$

$$J_2(t) = \int_0^t \Big(1\!\!1_{D_m}(s)\, F(x^m(s-m^{-1}),\, x^m(s-\delta(s)-m^{-1}),\, ds) - 1\!\!1_{D_n}(s)\, F(x^n(s-n^{-1}),\, x^n(s-\delta(s)-n^{-1}),\, ds) \Big).$$

Note

$$J_1(t) = \int_0^t 1\!\!1_{D_n^c}(s)\Big(F(x^m(s-m^{-1}),\, x^m(s-\delta(s)),\, ds) - F(x^n(s-n^{-1}),\, x^n(s-\delta(s)),\, ds) \Big)$$

$$+ \int_0^t \mathbb{1}_{D_m^c - D_n^c}(s)\, F(x^m(s-m^{-1}),\ x^m(s-\delta(s)),\ ds).$$

We can derive that

$$E\Big(\sup_{0 \le s \le t} |J_1(s)|^2 \Big)$$

$$\le 4(\beta T + 4) \int_0^t \mathbb{1}_{D_n^c}(s)$$

$$\times\ \kappa\Big(E|x^m(s-m^{-1}) - x^n(s-n^{-1})|^2 + E|x^m(s-\delta(s)) - x^n(s-\delta(s))|^2 \Big)\, dA(s)$$

$$+ 4K(\beta T + 4) \int_0^t \mathbb{1}_{D_m^c - D_n^c}(s) \Big[1 + E|x^m(s-m^{-1})|^2 + E|x^m(s-\delta(s))|^2 \Big]\, dA(s)$$

$$\le 4(\beta T + 4) \int_0^t \kappa\Big(E|x^m(s-m^{-1}) - x^n(s-n^{-1})|^2 \Big)\, dA(s)$$

$$+ 4(\beta T + 4) \int_0^t \kappa\Big(E\Big[\sup_{0 \le r \le s} |x^m(r) - x^n(r)|^2 \Big] \Big)\, dA(s)$$

$$+ 4\beta K(\beta T + 4)(1 + 2C_1)\, \mu\Big[[0, T] \cap (D_m^c - D_n^c) \Big],$$

where (5.5) has been used and μ is the Lebesgue measure on R_+. But,

$$\int_0^t \kappa\Big(E|x^m(s-m^{-1}) - x^n(s-n^{-1})|^2 \Big)\, dA(s)$$

$$\le 3 \int_0^t \kappa\Big(E|x^m(s-m^{-1}) - x^m(s)|^2 \Big)\, dA(s) + 3 \int_0^t \kappa\Big(E|x^m(s) - x^n(s)|^2 \Big)\, dA(s)$$

$$+ 3 \int_0^t \kappa\Big(E|x^n(s) - x^n(s-n^{-1})|^2 \Big)\, dA(s)$$

$$\le 3\beta m^{-1} \kappa(4C_1) + 3 \int_{m^{-1}}^T \kappa\Big(E|x^m(s-m^{-1}) - x^m(s)|^2 \Big)\, dA(s)$$

$$+ 3 \int_0^t \kappa\Big(E\Big[\sup_{0\le r\le s} |x^m(r) - x^n(r)|^2 \Big] \Big)\, dA(s)$$

$$+ 3\beta n^{-1}\kappa(4C_1) + 3 \int_{n^{-1}}^T \kappa\Big(E|x^n(s) - x^n(s-n^{-1})|^2 \Big)\, dA(s)$$

$$\le 12\beta\kappa(C_1)(m^{-1} + n^{-1}) + 3\beta T \Big[\kappa(C_2 m^{-1}) + \kappa(C_2 n^{-1}) \Big]$$

$$+ 3 \int_0^t \kappa\Big(E\Big[\sup_{0\le r\le s} |x^m(r) - x^n(r)|^2 \Big] \Big)\, dA(s).$$

Hence

$$E\Big(\sup_{0\le s\le t} |J_1(s)|^2 \Big)$$
$$\le 4(\beta T + 4) \Big(12\beta\kappa(C_1)(m^{-1} + n^{-1}) + 3\beta T \Big[\kappa(C_2 m^{-1}) + \kappa(C_2 n^{-1}) \Big] \Big)$$
$$+ 4\beta K(\beta T + 4)(1 + 2C_1)\, \mu\Big[[0, T] \cap (D_m^c - D_n^c) \Big]$$
$$+ 16(\beta T + 4) \int_0^t \kappa\Big(E\Big[\sup_{0\le r\le s} |x^m(r) - x^n(r)|^2 \Big] \Big)\, dA(s). \qquad (5.8)$$

Similarly, we can show that

$$E\Big(\sup_{0\le s\le t} |J_2(s)|^2 \Big)$$
$$\le 4(\beta T + 4) \Big(12\beta\kappa(C_1)(m^{-1} + n^{-1}) + 3\beta T \Big[\kappa(C_2 m^{-1}) + \kappa(C_2 n^{-1}) \Big] \Big)$$
$$+ 4\beta K(\beta T + 4)(1 + 2C_1)\, \mu\Big([0, T] \cap (D_n - D_m) \Big)$$
$$+ 12(\beta T + 4) \int_0^t \kappa\Big(E\Big[\sup_{0\le r\le s} |x^m(r) - x^n(r)|^2 \Big] \Big)\, dA(s)$$
$$+ 4(\beta T + 4) \int_0^t \mathbb{1}_{D_m}(s)\, \kappa\Big(E|x^m(s-\delta(s)-m^{-1}) - x^n(s-\delta(s)-n^{-1})|^2 \Big)\, dA(s).$$

But

$$\int_0^t \mathbb{1}_{D_m}(s)\, \kappa\Big(E|x^m(s-\delta(s)-m^{-1}) - x^n(s-\delta(s)-n^{-1})|^2 \Big)\, dA(s).$$

$$\le 24\beta m^{-1}\kappa(C_1) + 3\int_{2m^{-1}}^{T} 1\!\!1_{D_m}(s)\,\kappa\Big(E|x^m(s-\delta(s)-m^{-1}) - x^m(s-\delta(s))|^2\Big)\,dA(s)$$

$$+ 3\int_0^t 1\!\!1_{D_m}(s)\,\kappa\Big(E\Big[\sup_{0\le r\le s}|x^m(r) - x^n(r)|^2\Big]\Big)\,dA(s)$$

$$+ 24\beta n^{-1}\kappa(C_1) + 3\int_{2n^{-1}}^{T} 1\!\!1_{D_m}(s)\,\kappa\Big(E|x^n(s-\delta(s)) - x^n(s-\delta(s)-n^{-1})|^2\Big)\,dA(s)$$

$$\le 24\beta\kappa(C_1)(n^{-1} + m^{-1}) + 3\beta T\Big[\kappa(C_2 m^{-1}) + \kappa(C_2 n^{-1})\Big]$$

$$+ 3\int_0^t \kappa\Big(E\Big[\sup_{0\le r\le s}|x^m(r) - x^n(r)|^2\Big]\Big)\,dA(s).$$

We therefore obtain

$$E\Big(\sup_{0\le s\le t}|J_2(s)|^2\Big)$$

$$\le 4(\beta T + 4)\Big(36\beta\kappa(C_1)(m^{-1} + n^{-1}) + 6\beta T\Big[\kappa(C_2 m^{-1}) + \kappa(C_2 n^{-1})\Big]\Big)$$

$$+ 4\beta K(\beta T + 4)(1 + 2C_1)\,\mu\Big[[0, T]\cap(D_n - D_m)\Big]$$

$$+ 24(\beta T + 4)\int_0^t \kappa\Big(E\Big[\sup_{0\le r\le s}|x^m(r) - x^n(r)|^2\Big]\Big)\,dA(s). \tag{5.9}$$

We therefore see from (5.7)–(5.9) that for any $0 \le t \le T$,

$$E\Big(\sup_{0\le s\le t}|x^m(s) - x^n(s)|^2\Big)$$

$$\le 2E\Big(\sup_{0\le s\le t}|J_1(s)|^2\Big) + 2E\Big(\sup_{0\le s\le t}|J_2(s)|^2\Big)$$

$$\le \delta_{m,n} + 80(\beta T + 4)\int_0^t \kappa\Big(E\Big[\sup_{0\le r\le s}|x^m(r) - x^n(r)|^2\Big]\Big)\,dA(s), \tag{5.10}$$

where

$$\delta_{m,n} = 8(\beta T + 4)\Big(48\beta\kappa(C_1)(m^{-1} + n^{-1}) + 9\beta T\Big[\kappa(C_2 m^{-1}) + \kappa(C_2 n^{-1})\Big]\Big)$$

$$+ 16\beta K(\beta T + 4)(1 + 2C_1)\mu\Big[[0, T]\cap(D_n - D_m)\Big].$$

Using Theorem 1-7.2 we get

$$E\Big(\sup_{0\le t\le T}|x^m(t)-x^n(t)|^2\Big) \le Q^{-1}\Big(Q(\delta_{m,n})+80\beta T(\beta T+4)\Big) \tag{5.11}$$

provided $Q(\delta_{m,n})+80\beta T(\beta T+4)\in \mathrm{Dom}(Q^{-1})$, where

$$Q(u)=\int_1^u \frac{ds}{\kappa(s)} \qquad \text{for } u>0,$$

and Q^{-1} is the inverse function of Q. By the condition on κ we see $(-\infty, 0]\subset \mathrm{Dom}(Q^{-1})$, $Q(u)\to-\infty$ as $u\to 0$ and $Q^{-1}(u)\to 0$ as $u\to-\infty$. Also note $\delta_{m,n}\to 0$ as $m, n\to\infty$. So $Q(\delta_{m,n})\to-\infty$ as $m, n\to\infty$. Thus (5.11) gives

$$E\Big(\sup_{0\le t\le T}|x^m(t)-x^n(t)|^2\Big)\to 0 \qquad \text{as } m, n\to\infty.$$

Hence, $\{x^n(t): -\tau\le t\le T\}_{n\ge1}$ is a Cauchy sequence in $L^2(\Omega; C([-\tau, T], R^d))$ for any $T>0$. Therefore, there exists a continuous process $x(t)$ on $t\ge-\tau$ such that for any $T>0$, $\{x(t): -\tau\le t\le T\}\in L^2(\Omega; C([-\tau, T], R^d))$ and

$$\lim_{n\to\infty} E\Big(\sup_{-\tau\le t\le T}|x(t)-x^n(t)|^2\Big)=0. \tag{5.12}$$

Letting $n\to\infty$ in (5.5) gives

$$E\Big(\sup_{-\tau\le t\le T}|x(t)|^2\Big)\le C_1. \tag{5.13}$$

Step 4. Now we show $x(t)$ is a solution of equation (5.1). In fact, it is obvious that $x(t)$ is $\{\mathcal{F}_t\}$-adapted and $x(t)=\xi(t)$ on $-\tau\le t\le 0$. Moreover

$$E\Big|\int_0^t F(x(s), x(s-\delta(s)), ds) - \int_0^t \mathbb{1}_{D_n^c}(s)\, F(x^n(s-n^{-1}), x^n(s-\delta(s)), ds)$$

$$- \int_0^t \mathbb{1}_{D_n}(s)\, F(x^n(s-n^{-1}), x^n(s-\delta(s)-n^{-1}), ds)\Big|^2$$

$$\le 4(\beta t+4)\int_0^t \mathbb{1}_{D_n^c}(s)\,\kappa\Big(E|x(s)-x^n(s-n^{-1})|^2 + E|x(s-\delta(s))-x^n(s-\delta(s))|^2\Big)\,dA(s)$$

$$+ 4(\beta t+4)\int_0^t \mathbb{1}_{D_n}(s)\,\kappa\Big(E|x(s)-x^n(s-n^{-1})|^2 + E|x(s-\delta(s))-x^n(s-\delta(s)-n^{-1})|^2\Big)\,dA(s)$$

$$\leq 8(\beta t + 4)\int_0^t \kappa\Big(2E|x(s)-x^n(s)|^2 + 2E|x^n(s)-x^n(s-n^{-1})|^2 \Big)\, dA(s)$$

$$+ 12(\beta t + 4)\int_0^t \kappa\Big(E|x(s-\delta(s)) - x^n(s-\delta(s))|^2 \Big)\, dA(s)$$

$$+ 8(\beta t + 4)\int_0^t \mathbb{1}_{D_n}(s)\, \kappa\Big(E|x^n(s-\delta(s)) - x^n(s-\delta(s)-n^{-1})|^2 \Big)\, dA(s)$$

$$\to 0 \qquad \text{as } n \to \infty.$$

Consequently, we can let $n \to \infty$ in (5.2) to obtain that

$$x(t) = \xi(0) + \int_0^t F(x(s), x(s-\delta(s)), ds).$$

Step 5. In order to show the uniqueness, let $\bar{x}(t)$, $t \geq -\tau$ be another solution. Then we can derive that

$$E\Big(\sup_{0\leq s\leq t} |x(s) - \bar{x}(s)|^2 \Big) \leq 4(\beta t + 4)\int_0^t \kappa\Big(E\Big[\sup_{0\leq r\leq s} |x(r) - \bar{x}(r)|^2 \Big]\Big)\, dA(s)$$

for all $t \geq 0$, which implies, by Theorem 1-7.2, that

$$E\Big(\sup_{0\leq s\leq t} |x(s) - \bar{x}(s)|^2 \Big) = 0 \qquad \text{for all } t \geq 0.$$

So $x(t) = \bar{x}(t)$ a.s. for all $t \geq 0$. The proof is complete.

3.6. L^p-ESTIMATES

In Section 3.2 we obtained an L^p-estimate (i.e. Theorem 2.4) for the solution of the stochastic differential delay equation (2.1). In this section we shall establish L^p-estimates for the solution of a stochastic differential-functional equation. It should be pointed out that the technique employed in this section is better than that used in the proof of Theorem 2.4. It is therefore possible to adopt this technique to establish an even better estimate than Theorem 2.4, but the details are left to the reader.

Let $F(x, t) = (F_1(x, t), ..., F_d(x, t))^T$, $x \in R^m$ be a d-dimensional continuous C-semimartingale which has the decomposition $F(x, t) = M(x, t) + B(x, t)$ and the local characteristic $(a(x, y, t), b(x, t), A(t))$ belonging to the class $B^{0,\delta}$. Let $\tau > 0$ and D be an operator from $C([-\tau, 0]; R^d) \times [0, \infty)$ to R^m. Let $\xi(.)$ be a $C([-\tau, 0]; R^d)$-valued $\mathcal{F}_0$-measurable random variable. Consider a stochastic differential-functional equation

$$x(t) = \xi(0) + \int_0^t F(D(\hat{x}_s, s), ds) \qquad \text{on } t \geq 0 \tag{6.1}$$

with initial data $\hat{x}_0 = \xi$, where $\hat{x}_t = \{x(t + s): -\tau \leq s \leq 0\}$ which is $C([-\tau, 0]; R^d)$-valued. Assume the equation has a unique solution $x(t)$ on $t \geq -\tau$.

Theorem 6.1 Assume:

(i) $E\|\xi\|_\tau^p < \infty$ for some $p \geq 2$,

(ii) $|b(x, t)|^2 \leq K_1(1 + x|^2)$ and $a_{ii}(x, x, t) \leq K_1(1 + x|^2)$ a.s.

for all $x, y \in R^m$, $t \geq 0$ and $1 \leq i \leq d$,

(iii) $|D(u, t)|^2 \leq K_2(1 + \|u\|_\tau^2)$ for all $u \in C([-\tau, 0]; R^d)$, $t \geq 0$,

where $K_1 > 0$ and $K_2 > 0$. For each $u \geq 0$, define

$$\sigma_u = \inf \{t \geq 0: A(t) > u\}.$$

Then

$$E\Big(\sup_{0 \leq t \leq \sigma_u} |x(t)|^p\Big) \leq \Big[(2 + (p + C)u)E\|\xi\|_\tau^p + Cu\Big] e^{(p+C)u} \tag{6.2}$$

where

$$C = pK_1(1 + 2K_2)(1 + d(9p - 1)).$$

Proof. By Itô's formula,

$$|x(t)|^p = |\xi(0)|^p + p\int_0^t |x(s)|^{p-2} \sum_{i=1}^d x_i(s)\, b_i(D(\hat{x}_s, s), s)\, dA(s)$$

$$+ \frac{1}{2}\int_0^t \sum_{i,j=1}^d \Big[p|x(s)|^{p-2}\delta_{ij} + p(p - 2)|x(s)|^{p-4}\, x_i(s)\, x_j(s)\Big]$$

$$\times\, a_{ij}(D(\hat{x}_s, s), D(\hat{x}_s, s), s)\, dA(s)$$

$$+\ p\int_0^t |x(s)|^{p-2} \sum_{i=1}^d x_i(s)\, M_i(D(\hat{x}_s, s), ds) \tag{6.3}$$

for all $t \geq 0$, where $\delta_{ij} = 1$ if $i = j$ otherwise $= 0$. But by the hypotheses we compute

$$|x(s)|^{p-2} \sum_{i=1}^{d} x_i(s)\, b_i(D(\hat{x}_s, s), s)$$

$$\leq \frac{1}{2} |x(s)|^{p-2} \left[|x(s)|^2 + |b(D(\hat{x}_s, s), s)|^2 \right]$$

$$\leq \frac{1}{2} |x(s)|^{p-2} \left[|x(s)|^2 + K_1(1 + K_2(1+\|\hat{x}_s\|_\tau^2)) \right]$$

$$\leq \frac{1}{2} K_1(1 + K_2)|x(s)|^{p-2} + \frac{1}{2}(1 + K_1K_2)\|\hat{x}_s\|_\tau^p$$

$$\leq \frac{1}{2} K_1(1 + K_2) + \frac{1}{2} (1 + K_1 + 2K_1K_2)\|\hat{x}_s\|_\tau^p.$$

Also by Theorem 1-3.3,

$$|a_{ij}(x, x, s)| \leq (a_{ii}(x, x, s)\, a_{jj}(x, x, s))^{1/2}$$

and therefore

$$\sum_{i,j=1}^{d} \left[p|x(s)|^{p-2}\delta_{ij} + p(p-2)|x(s)|^{p-4} x_i(s)\, x_j(s) \right] a_{ij}(D(\hat{x}_s, s), D(\hat{x}_s, s), s)$$

$$\leq dp(p - 1)|x(s)|^{p-2} K_1 \left[1 + K_2(1 + \|\hat{x}_s\|_\tau^2) \right]$$

$$\leq dp(p - 1)K_1(1 + K_2) + dp(p - 1)K_1(1 + 2K_2)\|\hat{x}_s\|_\tau^p.$$

We then see from (6.3) that for any stopping time $0 \leq \rho \leq \sigma_u$

$$E\Big(\sup_{0 \leq t \leq \rho} |x(t)|^p \Big) \leq E\|\xi\|_\tau^p + \frac{1}{2} upK_1(1 + K_2)(1+d(p - 1))$$

$$+ \frac{1}{2} p\Big[1 + K_1(1 + 2K_2)(1 + d(p - 1)) \Big] \int_0^\rho \|\hat{x}_s\|_\tau^p \, dA(s)$$

$$+ p\, E\Big(\sup_{0 \leq t \leq \rho} \int_0^t |x(s)|^{p-2} \sum_{i=1}^{d} x_i(s)\, M_i(D(\hat{x}_s, s), ds) \Big). \tag{6.4}$$

But by the Burkholder-Davis-Gundy inequality (Theorem 1-2.6),

$$pE\Big(\sup_{0\le t\le\rho}\int_0^t |x(s)|^{p-2}\sum_{i=1}^d x_i(s)\, M_i(D(\hat{x}_s, s), ds)\Big)$$

$$\le 2\sqrt{2}p\, E\Big(\int_0^\rho |x(s)|^{2p-4}\sum_{i,j=1}^d x_i(s)\, x_j(s)\, a_{ij}(D(\hat{x}_s, s), D(\hat{x}_s, s), s)\, dA(s)\Big)^{1/2}$$

$$\le 2\sqrt{2}\, p\, E\Big(\int_0^\rho dK_1 |x(s)|^{2p-2}\big[1 + K_2(1 + \|\hat{x}_s\|_\tau^2)\big]\, dA(s)\Big)^{1/2}$$

$$\le 2\sqrt{2}\, p\, E\Big(\Big[\sup_{0\le t\le\rho}|x(t)|^p\Big]\int_0^\rho dK_1 |x(s)|^{p-2}\big[1 + K_2(1 + \|\hat{x}_s\|_\tau^2)\big]\, dA(s)\Big)^{1/2}$$

$$\le \frac{1}{2}E\Big(\sup_{0\le t\le\rho}|x(t)|^p\Big) + 4p^2\, E\int_0^\rho \Big(dK_1(1 + K_2) + dK_1(1 + 2K_2)\|\hat{x}_s\|_\tau^2\Big)\, dA(s)$$

$$\le \frac{1}{2}E\Big(\sup_{0\le t\le\rho}|x(t)|^p\Big) + 4p^2 u dK_1(1 + K_2)$$

$$+ 4p^2 dK_1(1 + 2K_2)\, E\int_0^\rho \|\hat{x}_s\|_\tau^2\, dA(s).$$

Therefore

$$E\Big(\sup_{0\le t\le\rho}|x(t)|^p\Big) \le 2E\|\xi\|_\tau^p + Cu + (p + C)\, E\int_0^\rho \|\hat{x}_s\|_\tau^p\, dA(s), \tag{6.5}$$

where

$$C = pK_1(1 + 2K_2)(1 + d(9p - 1)).$$

Note

$$\|\hat{x}_s\|_\tau^p \le \|\xi\|_\tau^p + \sup_{0\le t\le s}|x(t)|^p.$$

It follows from (6.5)

$$E\Big(\sup_{0\le t\le\rho}|x(t)|^p\Big) \le 2E\|\xi\|_\tau^p + Cu + (p + C)uE\|\xi\|_\tau^p$$

$$+ (p + C)\, E\int_0^\rho \Big[\sup_{0\le t\le s}|x(t)|^p\Big]\, dA(s).$$

By Theorem 1-7.11 we obtain

$$E\Big(\sup_{0\le t\le\sigma_u}|x(t)|^p\Big)\le\Big[(2+(p+C)u)E\|\xi\|_\tau^p+Cu\Big]e^{(p+C)u}$$

as required. The proof is complete.

Corollary 6.2 Assume the hypotheses of Theorem 6.1 hold. Assume furthermore that $A(t)\le\mu(t)$ a.s. for all $t\ge 0$, where $\mu: R_+\to R_+$ is nondecreasing. Then

$$E\Big(\sup_{0\le t\le T}|x(t)|^p\Big)\le\Big[(2+(p+C)\mu(T))E\|\xi\|_\tau^p+C\mu(T)\Big]e^{(p+C)\mu(T)} \tag{6.6}$$

for all $T>0$.

Proof. Since $\sigma_{\mu(T)}\ge T$,

$$E\Big(\sup_{0\le t\le T}|x(t)|^p\Big)\le E\Big(\sup_{0\le t\le\sigma_{\mu(T)}}|x(t)|^p\Big).$$

Then (6.6) follows from Theorem 6.1.

3.7. ALMOST SURELY ASYMPTOTIC BOUNDS OF SOLUTIONS

In this section we shall discuss the almost surely asymptotic bounds for solutions of various types of stochastic differential delay equations.

To start, we let $F(x, y, t)$, $x, y\in R^d$ be a d-dimensional continuous C-semimartingale such that it has the decomposition $F(x, y, t) = M(x, y, t) + B(x, y, t)$ and the local characteristic $(a(x, y, \bar{x}, \bar{y}, t), b(x, y, t), A(t))$ belonging to the class $B^{0,\delta}$. Let $\tau>0$ and $\delta(.): R_+\to[-\tau, 0]$. Let $\xi(.)$ be a $C([-\tau, 0]; R^d)$-valued $\mathcal{F}_0$-measurable random variable. We first study the almost surely asymptotic bound for a d-dimensional stochastic differential delay equation driven by $F(x, y, t)$ of the form

$$x(t)=\xi(0)+\int_0^t F(x(s), x(s-\delta(s)), ds)\qquad \text{on } t\ge 0 \tag{7.1}$$

with initial data $x(s)=\xi(s)$ on $-\tau\le s\le 0$. Assume the equation has a unique global solution $x(t)$ on $t\ge -\tau$.

Theorem 7.1 Assume there exist four positive constants c_1 - c_4 such that for all $x, y \in R^d$, $t\ge 0$,

$$(x, b(x, y, t)) \le c_1|x|^2 + c_2|y|^2 + c_3,$$

$$\sum_{i=1}^{d} a_{ii}(x, y, x, y, t) \le c_4,$$

(where (., .) denotes the inner product of R^d-vectors). Then

$$\lim_{t\to\infty} \frac{|x^*(t)|^2}{e^{2(c_1+c_2)A(t)} \log\log t} = 0 \quad \text{a.s.} \tag{7.2}$$

where $|x^*(t)|^2 = \sup\{|x(s)|^2 : 0 \le s \le t\}$.

Proof. By Itô's formula and the hypothesis we see that

$$e^{-2c_1A(t)} |x(t)|^2$$

$$= |x(0)|^2 + \int_0^t e^{-2c_1A(s)} \Big(-2c_1|x(s)|^2 + 2(x(s), b(x(s), x(s-\delta(s)), s))$$

$$+ \sum_{i=1}^{d} a_{ii}(x(s), x(s-\delta(s)), x(s), x(s-\sigma(s)), t) \Big) dA(s) + 2N(t)$$

$$\le |\xi(0)|^2 + (2c_3 + c_4)/2c_1$$

$$+ 2c_2 \int_0^t e^{-2c_1A(s)} |x(s-\delta(s))|^2 \, dA(s) + 2N(t) \tag{7.3}$$

for all $t \ge 0$, where

$$N(t) := \int_0^t e^{-2c_1A(s)} (x(s), M(x(s), x(s-\delta(s)), ds)).$$

Obviously, N(t) is a continuous real-valued martingale and its quadratic variation

$$\langle N(t)\rangle = \int_0^t e^{-4c_1A(s)} \sum_{i,j=1}^{d} x_i(s)\, x_j(s)\, a_{ij}(x(s), x(s-\delta), x(s), x(s-\delta), s)\, dA(s).$$

By Theorem 1-3.3 and the hypothesis,

$$|a_{ij}(x(s), x(s-\delta), x(s), x(s-\delta), s)|$$

$$\le (a_{ii}(x(s), x(s-\delta), x(s), x(s-\delta), s)\, a_{jj}(x(s), x(s-\delta), x(s), x(s-\delta), s))^{1/2}$$

$$\le c_4.$$

So

$$\langle N(t)\rangle \le c_4 \int_0^t e^{-4c_1A(s)} \sum_{i,j=1}^{d} |x_i(s)|\,|x_j(s)|\, dA(s)$$

$$\le dc_4 \int_0^t e^{-4c_1A(s)} |x(s)|^2\, dA(s).$$

Assign $p \ge 1$ arbitrarily and let $k = 1, 2, \ldots$. Applying Lemma 2-6.2 with $\theta = 2$, $g(t) = t$, $\tau_k = 2^{kp}$ and $\gamma_k = 1$ we see that for almost all $\omega \in \Omega$ there exists a random integer $k_o(\omega)$ such that for all $k \ge k_o$

$$N(t) \le \frac{1}{2}\langle N(t)\rangle + 2\log k$$

$$\le \frac{1}{2} dc_4 \int_0^t e^{-4c_1A(s)} |x(s)|^2\, dA(s) + 2\log k$$

on $0 \le t \le 2^{kp}$. Substituting this into (7.3) yields that for almost all $\omega \in \Omega$, if $k \ge k_o$ and $0 \le t \le 2^{kp}$, then

$$e^{-2c_1A(t)} |x(t)|^2$$

$$\le |\xi(0)|^2 + (2c_3 + c_4)/2c_1 + 4\log k + dc_4 \int_0^t e^{-4c_1A(s)} |x(s)|^2\, dA(s)$$

$$+ 2c_2 \int_0^t e^{-2c_1A(s)} |x(s-\delta(s))|^2\, dA(s) := z(t), \tag{7.4}$$

where z(t) is defined evidently. Note that z(t) is nondecreasing. We now let $A(t) = 0$ for $-\tau \le t \le 0$ and compute

$$\int_0^t e^{-2c_1A(s)} |x(s-\delta(s))|^2\, dA(s)$$

$$\le \int_0^{\tau} e^{-2c_1A(s-\delta(s))} |x(s-\delta(s))|^2\, dA(s) + \int_{\tau}^{t\vee\tau} e^{-2c_1A(s-\delta(s))} |x(s-\delta(s))|^2\, dA(s)$$

$$\leq \int_0^\tau \left[\|\xi\|_\tau^2 + z(s) \right] dA(s) + \int_\tau^{t\vee\tau} z(s)\, dA(s)$$

$$\leq A(\tau)\|\xi\|_\tau^2 + \int_0^t z(s)\, dA(s).$$

We therefore obtain from (7.4) that

$$z(t) \leq |\xi(0)|^2 + (2c_3 + c_4)/2c_1 + 4\log k + 2c_2 A(\tau)\|\xi\|_\tau^2$$

$$+ \int_0^t \left[2c_2 + dc_4\, e^{-2c_1 A(s)} \right] z(s)\, dA(s).$$

Applying Theorem 1-7.1 to the above inequality yields

$$z(t) \leq \left(|\xi(0)|^2 + (2c_3 + c_4)/2c_1 + 4\log k + 2c_2 A(\tau)\|\xi\|_\tau^2 \right)$$

$$\times \exp\left(\int_0^t \left[2c_2 + dc_4\, e^{-2c_1 A(s)} \right] dA(s) \right)$$

$$\leq \left(|\xi(0)|^2 + (2c_3 + c_4)/2c_1 + 4\log k + 2c_2 A(\tau)\|\xi\|_\tau^2 \right)$$

$$\times \exp\left[2c_2 A(t) + dc_4/2c_1 \right].$$

That is, we have showed that

$$e^{-2c_1 A(t)} |x(t)|^2 \leq \left(|\xi(0)|^2 + (2c_3+c_4)/2c_1 + 4\log k + 2c_2 A(\tau)\|\xi\|_\tau^2 \right)$$

$$\times \exp\left[2c_2 A(t) + dc_4/2c_1 \right]$$

for all $0 \leq t \leq 2^{kp}$, $k \geq k_o$ almost surely. Thus, if $2^{(k-1)p} \leq t \leq 2^{kp}$, $k \geq k_o$

$$\frac{|x(t)|^2}{e^{2(c_1+c_2)A(t)} \log\log t}$$

$$\leq \left(|\xi(0)|^2 + (2c_3+c_4)/2c_1 + 4\log k + 2c_2 A(\tau)\|\xi\|_\tau^2 \right)$$

$$\times \exp(dc_4/2c_1) \left(p \log(k-1) + \log\log 2 \right)^{-1},$$

which implies

$$\limsup_{t\to\infty} \frac{|x(t)|^2}{e^{2(c_1+c_2)A(t)} \operatorname{loglog} t} \le 4p^{-1} \exp(dc_4/2c_1) \quad \text{a.s.}$$

Since $p \ge 1$ is arbitrary,

$$\lim_{t\to\infty} \frac{|x(t)|^2}{e^{2(c_1+c_2)A(t)} \operatorname{loglog} t} = 0 \quad \text{a.s.}$$

and then the required (7.3) follows from Lemma 2-6.3. The proof is complete.

If $\delta(t) \equiv \tau$, equation (7.1) becomes

$$x(t) = \xi(0) + \int_0^t F(x(s),\ x(s-\tau),\ ds). \tag{7.6}$$

For this equation, we can obtain an even stronger asymptotic bound.

Theorem 7.2 In addition to the hypotheses of Theorem 7.1, assume

$$c_5 \le A(t) - A(t-\tau) \quad \text{for all } t \ge \tau \tag{7.7}$$

for some $c_5 \ge 0$. Then the solution of equation (7.6) satisfies

$$\lim_{t\to\infty} \frac{|x^*(t)|^2}{e^{\lambda A(t)} \operatorname{loglog} t} = 0 \quad \text{a.s.}$$

where $\lambda = 2\left[c_1 + c_2 e^{-2c_1c_5}\right]$.

Proof. Replace $\delta(s)$ by τ in (7.4) and compute

$$\int_0^t e^{-2c_1 A(s)} |x(s-\tau)|^2 \, dA(s)$$

$$\le \int_0^\tau |x(s-\tau)|^2 \, dA(s) + \int_\tau^{t\vee\tau} e^{-2c_1(A(s-\tau)+c_5)} |x(s-\tau)|^2 \, dA(s)$$

$$\le A(\tau)\|\xi\|_\tau^2 + e^{-2c_1c_5} \int_0^t z(s) \, dA(s).$$

Hence we obtain from (7.4) that

$$z(t) \le |\xi(0)|^2 + (2c_3+c_4)/2c_1 + 4\log k + 2c_2A(\tau)\|\xi\|_\tau^2$$

$$+ \int_0^t \left[2c_2e^{-2c_1c_5} + dc_4\, e^{-2c_1A(s)} \right] z(s)\, dA(s).$$

The remainder of the proof can be carried out in the same lines as the previous theorem. The proof is complete.

We finally discuss the almost surely asymptotic bound for a general stochastic differential-functional equation. Let $F(x, t) = (F_1(x, t), ..., F_d(x, t))^T$, $x \in R^m$ be a d-dimensional continuous C-semimartingale which has the decomposition $F(x, t) = M(x, t) + B(x, t)$ and the local characteristic $(a(x, y, t), b(x, t), A(t))$ belonging to the class $B^{0,\delta}$. Let $\tau > 0$ and D be an operator from $C([-\tau, 0]; R^d)\times[0, \infty)$ to R^m. Let $\xi(.)$ still be a $C([-\tau, 0]; R^d)$-valued $\mathcal{F}_0$-measurable random variable. Consider a stochastic differential-functional equation

$$x(t) = \xi(0) + \int_0^t F(D(\hat{x}_s, s), ds) \qquad \text{on } t \ge 0 \tag{7.8}$$

with initial data $\hat{x}_0 = \xi$, where $\hat{x}_t = \{x(t + s): -\tau \le s \le 0\}$ which is $C([-\tau, 0]; R^d)$-valued. Assume the equation has a unique solution $x(t)$ on $t \ge -\tau$.

Theorem 7.3 Assume there exist seven nonnegative constants $c_1 - c_7$ such that

(i) $|D(u, t)|^2 \le c_1\|u\|_\tau^2 + c_2$ for all $u \in C([-\tau, 0]; R^d)$, $t \ge 0$,

(ii) $|b(x, t)|^2 \le c_3|x|^2 + c_4$ and $\sum_{i=1}^{d} a_{ii}(x, x, t) \le c_5$ a.s.

all $x \in R^m$, $t \ge 0$,

(iii) $A(t) \le c_6t + c_7$ a.s. for all $t \ge 0$.

Then the solution of equation (7.8) satisfies

$$\limsup_{t\to\infty} \frac{|x^*(t)|^2}{e^{\lambda t}\, t \log\log t} \le 2dc_5c_6\exp\left(1 + 2c_7\sqrt{c_1c_3} \right) \quad \text{a.s.} \tag{7.8}$$

where $\lambda = 2c_6\sqrt{c_1c_3}$ and $|x^*(t)|^2 = \sup\{|x(s)|^2: 0 \le s \le t\}$.

Proof. By Itô's formula,

$$|x(t)|^2 = |x(0)|^2 + \int_0^t 2(x(s), b(D(\hat{x}_s, s), s))\, dA(s)$$

$$+ \int_0^t \sum_{i=1}^d a_{ii}(D(\hat{x}_s, s), D(\hat{x}_s, s), s)\, dA(s) + 2N(t) \tag{7.9}$$

for all $t \geq 0$, where

$$N(t) := \int_0^t (x(s), M(D(\hat{x}_s, s), ds)).$$

Note for any $\varepsilon > 0$,

$$2|\,(x(s), b(D(\hat{x}_s, s), s))\,| \leq \varepsilon |x(s)|^2 + \varepsilon^{-1} |b(D(\hat{x}_s, s), s))|^2.$$

By hypotheses (i) and (ii)

$$2|\,(x(s), b(D(\hat{x}_s, s), s))\,| \leq \varepsilon |x(s)|^2 + \varepsilon^{-1} \left[\, c_3(c_1 \|\hat{x}_s\|_\tau^2 + c_2) + c_4 \,\right].$$

Choosing $\varepsilon = \sqrt{c_1 c_3}$ we obtain that

$$2|\,(x(s), b(D(\hat{x}_s, s), s))\,| \leq 2\sqrt{c_1 c_3}\, \|\hat{x}_s\|_\tau^2 + C,$$

where $C = (c_2 c_3 + c_4)/\sqrt{c_1 c_3}$. By this estimate and the hypotheses (ii) and (iii) we see from (7.9) that

$$|x(t)|^2 \leq |\xi(0)|^2 + (C + c_5)(c_6 t + c_7) + 2N(t) + 2\sqrt{c_1 c_3} \int_0^t \|\hat{x}_s\|_\tau^2\, dA(s) \tag{7.10}$$

for all $t \geq 0$. On the other hand, let $\theta > 1$ and $\alpha > 0$ be arbitrary and $k = 1, 2, \ldots$. By Lemma 2-6.2 with $g(t) = t$, $\tau_k = \theta^k$ and $\gamma_k = \alpha\theta^{-k}$, we see that for almost all $\omega \in \Omega$ there exists a random integer $k_o(\omega)$ such that for all $k \geq k_o$

$$N(t) \leq \frac{1}{2} \alpha\theta^{-k} \langle N(t) \rangle + \alpha^{-1}\theta^{k+1} \log k \qquad \text{for } 0 \leq t \leq \theta^k.$$

But it is easy to check

$$\langle N(t) \rangle \leq d c_5 \int_0^t |x(s)|^2\, dA(s).$$

So

$$N(t) \le \frac{1}{2}\alpha d c_5 \theta^{-k} \int_0^t |x(s)|^2 \, dA(s) + \alpha^{-1}\theta^{k+1} \log k$$

for all $0 \le t \le \theta^k$, $k \ge k_o$ almost surely. Substituting this to (7.10) gives

$$|x(t)|^2 \le |\xi(0)|^2 + (C+c_5)(c_6 t + c_7) + 2\alpha^{-1}\theta^{k+1} \log k$$

$$+ \alpha d c_5 \theta^{-k} \int_0^t |x(s)|^2 \, dA(s) + 2\sqrt{c_1 c_3} \int_0^t \|\hat{x}_s\|_\tau^2 \, dA(s)$$

for all $0 \le t \le \theta^k$, $k \ge k_o$ almost surely. This implies

$$|x^*(t)|^2 \le |\xi(0)|^2 + (C + c_5)(c_6 \theta^k + c_7) + 2\alpha^{-1}\theta^{k+1} \log k$$

$$+ 2(c_6 \tau + c_7)\sqrt{c_1 c_3}\, \|\xi\|_\tau^2 + \left[\alpha d c_5 \theta^{-k} + 2\sqrt{c_1 c_3} \right] \int_0^t |x^*(s)|^2 \, dA(s)$$

for all $0 \le t \le \theta^k$, $k \ge k_o$ almost surely. By Theorem 1–7.1 we then obtain

$$|x^*(t)|^2 \le \Big(|\xi(0)|^2 + (C + c_5)(c_6 \theta^k + c_7)$$

$$+ 2\alpha^{-1}\theta^{k+1} \log k + 2(c_6 \tau + c_7)\sqrt{c_1 c_3}\, \|\xi\|_\tau^2 \Big)$$

$$\times \exp\Big((c_6 t + c_7) \left[\alpha d c_5 \theta^{-k} + 2\sqrt{c_1 c_3} \right] \Big)$$

for all $0 \le t \le \theta^k$, $k \ge k_o$ almost surely. Particularly, for almost all $\omega \in \Omega$, if $\theta^{k-1} \le t \le \theta^k$, $k \ge k_o$,

$$|x^*(t)|^2 \,/\, \exp\left[2c_6 t \sqrt{c_1 c_3} \right] t \log\log t$$

$$\le \Big(|\xi(0)|^2 + (C + c_5)(c_6 \theta^k + c_7) + 2\alpha^{-1}\theta^{k+1} \log k + 2(c_6 \tau + c_7)\sqrt{c_1 c_3}\, \|\xi\|_\tau^2 \Big)$$

$$\times \exp\Big((c_6 \theta^k + c_7)\alpha d c_5 \theta^{-k} + 2c_7\sqrt{c_1 c_3} \Big) \,/\, \theta^{k-1} (\log(k-1) + \log\log\theta).$$

Hence

$$\limsup_{t\to\infty} |x^*(t)|^2 \,/\, \exp\left[2c_6 t \sqrt{c_1 c_3} \right] t \log\log t$$

$$\le 2\alpha^{-1}\theta^2 \exp\Big(\alpha d c_5 c_6 \theta + 2c_7\sqrt{c_1 c_3} \Big) \qquad \text{a.s.}$$

Letting $\theta \to 1$ and choosing $\alpha = (dc_5c_6)^{-1}$ we get

$$\limsup_{t\to\infty} |x^*(t)|^2 \Big/ \exp\Big[\,2c_6 t\sqrt{c_1c_3}\,\Big]\, t \log\log t$$

$$\le 2dc_5c_6 \exp\Big(\,1 + 2c_7\sqrt{c_1c_3}\,\Big) \qquad \text{a.s.}$$

as required. The proof is complete.

We finally give two examples to illustrate the theory and also to close this chapter.

Example 7.1 Let $\tau > 0$ and $\delta(.) : R_+ \to [-\tau, 0]$. Let $\xi(.)$ be a $C([-\tau, 0]; R^d)$-valued $\mathcal{F}_0$-measurable random variable. Consider a d-dimensional Itô delay equation

$$dx(t) = f(x(t), x(t-\delta(t)), t)\, dt$$

$$+ \sum_{k=1}^{\infty} g_k(x(t), x(t-\delta(t)), t)\, dw_k(t) \qquad \text{on } t \ge 0, \tag{7.11}$$

with initial data $x(t) = \xi(t)$ on $-\tau \le t \le 0$, where $\{w_k: k \ge 1\}$ is a sequence of independent of Brownian motions, $f = (f_1, ..., f_d)^T$ and $g_k = (g_{k1}, ..., g_{kd})^T$ are all R^d-valued functions defined on $R^{2d} \times R_+$. Suppose that the equation has a unique solution $x(t)$ on $t \ge -\tau$. Assume

$$\sum_{k=1}^{\infty} |g_k(x, y, t)|^2 \le K < \infty \quad \text{for all } (x, y, t) \in R^{2d} \times R_+. \tag{7.12}$$

Also assume that f satisfies the linear growth condition, that is

$$|f(x, y, t)|^2 \le c(1 + |x|^2 + |y|^2) \quad \text{for all } (x, y, t) \in R^{2d} \times R_+. \tag{7.13}$$

Note that for any $\varepsilon > 0$,

$$(x, f(x, y, t)) \le |x|\,|f(x, y, t)| \le 2\Big[\,|x|/2\sqrt{\varepsilon}\,\Big]\Big[\sqrt{\varepsilon}\,|f(x, y, t)|\,\Big]$$

$$\le (4\varepsilon)^{-1}|x|^2 + \varepsilon|f(x, y, t)|^2 \le (4\varepsilon)^{-1}|x|^2 + \varepsilon c(1 + |x|^2 + |y|^2)$$

$$\le ((4\varepsilon)^{-1} + \varepsilon c)|x|^2 + \varepsilon c|y|^2 + \varepsilon c.$$

By Theorem 7.1, we have

$$\lim_{t\to\infty} \frac{|x^*(t)|^2}{e^{2((4\varepsilon)^{-1}+2\varepsilon c)} \log\log t} = 0 \qquad \text{a.s.}$$

In particular, choose $\varepsilon = 1/2\sqrt{2c}$. Then

$$\lim_{t\to\infty} \frac{|x^*(t)|^2}{e^{2t\sqrt{2c}}\ \log\log t} = 0 \quad \text{a.s.} \tag{7.14}$$

Moreover, if $\delta(t) = \tau$, then equation (7.11) becomes

$$dx(t) = f(x(t), x(t-\tau), t)\, dt + \sum_{k=1}^{\infty} g_k(x(t), x(t-\tau), t)\, dw_k(t). \tag{7.15}$$

For this delay equation, of course (7.14) holds. However, we can use Theorem 7.2 to obtain an even better asymptotic bound. In fact, we have

$$\lim_{t\to\infty} \frac{|x^*(t)|^2}{e^{\lambda t}\ \log\log t} = 0 \quad \text{a.s.} \tag{7.16}$$

where

$$\lambda = \lambda(\varepsilon) = 2\Big((4\varepsilon)^{-1} + \varepsilon c + \varepsilon c e^{-2\tau((4\varepsilon)^{-1}+\varepsilon c)} \Big).$$

Let $\varepsilon = 1/2\sqrt{2c}$ as before. Then

$$\lambda = \sqrt{c/2}\,\Big(3 + \exp\big[-3\tau\sqrt{c/2}\,\big]\Big) < 2\sqrt{2c}.$$

Of course this λ is not optimal yet.

Example 7.2 Let τ, $\xi(.)$ and $\{w_k\!: k\geq 1\}$ be the same as in Example 7.1. Consider a d-dimensional Itô differential-functional equation

$$dx(t) = f(\hat{x}(t), t)\, dt + \sum_{k=1}^{\infty} g_k(\hat{x}(t), t)\, dw_k(t) \qquad \text{on } t \geq 0, \tag{7.11}$$

with initial data $\hat{x}_0 = \xi$, where $f = (f_1, ..., f_d)^T$ and $g_k = (g_{k1}, ..., g_{kd})^T$ are all R^d-valued functionals defined on $C([-\tau, 0]; R^d)\times R_+$. Suppose that the equation has a unique solution $x(t)$ on $t \geq -\tau$. Assume

$$\sum_{k=1}^{\infty} |g_k(u, t)|^2 \leq K < \infty$$

and

$$|f(u, t)|^2 \leq c\|u\|_\tau^2 + \rho$$

for all $(u, t) \in C([-\tau, 0]; R^d) \times R_+$, where $c > 0$ and $\rho \geq 0$. Then by Theorem 7.3 with a little modification we deduce (or more precisely we can show in the same way as Theorem 7.3) that

$$\limsup_{t\to\infty} \frac{|x^*(t)|^2}{e^{2t\sqrt{c}}\, t \log\log t} \le 2edK \quad \text{a.s.}$$

Also we can apply Corollary 6.2 to obtain an L^p-estimate ($p \ge 2$)

$$E\Big(\sup_{0\le s\le t} |x(s)|^p \Big) \le \Big[(2 + (p + C)t)\, E\|\xi\|_\tau^p + Ct \Big] e^{(p+C)t}$$

for all $t \ge 0$, where

$$C = p\bar{c}(1 + 2\bar{c})(1 + d(9p - 1)) \quad \text{and} \quad \bar{c} = c \vee \rho \vee K,$$

provided $E\|\xi\|_\tau^p < \infty$.

CHAPTER 4

Exponential Stability of Stochastic Differential Equations

4.1. INTRODUCTION

Lyapunov functions play an important role in stability analysis of stochastic differential equations. Bucy (1965) discovered that diffusion processes along Lyapunov functions were supermartingales, which gave surprisingly simple sufficient criteria for stability in probability and for stability of moments. Almost sure exponential stability was considered by Has'minskii (1967) for linear stochastic differential equations with Brownian motions as integrators using Lyapunov exponents and ergodic theory. Arnold et al. (1984) proved that the almost sure Lyapunov exponent of a linear stochastic system was the derivative (at zero) of the pth moment Lyapunov exponents $g(p)$, $p \in R$. While the theory was developed for hypoelliptic diffusion processes, where the top Lyapunov exponent was realized with probability 1, the multiplicative ergodic theorem of Oseledec (1968) allowed the analysis of all exponents for a stochastic flow, leading to a detailed analysis of the dynamics of random systems (see, e.g. Arnold and Crauel (1991)). Mohammed (1990) obtained the first result in this direction for linear stochastic delay equations.

In this chapter we shall take a different route: making use of stochastic Lyapunov functions and the exponential martingale inequality, we obtain upper estimates for the almost sure Lyapunov exponent of the solution of a stochastic differential equation driven by a nonlinear integrator, which lead to the sufficient criteria for almost sure exponential stability. The pth moment exponential stability will also be studied via stochastic Lyapunov functions. For illustration, our results will be applied to the study of stochastic stabilization and destabilization, stochastic flows and stochastic oscillators.

4.2. ALMOST SURE EXPONENTIAL STABILITY

In this section we shall discuss the almost sure exponential stability for a d-dimensional stochastic differential equation driven by a nonlinear integrator.

Throughout Sections 4.2 and 4.3, we let $F(x, t)$, $x \in R^d$ be a d-dimensional continuous C-semimartingale which has the local characteristic $(a(x, y, t), b(x, t), A(t))$ belonging to the class $B^{0,\delta}$ and the decomposition $F(x, t) = B(x, t) + M(x, t)$, where $a(x, y, t) = (a_{ij})_{d\times d}$, $b(x, t) = (b_1, ..., b_d)^T$, $M(x, t) = (M_1, ..., M_d)^T$ and $B(x, t) = (B_1, ..., B_d)^T$. Consider a d-dimensional stochastic differential equation driven by $F(x, t)$ of the form

$$x(t) = \xi + \int_{t_o}^{t} F(x(s), ds) \qquad \text{on } t \geq t_o, \tag{2.1}$$

where ξ is an $\mathcal{F}_{t_o}$-measurable R^d-valued random variable. Assume that equation (2.1) has a unique solution which is denoted by $x(t; t_o, \xi)$. Let us first give the definition of almost sure exponential stability.

Definition 2.1 Equation (2.1) is said to be *almost surely exponential stable* if there exists a $\lambda > 0$ such that for each pair of t_o and ξ there is a finite random variable β such that

$$|x(t; t_o, \xi)| \leq \beta e^{-\lambda(t-t_o)} \qquad \text{a.s. for all } t \geq t_o, \tag{2.2}$$

or if

$$\limsup_{t\to\infty} \frac{1}{t} \log|x(t; t_o, \xi)| \leq -\lambda \qquad \text{a.s.} \tag{2.3}$$

The left hand side of (2.3) is called the *almost sure Lyapunov exponent* of the solution.

We shall introduce some new notations. Let D be an open subset of $R^d \times R_+$. Denote by $C^{2,1}(D; R)$ the family of all real-valued functions $V(x, t)$ defined on D which are continuously twice differentiable in x and once differentiable in t. Introduce two operators L and H acting on $C^{2,1}(D; R)$-valued functions as follows:

$$LV(x, A(t)) = \frac{\partial}{\partial t} V(x, A(t)) + \sum_{i=1}^{d} b_i(x, t) \frac{\partial}{\partial x_i} V(x, A(t))$$

$$+ \frac{1}{2} \sum_{i,j=1}^{d} a_{ij}(x, x, t) \frac{\partial^2}{\partial x_i \partial x_j} V(x, A(t))$$

and

$$HV(x, A(t)) = \sum_{i,j=1}^{d} a_{ij}(x, x, t) \frac{\partial}{\partial x_i} V(x, A(t)) \frac{\partial}{\partial x_j} V(x, A(t)).$$

We shall need a standing hypothesis in the first half of this section although this restriction will be removed later on.

(H2.1) $x(t; t_o, \xi) \neq 0$ for all $t \geq t_o$ a.s. provided $\xi \neq 0$ a.s.

Before we start to discuss the stability, let us give a sufficient condition for (H2.1) to be satisfied.

Proposition 2.1 Assume for any $n > 0$ there exists a $K_n > 0$ such that

$$|b(x, t)|^2 + \sum_{i=1}^{d} a_{ii}(x, x, t) \leq K_n |x|^2 \qquad \text{if } |x| \leq n \text{ and } t \geq 0.$$

Then (H2.1) is satisfied.

Proof. Obviously we only need to prove that $x(t; t_o, \xi) \neq 0$ for all $t \geq t_o$ a.s. if $t_o \geq 0$ and $\xi \in R^d - \{0\}$. Suppose it were false, then there exists a pair of t_o and ξ ($\neq 0$) such that $P(\tau < \infty) > 0$, where τ is the first time of zero, i.e.

$$\tau = \inf\{t \geq t_o: |x(t; t_o, \xi)| = 0\}.$$

Hence one can find a pair of $T > t_o$ and $n > 0$ large enough such that $P(B) > 0$, where

$$B = \{\omega : \tau \leq T \text{ and } |x(t; t_o, \xi)| \leq n - 1 \text{ for all } t_o \leq t \leq \tau\}.$$

Let $V(x, t) = |x|^{-1}$. It is easy to check

$$LV(x, A(t)) \leq \mu V(x, A(t)) \qquad \text{if } 0 < |x| \leq n, t \geq 0,$$

where μ is a positive constant dependent of only K_n. Write $x(t) = x(t; t_o, \xi)$ simply and define

$$\tau_\varepsilon = \inf\{t \geq t_o: |x(t)| \leq \varepsilon \text{ or } |x(t)| \geq n\}$$

for each $0 < \varepsilon < |\xi|$. Then by Itô's formula,

$$E\Big[e^{-\mu(A(\tau_\varepsilon \wedge T) - A(t_o))} V(x(\tau_\varepsilon \wedge T), A(\tau_\varepsilon \wedge T)) \Big]$$

$$= V(\xi, t_o) + E \int_{t_o}^{\tau_\varepsilon \wedge T} e^{-\mu(A(s)-A(t_o))} \big[-\mu V(x(s), A(s)) + LV(x(s), A(s))\big] \, dA(s)$$

$$\leq |\xi|^{-1}.$$

Note that if $\omega \in B$, then $\tau_\varepsilon \leq T$ and $|x(\tau_\varepsilon)| = \varepsilon$. We therefore see from the above inequality that

$$E\Big[e^{-\mu(A(T)-A(t_o))} \varepsilon^{-1} \mathbb{1}_B \Big] \leq |\xi|^{-1}$$

i.e.

$$E\Big[e^{-\mu(A(T)-A(t_o))} \mathbb{1}_B \Big] \leq \varepsilon |\xi|^{-1}.$$

Letting $\varepsilon \to 0$ we obtain

$$E\Big[e^{-\mu(A(T)-A(t_o))} \mathbb{1}_B \Big] = 0.$$

We must then have $P(B) = 0$, but this contradicts the definition of B. The proof is complete.

The proposition particularly tells us that (H2.1) is satisfied if $b(x, t)$ and $a(x, y, t)$ are locally Lipschitz continuous and $b(0, t) \equiv 0$, $a(0, 0, t) \equiv 0$. But generally speaking, the local Lipschitz continuity is required in order to have a unique solution, and both $b(0, t) \equiv 0$ and $a(0, 0, t) \equiv 0$ can be assumed to be true in the study of stability of equilibrium states as we did in the theory of stability of ordinary differential equation. In other words, (H2.1) will be satisfied in many interesting cases.

We now can start to study the exponential stability, the main topic of this book.

Theorem 2.1 Let (H2.1) hold. Assume furthermore that there exists a function $V(x, t) \in C^{2,1}((R^d - \{0\}) \times R_+; R)$ and positive constants p, λ, σ such that

(i) $|x|^p \leq V(x, t)$ and $LV(x, A(t)) \leq -\lambda V(x, A(t))$

for $x \neq 0$, $t \geq 0$, and

(ii) $\liminf_{t\to\infty} \frac{A(t)}{t} \geq \sigma$ a.s.

Then

$$\limsup_{t\to\infty} \frac{1}{t} \log|x(t; t_o, \xi)| \leq -\lambda\sigma/p \quad \text{a.s.} \tag{2.4}$$

for all $t_o \geq 0$ and $\xi \neq 0$ a.s.

Proof. Fix t_o and ξ arbitrarily and write $x(t)$ for $x(t; t_o, \xi)$ simply. Itô's formula gives

$$\log V(x(t), A(t)) = \log V(\xi, A(t_o)) + N(t)$$

$$+ \int_{t_o}^{t} \left(\frac{LV(x(s), A(s))}{V(x(s), A(s))} - \frac{1}{2} \frac{HV(x(s), A(s))}{V(x(s), A(s))^2} \right) dA(s)$$

$$\leq \log V(\xi, t_o) + N(t) - \lambda A(t) - \frac{1}{2} \int_{0}^{t} \frac{HV(x(s), A(s))}{V(x(s), A(s))^2}\, dA(s), \tag{2.5}$$

where

$$N(t) = \int_{t_o}^{t} \frac{1}{V(x(s), A(s))} \sum_{i=1}^{d} \frac{\partial}{\partial x_i} V(x(s), A(s))\, M_i(x(s), ds)$$

which is a continuous real-valued martingale. Let $k = 1, 2, \ldots$. Applying Lemma 2-6.2 with $g(t) \equiv t$, $\tau_k = k$, $\gamma_k = 1$ and $\theta = 2$, we deduce that, for almost all $\omega \in \Omega$, there exists an integer $k_o(\omega)$ such that for all $k \geq k_o$

$$N(t) \leq 2 \log k + \frac{1}{2} \langle N(t) \rangle, \qquad t_o \leq t \leq t_o + k.$$

Note

$$\langle N(t) \rangle = \int_{t_o}^{t} \frac{HV(x(s), A(s))}{V(x(s), A(s))^2}\, dA(s).$$

So

$$N(t) \leq 2 \log k + \frac{1}{2} \int_{t_o}^{t} \frac{HV(x(s), A(s))}{V(x(s), A(s))^2}\, dA(s)$$

for all $t_o \leq t \leq t_o + k$, $k \geq k_o$ almost surely. Substituting this into (2.5) gives

$$\log V(x(t), A(t)) \le \log V(\xi, A(t_o)) - \lambda A(t) + 2\log k$$

for all $t_o \le t \le t_o + k$, $k \ge k_o$ almost surely. This along with condition (i) implies

$$\frac{1}{t}\log|x(t)| \le \frac{1}{pt}\log V(x(t), A(t)) \le \frac{1}{pt}\Big(\log V(\xi, A(t_o)) - \lambda A(t) + 2\log k\Big)$$

$$\le \frac{1}{pt}\Big(\log V(\xi, A(t_o)) - \lambda A(t)\Big) + 2\log k/p(t_o + k - 1)$$

for $t_o + k - 1 \le t \le t_o + k$, $k \ge k_o$ almost surely. By condition (ii) we then derive

$$\limsup_{t\to\infty}\frac{1}{t}\log|x(t)| \le -\frac{\lambda}{p}\liminf_{t\to\infty}\frac{A(t)}{t} \le -\lambda\sigma/p \qquad \text{a.s.}$$

as required. The proof is complete.

The following theorem will play a very important role in the study of stochastic stabilization (see Section 4.4 below).

Theorem 2.2 Let (H2.1) hold. Let $V \in C^{2,1}((R^d - \{0\}) \times R_+; R)$ and $\psi(t)$ be a nonnegative predictable process, and let $p > 0$, $\rho \ge 0$, $\sigma > 0$, $\lambda < \rho/2$. Assume that for all $x \ne 0$ and $t \ge 0$,

(i) $|x|^p \le V(x, t)$,

(ii) $LV(x, A(t)) \le \lambda\, \psi(t)\, V(x, A(t))$,

(iii) $HV(x, A(t)) \ge \rho\, \psi(t)\, (V(x, A(t)))^2$,

(iv) $\liminf_{t\to\infty}\frac{1}{t}\int_0^t \psi(s)dA(s) \ge \sigma$ a.s.

Then the solution of equation (2.1) satisfies

$$\limsup_{t\to\infty}\frac{1}{t}\log|x(t; t_o, \xi)| \le -\frac{1}{p}\sigma(\rho/2 - \lambda) \qquad \text{a.s.} \tag{2.6}$$

whenever $t_o \ge 0$ and $\xi \ne 0$ a.s.

Proof. Again fix t_o and ξ arbitrarily and write $x(t)$ for $x(t; t_o, \xi)$ simply. It is easy to deduce by Itô's formula and condition (ii) that

$$\log V(x(t), A(t)) \le \log V(\xi, A(t_o)) + N(t)$$

$$+ \lambda \int_{t_o}^{t} \psi(s)\, dA(s) - \frac{1}{2} \int_{t_o}^{t} \frac{HV(x(s), A(s))}{V(x(s), A(s))^2}\, dA(s), \tag{2.7}$$

where

$$N(t) = \int_{t_o}^{t} \frac{1}{V(x(s), A(s))} \sum_{i=1}^{d} \frac{\partial}{\partial x_i} V(x(s), A(s))\, M_i(x(s), ds)$$

the same as before. Let $k = 1, 2, \ldots$ and assign $0 < \varepsilon < 1 - 2\lambda/\rho$ arbitrarily. In view of Lemma 2-6.2 with $g(t) \equiv t$, $\tau_k = k$, $\gamma_k = \varepsilon$ and $\theta = 2$, we derive that, for almost all $\omega \in \Omega$, there exists an integer $k_o(\omega)$ such that for all $k \geq k_o$

$$N(t) \leq 2\varepsilon^{-1} \log k + \frac{1}{2} \varepsilon \langle N(t) \rangle$$

$$\leq 2\varepsilon^{-1} \log k + \frac{1}{2} \varepsilon \int_{t_o}^{t} \frac{HV(x(s), A(s))}{V(x(s), A(s))^2}\, dA(s)$$

for all $t_o \leq t \leq t_o + k$. Substituting this into (2.7) and using condition (iii) one obtains

$$\log V(x(t), A(t)) \leq \log V(\xi, A(t_o)) + 2\varepsilon^{-1} \log k$$

$$- \left[\frac{1}{2} (1-\varepsilon)\rho - \lambda \right] \int_{t_o}^{t} \psi(s)\, dA(s)$$

for all $t_o \leq t \leq t_o + k$, $k \geq k_o$ almost surely. This along with condition (i) implies

$$\frac{1}{t} \log|x(t)| \leq \frac{1}{pt} \log V(x(t), A(t))$$

$$\leq \frac{1}{pt} \left(\log V(\xi, A(t_o)) + 2\varepsilon^{-1} \log k - \left[\frac{1}{2} (1-\varepsilon)\rho - \lambda \right] \int_{t_o}^{t} \psi(s)\, dA(s) \right)$$

$$\leq \frac{1}{pt} \left(\log V(\xi, A(t_o)) - \left[\frac{1}{2} (1 - \varepsilon)\rho - \lambda \right] \int_{t_o}^{t} \psi(s)\, dA(s) \right) + \frac{2 \log k}{\varepsilon p (t_o + k - 1)}$$

for $t_o + k - 1 \leq t \leq t_o + k$, $k \geq k_o$ almost surely. Consequently

$$\limsup_{t \to \infty} \frac{1}{t} \log|x(t)| \leq -\frac{1}{p} \left[\frac{1}{2} (1 - \varepsilon)\rho - \lambda \right] \liminf_{t \to \infty} \frac{1}{t} \int_{t_o}^{t} \psi(s)\, dA(s)$$

$$\leq -\frac{1}{p}\sigma\left[\frac{1}{2}(1-\varepsilon)\rho-\lambda\right] \qquad \text{a.s.}$$

where condition (iv) has also been used. Letting $\varepsilon \to 0$ we obtain the required (2.6). The proof is complete.

The following theorem gives a sufficient criterion for a stochastic system which is almost surely exponentially unstable. This theorem is useful in the application of destabilization by noise.

Theorem 2.3 Let (H2.1) hold. Let $V \in C^{2,1}((R^d - \{0\}) \times R_+; R)$ and $\psi(t)$ be a nonnegative predictable process, and let $p > 0$, $\rho > 0$, $\sigma > 0$, $\lambda > \rho/2$. Assume that for all $x \neq 0$ and $t \geq 0$,

(i) $|x|^p \geq V(x, t)$,

(ii) $LV(x, A(t)) \geq \lambda\, \psi(t)\, V(x, A(t))$,

(iii) $HV(x, A(t)) \leq \rho\, \psi(t)\, (V(x, A(t)))^2$,

(iv) $\liminf\limits_{t\to\infty} \frac{1}{t}\int_0^t \psi(s)dA(s) \geq \sigma$ a.s.

Then the solution of equation (2.1) satisfies

$$\liminf_{t\to\infty} \frac{1}{t}\log|x(t; t_o, \xi)| \geq \frac{1}{p}\sigma(\lambda - \rho/2) \qquad \text{a.s.}$$

whenever $t_o \geq 0$ and $\xi \neq 0$ a.s. In this case, equation (2.1) is said to be *almost surely exponentially unstable.*

Proof. By Itô's formula and conditions (ii) and (iii) one can derive

$$\log V(x(t), A(t)) \geq \log V(\xi, A(t_o)) + N(t) + (\lambda - \rho/2)\int_{t_o}^t \psi(s)\, dA(s),$$

where $N(t)$ is the same as before. Let $k = 1, 2, \ldots$ and assign $0 < \varepsilon < 2\lambda/\rho - 1$ arbitrarily. Applying Lemma 2-6.2 to the martingale $-N(t)$ we deduce that, for almost all $\omega \in \Omega$, there exists an integer $k_o(\omega)$ such that for all $k \geq k_o$

$$-N(t) \leq 2\varepsilon^{-1}\log k + \frac{1}{2}\varepsilon\rho\int_{t_o}^t \psi(s)\, dA(s), \qquad t_o \leq t \leq t_o + k.$$

Therefore

$$\log V(x(t), A(t)) \geq \log V(\xi, A(t_o)) - 2\varepsilon^{-1}\log k$$

$$+ (\lambda-(1+\varepsilon)\rho/2)\int_{t_o}^{t}\psi(s)\, dA(s)$$

for all $t_o \leq t \leq t_o + k$, $k \geq k_o$ almost surely. This, together with conditions (i) and (iv), implies

$$\liminf_{t\to\infty} \frac{1}{t}\log|x(t; t_o, \xi)| \geq \frac{1}{p}\sigma(\lambda-(1+\varepsilon)\rho/2) \quad \text{a.s.}$$

Since ε is arbitrary, the conclusion follows and the proof is complete.

In what follows we shall remove hypothesis (H2.1). However it should be pointed out that the following results are not the generalizations of the previous ones, but cover different cases.

Theorem 2.4 Let $V \in C^{2,1}(R^d \times R_+; R)$ and $\mu(t)$ be a polynomial with positive coefficients, and let p, λ, σ be positive constants. Assume that for all $x \in R^d$ and $t \geq 0$,

(i) $|x|^p \leq V(x, t)$,

(ii) $LV(x, A(t)) \leq -\lambda V(x, A(t)) + \mu(t)\, e^{-\lambda A(t)}$,

(iii) $HV(x, A(t)) \leq \mu(t)\, e^{-\lambda A(t)}\, V(x, A(t))$,

(iv) $A(t) \leq \mu(t)$ and $\liminf_{t\to\infty} \frac{A(t)}{t} \geq \sigma$ a.s.

Then the solution of equation (2.1) satisfies that, for all t_o and ξ,

$$\limsup_{t\to\infty} \frac{1}{t}\log|x(t; t_o, \xi)| \leq -\lambda\sigma/p \quad \text{a.s.}$$

Proof. By Itô's formula and condition (ii),

$$e^{\lambda A(t)}\, V(x(t), A(t)) \leq e^{\lambda A(t_o)}\, V(\xi, A(t_o)) + (\mu(t))^2 + N(t), \tag{2.8}$$

where

$$N(t) = \int_{t_o}^{t} e^{\lambda A(s)} \sum_{i=1}^{d} \frac{\partial}{\partial x_i} V(x(s), A(s))\, M_i(x(s), ds).$$

Let $q/2$ be the degree of the polynomial $\mu(.)$ and $k = 1, 2, \ldots$. Assign $\theta > 1$ arbitrarily. An application of Lemma 2-6.2 with $g(t) \equiv t$, $\tau_k = \theta^k$, $\gamma_k = \theta^{-qk}$ yields that, for almost all $\omega \in \Omega$, there exists an integer $k_o(\omega)$ such that for all $k \ge k_o$

$$N(t) \le \theta^{qk+1} \log k + \frac{1}{2}\theta^{-qk} \langle N(t) \rangle$$

$$\le \theta^{qk+1} \log k + \frac{1}{2}\theta^{-qk} \int_{t_o}^{t} e^{2\lambda A(s)} HV(x(s), A(s))\, dA(s)$$

$$\le \theta^{qk+1} \log k + \frac{1}{2}\theta^{-qk} \int_{t_o}^{t} \mu(s)\, e^{\lambda A(s)} V(x(s), A(s))\, dA(s)$$

for $t_o \le t \le t_o + \theta^k$, where condition (iii) has been used. We therefore see from (2.8) that

$$e^{\lambda A(t)} V(x(t), A(t)) \le e^{\lambda A(t_o)} V(\xi, A(t_o)) + (\mu(t_o + \theta^k))^2$$

$$+ \theta^{qk+1} \log k + \frac{1}{2}\theta^{-qk} \int_{t_o}^{t} \mu(s)\, e^{\lambda A(s)} V(x(s), A(s))\, dA(s)$$

for all $t_o \le t \le t_o + \theta^k$, $k \ge k_o$ almost surely. By Theorem 1-7.1,

$$e^{\lambda A(t)} V(x(t), A(t)) \le \left[e^{\lambda A(t_o)} V(\xi, A(t_o)) + (\mu(t_o + \theta^k))^2 + \theta^{qk+1} \log k \right]$$

$$\times \exp\left(\frac{1}{2}\theta^{-qk} (\mu(t_o + \theta^k))^2 \right) \le C\, \theta^{qk+1} \log k$$

for all $t_o \le t \le t_o + \theta^k$, $k \ge k_o$ almost surely, where C is a finite random variable independent of k. Consequently, if $t_o + \theta^{k-1} \le t \le t_o + \theta^k$, $k \ge k_o$

$$\frac{e^{\lambda A(t)} V(x(t), A(t))}{(t-t_o)^q \log\log(t-t_o)} \le C\, \theta^{q+1} \log k/(\log(k-1) + \log\log\theta) \qquad \text{a.s.}$$

which implies

$$\limsup_{t\to\infty} \frac{e^{\lambda A(t)} V(x(t), A(t))}{(t-t_o)^q \log\log(t-t_o)} \le C\, \theta^{q+1} \qquad \text{a.s.}$$

Since $\theta > 1$ is arbitrary,

$$\limsup_{t\to\infty} \frac{e^{\lambda A(t)} V(x(t), A(t))}{(t-t_o)^q \log\log(t-t_o)} \le C \qquad \text{a.s.}$$

Finally, by conditions (i) and (iv),

$$\limsup_{t\to\infty} \frac{1}{t} \log|x(t)| \le \limsup_{t\to\infty} \frac{1}{pt} \log V(x(t), A(t))$$

$$= \limsup_{t\to\infty} \frac{1}{pt} \log\Big(e^{-\lambda A(t)} \frac{e^{\lambda A(t)} V(x(t), A(t))}{(t-t_o)^q \log\log(t-t_o)} (t-t_o)^q \log\log(t-t_o) \Big)$$

$$\le -(\lambda/p) \liminf_{t\to\infty} \frac{A(t)}{t} \le -\lambda\sigma/p \qquad \text{a.s.}$$

as required. The proof is complete.

Theorem 2.5 Let $V \in C^{2,1}(R^d \times R_+; R)$ and $\psi(t)$ be a nonnegative predictable process. Let p, λ, η, q, σ be positive constants and $\alpha \in [0, 1)$. Assume for $x \in R^d$, $t \ge 0$,

(i) $e^{\lambda t} |x|^p \le V(x, t)$,

(ii) $LV(x, A(t)) + \eta(1 + t)^{-q} HV(x, A(t)) \le \psi(t) \big[1 + (V(x, A(t)))^\alpha \big]$,

(iii) $\liminf_{t\to\infty} \frac{A(t)}{t} \ge \sigma$,

(iv) $\limsup_{t\to\infty} \frac{1}{t} \log\Big(\int_0^t \psi(s)\, dA(s) \Big) \le 0.$

Then the solution of equation (2.1) satisfies

$$\limsup_{t\to\infty} \frac{1}{t} \log|x(t; t_o, \xi)| \le -\lambda\sigma/p \qquad \text{a.s.}$$

Proof. Again fix any t_o and ξ, and write $x(t; t_o, \xi) - x(t)$ simply. By Itô's formula,

$$V(x(t), A(t)) = V(\xi, A(t_o)) + N(t) + \int_{t_o}^t LV(x(s), A(s))\, dA(s), \tag{2.9}$$

where

$$N(t) = \int_{t_o}^t \sum_{i=1}^d \frac{\partial}{\partial x_i} V(x(s), A(s))\, M_i(x(s), ds).$$

Let k = 1, 2, ... An application of Lemma 2-6.2 with $g(t) \equiv t$, $\gamma_k = 2\eta(1 + t_o + 2^k)^{-q}$, $\tau_k = 2^k$ and $\theta = 2$ yields that, for almost all $\omega \in \Omega$, there exists an integer $k_o(\omega)$ such that for all $k \ge k_o$

$$N(t) \le \frac{1}{\eta} (1 + t_o + 2^k)^q \log k + \eta(1 + t_o + 2^k)^{-q} \int_{t_o}^t HV(x(s), A(s))\, dA(s)$$

for $t_o \le t \le t_o + 2^k$. Substituting this into (2.9) and using condition (ii) one obtains

$$V(x(t), A(t)) \le V(\xi, A(t_o)) + \frac{1}{\eta}(1 + t_o + 2^k)^q \log k$$

$$+ \eta(1 + t_o + 2^k)^{-q} \int_{t_o}^{t} HV(x(s), A(s))\, dA(s) + \int_{t_o}^{t} LV(x(s), A(s))\, dA(s)$$

$$\le V(\xi, A(t_o)) + \frac{1}{\eta}(1 + t_o + 2^k)^q \log k$$

$$+ \int_{t_o}^{t} \Big(\eta(1 + s)^{-q} HV(x(s), A(s)) + V\, L(x(s), A(s)) \Big)\, dA(s)$$

$$\le V(\xi, A(t_o)) + \frac{1}{\eta}(1 + t_o + 2^k)^q \log k$$

$$+ \int_{t_o}^{t} \psi(s) \Big[1 + (V(x(s), A(s))^\alpha \Big]\, dA(s)$$

for all $t_o \le t \le t_o + 2^k$, $k \ge k_o$ almost surely. Now applying Corollary 1-7.5 we derive

$$V(x(t), A(t)) \le$$

$$\le \Bigg(\Big[V(\xi, A(t_o)) + \frac{1}{\eta}(1 + t_o + 2^k)^q \log k + \int_{t_o}^{t_o+2^k} \psi(s)\, dA(s) \Big]^{1-\alpha}$$

$$+ (1 - \alpha) \int_{t_o}^{t_o+2^k} \psi(s)\, dA(s) \Bigg)^{1/(1-\alpha)}$$

for all $t_o \le t \le t_o + 2^k$, $k \ge k_o$ almost surely. Now let $\varepsilon > 0$ be arbitrary and by condition (iv) one sees that there exists a random integer $k_1 = k_1(\omega)$ such that for all $k \ge k_1$,

$$\int_{t_o}^{t_o+2^k} \psi(s)\, dA(s) \le e^{\varepsilon(t_o+2^k)} \quad \text{a.s.}$$

Therefore, if $t_o + 2^{k-1} \le t \le t_o + 2^k$, $k \ge k_o \vee k_1$,

$$\frac{1}{t} \log\Big[e^{\lambda A(t)} |x(t)|^p \Big] \le \frac{1}{t} \log V(x(t), A(t))$$

$$\leq ((1-\alpha)(t_o + 2^{k-1}))^{-1} \log\Big(\Big[V(\xi, A(t_o))$$

$$+ \frac{1}{\eta}(1 + t_o + 2^k)^q \log k + e^{\varepsilon(t_o+2^k)} \Big]^{1-\alpha} + (1-\alpha)\, e^{\varepsilon(t_o+2^k)} \Big),$$

where condition (i) has been used too. This immediately implies

$$\limsup_{t\to\infty} \frac{1}{t} \log\Big[e^{\lambda A(t)} |x(t)|^p \Big] \leq 2\varepsilon/(1-\alpha) \qquad \text{a.s.}$$

Since $\varepsilon > 0$ is arbitrary,

$$\limsup_{t\to\infty} \frac{1}{t} \log\Big[e^{\lambda A(t)} |x(t)|^p \Big] \leq 0 \qquad \text{a.s.}$$

Finally

$$\limsup_{t\to\infty} \frac{1}{t} \log|x(t)| = \limsup_{t\to\infty} \frac{1}{pt} \log\Big[e^{-\lambda A(t)} e^{\lambda A(t)} |x(t)|^p \Big]$$

$$\leq -(\lambda/p) \liminf_{t\to\infty} \frac{A(t)}{t} \leq -\lambda\sigma/p \qquad \text{a.s.}$$

as required. The proof is complete.

Theorem 2.6 Let $V \in C^{2,1}(R^d \times R_+; R)$ and let $\psi_1(t)$, $\psi_2(t)$ be two nonnegative predictable processes. Let $p, \lambda, \eta, q, \sigma$ be positive constants and $0 \leq \alpha < 1$. Assume that for all $x \in R^d$ and $t \geq 0$,

(i) $e^{\lambda t} |x|^p \leq V(x, t)$,

(ii) $LV(x, A(t)) + \eta e^{-qt} HV(x, A(t)) \leq \psi_1(t) + \psi_2(t) (V(x, A(t)))^\alpha$,

(iii) $\displaystyle \liminf_{t\to\infty} \frac{A(t)}{t} \geq \sigma$,

(iv) $\displaystyle \limsup_{t\to\infty} \frac{1}{t} \log\Big(\int_0^t \psi_1(s)\, dA(s) \Big) \leq q$,

$$\limsup_{t\to\infty} \frac{1}{t} \log\Big(\int_0^t \psi_2(s)\, dA(s) \Big) \leq (1-\alpha)q.$$

Then the solution of equation (2.1) satisfies

$$\limsup_{t\to\infty} \frac{1}{t} \log|x(t; t_o, \xi)| \leq -(\lambda\sigma - q)/p \qquad \text{a.s.}$$

Proof. Note (2.9) still holds. Let k = 1, 2, ... Again making use of of Lemma 2–6.2 with $g(t) \equiv t$, $\gamma_k = 2\eta e^{-q(t_o+k)}$, $\tau_k = k$ and $\theta = 2$ one can derives that, for almost all $\omega \in \Omega$, there exists an integer $k_o(\omega)$ such that for all $k \geq k_o$

$$N(t) \leq \frac{1}{\eta} e^{q(t_o+k)} \log k + \eta e^{-q(t_o+k)} \int_{t_o}^{t} HV(x(s), A(s))\, dA(s)$$

for $t_o \leq t \leq t_o + k$. Substituting this into (2.9) and using condition (ii) one obtains

$$V(x(t), A(t)) \leq V(\xi, A(t_o)) + \frac{1}{\eta} e^{q(t_o+k)} \log k$$

$$+ \int_{t_o}^{t} \left[\psi_1(s) + \psi_2(s) (V(x, A(s)))^{\alpha} \right] dA(s)$$

for $t_o \leq t \leq t_o + k$, $k \geq k_o$ almost surely. So, by Corollary 1–7.5,

$$V(x(t), A(t)) \leq \Bigg(\Big[V(\xi, A(t_o)) + \frac{1}{\eta} e^{q(t_o+k)} \log k + \int_{t_o}^{t_o+k} \psi_1(s)\, dA(s) \Big]^{1-\alpha}$$

$$+ (1 - \alpha) \int_{t_o}^{t_o+k} \psi_2(s)\, dA(s) \Bigg)^{1/(1-\alpha)}$$

for $t_o \leq t \leq t_o + k$, $k \geq k_o$ almost surely. Let $\varepsilon > 0$ be arbitrary. By condition (iv) there exists a random integer $k_1 = k_1(\omega)$ such that

$$\int_{t_o}^{t_o+k} \psi_1(s)\, dA(s) \leq e^{(q+\varepsilon)(t_o+k)}$$

and

$$\int_{t_o}^{t_o+k} \psi_2(s)\, dA(s) \leq e^{(1-\alpha)(q+\varepsilon)(t_o+k)}$$

for all $k \geq k_1$ almost surely. Hence, if $t_o + k - 1 \leq t \leq t_o + k$, $k \geq k_o \vee k_1$,

$$V(x(t), A(t)) \leq \Bigg(\Big[V(\xi, A(t_o)) + \frac{1}{\eta} e^{q(t_o+k)} \log k + e^{(q+\varepsilon)(t_o+k)} \Big]^{1-\alpha}$$

$$+ (1 - \alpha) e^{(1-\alpha)(q+\varepsilon)(t_o+k)} \Bigg)^{1/(1-\alpha)} \quad \text{a.s.}$$

which implies immediately that

$$\limsup_{t\to\infty} \frac{1}{t} \log V(x(t), A(t)) \le q + \varepsilon \qquad \text{a.s.}$$

Letting $\varepsilon \to 0$ gives

$$\limsup_{t\to\infty} \frac{1}{t} \log V(x(t), A(t)) \le q \qquad \text{a.s.}$$

Finally

$$\limsup_{t\to\infty} \frac{1}{t} \log|x(t)| \le \limsup_{t\to\infty} \frac{1}{pt} \log\left[e^{-\lambda A(t)} V(x(t), A(t)) \right]$$

$$\le -(\lambda/p) \liminf_{t\to\infty} \frac{A(t)}{t} + q/p \le -(\lambda\sigma - q)/p \qquad \text{a.s.}$$

as required. The proof is complete.

As an application, let us consider a d-dimensional Itô equation

$$dx(t) = f(x(t), t)\, dt + \sum_{k=1}^{\infty} g_k(x(t), t)\, dw_k(t) \qquad \text{on } t \ge t_o \tag{2.10}$$

with initial data $x(t_o) = \xi$, where $\{w_k(t): k \ge 1\}$ is a sequence of independent Brownian motions, and $f(x, t) = (f_1, ..., f_d)^T$, $g_k(x, t) = (g_{1k}, ..., g_{dk})^T$ are all d-dimensional adapted processes with parameter $x \in R^d$. Assume the equation has a unique solution which is still denoted by $x(t; t_o, \xi)$. In addition, assume $f(0, t) = g_k(0, t) \equiv 0$ and all $f(x, t)$, $g_k(x, t)$ are locally Lipschitz continuous. Then $x(t; t_o, 0) \equiv 0$ obviously, and also by Proposition 2.1, $x(t; t_o, \xi) \ne 0$ a.s. if $\xi \ne 0$ a.s. For this equation, the operators L and H take the forms:

$$LV(x, t) = \frac{\partial}{\partial t} V(x, t) + \sum_{i=1}^{d} b_i(x, t) \frac{\partial}{\partial x_i} V(x, t)$$

$$+ \frac{1}{2} \sum_{i,j=1}^{d} \sum_{k=1}^{\infty} g_{ik}(x, t)\, g_{jk}(x, t) \frac{\partial^2}{\partial x_i \partial x_j} V(x, t)$$

and

$$HV(x, t) = \sum_{i,j=1}^{d} \sum_{k=1}^{\infty} g_{ik}(x, t)\, g_{jk}(x, t) \frac{\partial}{\partial x_i} V(x, t) \frac{\partial}{\partial x_j} V(x, t).$$

The following corollaries follow from Theorems 2.2–2.6 respectively.

Corollary 2.7 Let $V \in C^{2,1}((R^d - \{0\}) \times R_+; R)$ and $p > 0$, $\rho \ge 0$, $\lambda < \rho/2$. Assume that for all $x \neq 0$ and $t \geq 0$,

(i) $|x|^p \leq V(x, t)$,

(ii) $LV(x, t) \leq \lambda V(x, t)$,

(iii) $HV(x, t) \geq \rho\,(V(x, t))^2$.

Then the solution of equation (2.10) satisfies

$$\limsup_{t\to\infty} \frac{1}{t} \log|x(t; t_o, \xi)| \leq -\frac{1}{p}(\rho/2 - \lambda) \qquad \text{a.s.}$$

Corollary 2.8 Let $V \in C^{2,1}((R^d - \{0\}) \times R_+; R)$ and $p > 0$, $\rho > 0$, $\lambda > \rho/2$. Assume that for all $x \neq 0$ and $t \geq 0$,

(i) $|x|^p \geq V(x, t)$,

(ii) $LV(x, t) \geq \lambda V(x, t)$,

(iii) $HV(x, t) \leq \rho\,(V(x, t))^2$.

Then the solution of equation (2.10) satisfies

$$\liminf_{t\to\infty} \frac{1}{t} \log|x(t; t_o, \xi)| \geq \frac{1}{p}(\lambda - \rho/2) \qquad \text{a.s.}$$

Corollary 2.9 Let $V \in C^{2,1}(R^d \times R_+; R)$ and $\mu(t)$ be a polynomial with positive coefficients. Let $p > 0$ and $\lambda > 0$. Assume that for all $x \in R^d$ and $t \geq 0$,

(i) $|x|^p \leq V(x, t)$,

(ii) $LV(x, t) \leq -\lambda V(x, t) + \mu(t)\, e^{-\lambda t}$,

(iii) $HV(x, t) \leq \mu(t)\, e^{-\lambda t}\, V(x, t)$.

Then the solution of equation (2.10) satisfies

$$\limsup_{t\to\infty} \frac{1}{t} \log|x(t; t_o, \xi)| \leq -\lambda/p \quad \text{a.s.}$$

Corollary 2.10 Let $V \in C^{2,1}(R^d \times R_+; R)$ and $\psi(t)$ be a nonnegative predictable process. Let p, λ, η, q be positive constants and $0 \leq \alpha < 1$. Assume that for all $x \in R^d$ and $t \geq 0$,

(i) $e^{\lambda t}\, |x|^p \leq V(x, t)$,

(ii) $LV(x, t) + \eta(1 + t)^{-q}\, HV(x, t) \leq \psi(t)\left[\, 1 + (V(x, t))^{\alpha} \,\right]$,

$$\text{(iii)}\quad \limsup_{t\to\infty} \frac{1}{t} \log\Big(\int_0^t \psi(s)\, ds \Big) \le 0.$$

Then the solution of equation (2.10) satisfies

$$\limsup_{t\to\infty} \frac{1}{t} \log|x(t; t_o, \xi)| \le -\lambda/p \qquad \text{a.s.}$$

Corollary 2.11 Let $V \in C^{2,1}(R^d \times R_+; R)$ and let $\psi_1(t)$, $\psi_2(t)$ be two nonnegative predictable processes. Let p, λ, η, q be positive constants and $0 \le \alpha < 1$. Assume that for all $x \in R^d$ and $t \ge 0$,

$$\text{(i)}\quad e^{\lambda t}|x|^p \le V(x, t),$$

$$\text{(ii)}\quad LV(x, t) + \eta e^{-qt} HV(x, t) \le \psi_1(t) + \psi_2(t)\,(V(x, t))^\alpha,$$

$$\text{(iii)}\quad \limsup_{t\to\infty} \frac{1}{t} \log\Big(\int_0^t \psi_1(s)\, ds \Big) \le q,$$

$$\limsup_{t\to\infty} \frac{1}{t} \log\Big(\int_0^t \psi_2(s)\, ds \Big) \le (1 - \alpha)q.$$

Then the solution of equation (2.10) satisfies

$$\limsup_{t\to\infty} \frac{1}{t} \log|x(t; t_o, \xi)| \le -(\lambda - q)/p \qquad \text{a.s.}$$

4.3. Pth MOMENT EXPONENTIAL STABILITY

In this section we still consider the stochastic equation

$$x(t) = \xi + \int_{t_o}^t F(x(s), ds) \qquad \text{on } t \ge t_o. \tag{3.1}$$

Throughout this section we let $p > 0$ unless specified and ξ be an $\mathcal{F}_{t_o}$-measurable R^d-valued random variable such that $E|\xi|^p < \infty$. Let us first give the definition of pth moment exponential stability.

Definition 3.1 Equation (3.1) is said to be *pth moment exponentially stable* if there exists a pair of positive constants λ and C such that

$$E|x(t; t_o, \xi)|^p \le CE|\xi|^p e^{-\lambda(t-t_o)}, \quad t \ge t_o \tag{3.2}$$

holds for any t_o and ξ. In this case

$$\limsup_{t\to\infty} \frac{1}{t} \log(E|x(t; t_o, \xi)|^p) \le -\lambda. \tag{3.3}$$

The left hand side of (3.3) is called the *pth moment Lyapunov exponent* of the solution. When $p = 2$ it is usually called the *exponential stability in mean square.*

Generally speaking, the pth moment exponential stability and the almost sure exponential stability do not imply each other and additional conditions are required in order to deduce one from the other. The following theorem gives the additional conditions under which the pth moment exponential stability implies the almost sure exponential stability.

Theorem 3.1 Assume that $p \ge 2$, $A(t) \equiv t$ and

$$|b(x, t)|^2 \vee \text{trace } a(x, x, t) \le K|x|^2 \quad \text{for all } x \in R \text{ and } t \ge 0, \tag{3.4}$$

where $K > 0$. Then the pth moment exponential stability of equation (3.1) implies the almost sure exponential stability.

Proof. Assume (3.2) holds. Fix any t_o and ξ, and write $x(t; t_o, \xi) = x(t)$ simply. We first claim that there exists a constant c_1 such that

$$E\Big(\sup_{t_o+k-1\le t\le t_o+k} |x(t)|^p\Big) \le c_1 E|\xi|^p e^{-\lambda(k-1)}, \qquad k = 1, 2, \ldots \tag{3.5}$$

In fact, by Hölder's inequality and Burkholder-Davis-Gundy's inequality we can derive that

$$E\Big(\sup_{t_o+k-1\le t\le t_o+k} |x(t)|^p\Big)$$

$$\le 3^{p-1} E|x(t_o + k - 1)|^p + 3^{p-1} E \int_{t_o+k-1}^{t_o+k} |b(x(s), s)|^p ds$$

$$+ 3^{p-1} d^{(p/2-1)} C_p E\Big(\int_{t_o+k-1}^{t_o+k} \text{trace } a(x(s), x(s), s) ds\Big)^{p/2},$$

where C_p is the universal constant given by Theorem 1-2.6. Hence, by (3.2) and (3.4),

$$E\Big(\sup_{t_o+k-1\le t\le t_o+k} |x(t)|^p\Big) \le 3^{p-1}\, CE|\xi|^p\, e^{-\lambda(k-1)}$$

$$+ K^{p/2}\, 3^{p-1}\Big[1 + d^{(p/2-1)}\, C_p\Big] CE|\xi|^p \int_{t_o+k-1}^{t_o+k} e^{-\lambda(s-t_o)}\, ds$$

$$\le C3^{p-1}\Big(1 + K^{p/2}\Big[1 + d^{(p/2-1)}\, C_p\Big]\Big) E|\xi|^p\, e^{-\lambda(k-1)}$$

and (3.5) follows. Let $\varepsilon \in (0, \lambda)$ be arbitrary. It then follows from (3.5) that

$$P\Big(\omega: \sup_{t_o+k-1\le t\le t_o+k} |x(t)|^p > e^{-(\lambda-\varepsilon)(k-1)}\Big)$$

$$\le e^{(\lambda-\varepsilon)(k-1)}\, E\Big(\sup_{t_o+k-1\le t\le t_o+k} |x(t)|^p\Big)$$

$$\le c_1\, E|\xi|^p\, e^{-\varepsilon(k-1)}.$$

By the Borel-Cantelli lemma we see that for almost all $\omega \in \Omega$,

$$\sup_{t_o+k-1\le t\le t_o+k} |x(t)|^p \le e^{-(\lambda-\varepsilon)(k-1)} \tag{3.6}$$

holds for all but finitely many k. Hence there exists $k_o(\omega)$, for all $\omega \in \Omega$ excluding a P-null set, for which (3.6) holds whenever $k \ge k_o(\omega)$. Consequently

$$\frac{1}{t}\log|x(t)| \le \frac{1}{pt}\log|x(t)|^p \le -(\lambda - \varepsilon)(k-1)/p(t_o + k - 1)$$

if $t_o + k - 1 \le t \le t_o + k$, $k \ge k_o(\omega)$ almost surely. This implies

$$\limsup_{t\to\infty} \frac{1}{t}\log|x(t)| \le -(\lambda - \varepsilon)/p \qquad \text{a.s.}$$

Since $\varepsilon > 0$ is arbitrary,

$$\limsup_{t\to\infty} \frac{1}{t}\log|x(t)| \le -\lambda/p \qquad \text{a.s.}$$

That is, the equation is almost surely exponential stable and its almost sure Lyapunov exponent is not greater than $-\lambda/p$. The proof is complete.

We shall now give a sufficient criterion for the pth moment exponential stability via the Lyapunov function.

Theorem 3.2 Let $V \in C^{2,1}(R^d \times R_+; R)$ and p, α, β, λ, σ, δ be positive constants. Assume:

(i) $\alpha|x|^p \le V(x, t) \le \beta|x|^p, \quad x \in R^d, t \ge 0$

(ii) $LV(x, A(t)) \le -\lambda V(x, A(t)), \quad x \in R^d, t \ge 0$

(iii) $A(t) - A(s) \ge \sigma(t - s) - \delta, \quad 0 \le s < t < \infty.$

Then equation (3.1) is pth moment exponentially stable and the pth moment Lyapunov exponent is less than or equal to $-\lambda\sigma$.

Proof. By Itô's formula and hypothesis (ii),

$$E\left[e^{\lambda(A(t)-A(t_o))} V(x(t), A(t))\right] \le EV(\xi, A(t_o))$$

$$+ E\int_{t_o}^{t} e^{\lambda(A(s)-A(t_o))}\left[\lambda V(x(s), A(s)) + LV(x(s), A(s))\right] dA(s)$$

$$\le EV(\xi, A(t_o)).$$

Using conditions (i) and (iii), one then sees

$$E\left[e^{\lambda\sigma(t-t_o)-\lambda\delta}\,\alpha|x(t)|^p\right] \le \beta\, E|\xi|^p,$$

that is

$$E|x(t)|^p \le \frac{\beta e^{\lambda\delta}}{\alpha}\, E|\xi|^p\, e^{-\lambda\sigma(t-t_o)}.$$

The proof is complete.

In the same way we can prove the following theorem which gives a sufficient criterion for a stochastic system which is pth moment exponentially unstable.

Theorem 3.3 Let $V \in C^{2,1}(R^d \times R_+; R)$ and p, α, β, λ, σ, δ be positive constants. Assume:

(i) $\alpha|x|^p \le V(x, t) \le \beta|x|^p, \quad x \in R^d, t \ge 0$

(ii) $LV(x, A(t)) \ge \lambda V(x, A(t)), \quad x \in R^d, t \ge 0$

(iii) $A(t) - A(s) \ge \sigma(t - s) - \delta, \quad 0 \le s < t < \infty.$

Then

$$E|x(t; t_o, \xi)|^p \geq \frac{\alpha e^{-\lambda\delta}}{\beta} \; E|\xi|^p \, e^{\lambda\sigma(t-t_o)}.$$

In this case, equation (3.1) is said to be *pth moment exponentially unstable.*

As in the previous section, we can apply the above two theorems to equation (2.10) to obtain the following useful corollaries.

Corollary 3.4 Let $V \in C^{2,1}(R^d \times R_+; R)$ and let p, α, β, λ be positive constants. Assume for all $x \in R^d$ and $t \geq 0$

(i) $\alpha|x|^p \leq V(x, t) \leq \beta|x|^p$,

(ii) $LV(x, t) \leq -\lambda V(x, t)$,

where L is the operator associated with equation (2.10) (see page 160). Then equation (2.10) is pth moment exponentially stable and the pth moment Lyapunov exponent is not greater than $-\lambda$.

Corollary 3.5 Let $V \in C^{2,1}(R^d \times R_+; R)$ and let p, α, β, λ be positive constants. Assume for all $x \in R^d$ and $t \geq 0$:

(i) $\alpha|x|^p \leq V(x, t) \leq \beta|x|^p$,

(ii) $LV(x, t) \geq \lambda V(x, t)$.

Then

$$E|x(t; t_o, \xi)|^p \geq \frac{\alpha}{\beta} \; E|\xi|^p \, e^{\lambda(t-t_o)}.$$

and hence equation (2.10) is pth moment exponentially unstable.

4.4. STOCHASTIC STABILIZATION AND DESTABILIZATION

It has been observed that noise can destabilize a stable system. For example, suppose that a given exponentially stable system

$$\dot{x}(t) = -x(t) \quad \text{on } t \geq 0, \quad x(0) = x_o \in R \tag{4.1}$$

is perturbed by noise and the stochastically perturbed system is described by a one-dimensional Itô equation

$$dy(t) = -y(t)dt + \varepsilon dw(t) \quad \text{on } t \geq 0, \quad y(0) = x_o, \tag{4.2}$$

where $\varepsilon > 0$ and w(t) is a real-valued Brownian motion. Obviously, the stochastic perturbation does not change the mean value of the solution in the sense $Ey(t) = x(t)$. However, by Example 2-6.4,

$$\limsup_{t\to\infty} \frac{1}{\log t} \Big(\sup_{0\le s\le t} |y(s)|^2 \Big) = \varepsilon^2 \qquad \text{a.s.}$$

which implies

$$\limsup_{t\to\infty} |y(t)| = \infty \qquad \text{a.s.}$$

So the perturbed system (4.2) is no longer almost surely exponentially stable. In other words, the stochastic perturbation has destabilized the exponentially stable system (4.1).

On the other hands, it has also been observed that noise can have a stabilizing effect as well. For instance, consider an unstable system

$$\dot{x}(t) = x(t) \quad \text{on } t \ge 0, \;\; x(0) = x_o \in R. \tag{4.3}$$

Perturb this system by noise and say the perturbed system has the form

$$dy(t) = y(t)\,dt + 2y(t)\,dw(t) \quad \text{on } t \ge 0, \;\; y(0) = x_o \in R. \tag{4.4}$$

Again $Ey(t) = x(t)$. Note that equation (4.4) has the explicit solution

$$y(t) = x_o\, e^{-t+2w(t)},$$

which yields immediately that

$$\limsup_{t\to\infty} \frac{1}{t} \log|y(t)| = -1 \qquad \text{a.s.}$$

whenever $x_o \ne 0$. So the noise has stabilized the unstable system (4.3) without changing the mean value of the solution.

In this section we shall use the theory obtained in Section 4.2 to establish more general results on stochastic stabilization and destabilization.

Theorem 4.1 Let $A : R_+ \to R^{d\times d}$ be bounded and set $\|A\| = \sup\{\|A(t)\| : t \ge 0\}$. Let w(t) be a real-valued Brownian motion and σ be a constant such that $\sigma^2 > 2\|A\|$. Then the stochastic differential equation

$$dx(t) = A(t)\,x(t)\,dt + \sigma x(t)\,dw(t) \tag{4.5}$$

is almost surely exponentially stable and its almost sure Lyapunov exponent is not greater than $-(\sigma^2/2 - \|A\|)$. We therefore conclude that any linear system $\dot{x}(t) = A(t)\,x(t)$ can be stabilized by Brownian motion without changing the mean value of the solution provided $A(t)$ is bounded.

Proof. We use Corollary 2.7 to prove the theorem. Let $V(x, t) = |x|^2$. Then

$$LV(x, t) = 2x^T A(t)x + \sigma^2 |x|^2 \le (\sigma^2 + 2\|A\|)\,|x|^2,$$

$$HV(x, t) = 4\sigma^2 |x|^4.$$

So by Corollary 2.7,

$$\limsup_{t\to\infty} \frac{1}{t} \log|x(t)| \le -(2\sigma^2 - \sigma^2 - 2\|A\|)/2 = -(\sigma^2/2 - \|A\|) \qquad \text{a.s.}$$

The proof is complete.

Theorem 4.2 Let $d \ge 3$ and $(w_1(t), \ldots, w_d(t))$ be a d-dimensional Brownian motion. Let $A : R_+ \to R^{d\times d}$ be bounded and set $\|A\| = \sup\{\|A(t)\|: t \ge 0\}$. Let σ be a constant such that $\sigma^2 > 6\|A\|$. Then for the stochastic differential equation

$$dx(t) = A(t)\,x(t)\,dt + \sigma \begin{bmatrix} x_2(t)dw_1(t) \\ \vdots \\ x_d(t)dw_{d-1}(t) \\ x_1(t)dw_d(t) \end{bmatrix} \qquad \text{on } t \ge t_o \tag{4.6}$$

with initial data $x(t_o) = \xi \ne 0$ a.s., we have

$$\liminf_{t\to\infty} \frac{1}{t} \log|x(t)| \ge \sigma^2/6 - \|A\| \qquad \text{a.s.}$$

That is equation (4.6) is almost surely exponentially unstable.

Proof. We use Corollary 2.8 to prove the theorem. Let $V(x, t) = |x|^2$. Then

$$LV(x, t) = 2x^T A(t)x + \sigma^2 |x|^2 \ge (\sigma^2 - 2\|A\|)\,|x|^2$$

and

$$HV(x, t) = 4\sigma^2 \sum_{i=1}^{d} x_i^2\, x_{i+1}^2,$$

where we let $x_{d+1} = x_1$. But

$$\sum_{i=1}^{d} x_i^2 x_{i+1}^2 \le \sum_{i=1}^{d} (x_i^4 + x_{i+1}^4)/2 = \sum_{i=1}^{d} x_i^4$$

and then

$$3\sum_{i=1}^{d} x_i^2 x_{i+1}^2 \le 2\sum_{i=1}^{d} x_i^2 x_{i+1}^2 + \sum_{i=1}^{d} x_i^4 \le |x|^4.$$

So

$$HV(x, t) \le \frac{4}{3}\sigma^2 |x|^4.$$

Thus by Corollary 2.8,

$$\liminf_{t\to\infty} \frac{1}{t} \log|x(t)| \ge (\sigma^2 - 2\|A\| - 2\sigma^2/3)/2 = \sigma^2/6 - \|A\| \qquad \text{a.s.}$$

The proof is complete.

Theorem 4.3 Let $d = 2$ and $A : R_+ \to R^{d\times d}$ be bounded. Let $w(t)$ be a real-valued Brownian motion and σ be a constant such that $\sigma^2 > 2\|A\|$, where $\|A\| = \sup\{\|A(t)\| : t\ge 0\}$. Then for the stochastic differential equation

$$dx(t) = A(t)\, x(t)\, dt + \sigma \begin{bmatrix} x_2(t) \\ -x_1(t) \end{bmatrix} dw(t) \quad \text{on } t \ge t_o \tag{4.7}$$

with initial data $x(t_o) = \xi \ne 0$ a.s., we have

$$\liminf_{t\to\infty} \frac{1}{t} \log|x(t)| \ge \sigma^2/2 - \|A\| \qquad \text{a.s.}$$

That is equation (4.7) is almost surely exponentially unstable.

Proof. Let $V(x, t) = |x|^2$. Then

$$LV(x, t) = 2x^T A(t)x + \sigma^2 |x|^2 \ge (\sigma^2 - 2\|A\|)\, |x|^2,$$

$$HV(x, t) = 4\sigma^2 \left(x_1^2 x_2^2 - x_1^2 x_2^2 - x_1^2 x_2^2 + x_1^2 x_2^2 \right) = 0.$$

Thus by Corollary 2.8,

$$\liminf_{t\to\infty} \frac{1}{t} \log|x(t)| \ge (\sigma^2 - 2\|A\|)/2 = \sigma^2/2 - \|A\| \qquad \text{a.s.}$$

The proof is complete.

In view of Theorems 4.2 and 4.3 we obtain the following general result.

Theorem 4.4 Any linear system $\dot{x}(t) = A(t)x(t)$ can be destabilized by Brownian motion without changing the mean value of the solution provided A(t) is bounded and the dimension of the state space is not less than 2.

Before we carry on to cope with non-linear systems, let us look at a one-dimensional linear stochastic differential equation

$$dx(t) = bx(t) + \sum_{i=1}^{m} \sigma_i x(t)\, dw_i(t) \qquad \text{on } t \geq t_o \tag{4.8}$$

with initial data $x(t_o) = \xi \neq 0$ a.s., where $(w_1, ..., w_m)$ is an m-dimensional Brownian motion and b, σ_i $(1 \leq i \leq m)$ are all constants. It is known that the equation has the unique explicit solution

$$x(t) = x_o \exp\Big(\Big[b - \frac{1}{2}\sum_{i=1}^{m} \sigma_i^2 \Big] (t - t_o) + \sum_{i=1}^{m} \sigma_i(w_i(t) - w_i(t_o)) \Big).$$

So

$$\lim_{t\to\infty} \frac{1}{t} \log|x(t)| = b - \frac{1}{2}\sum_{i=1}^{m} \sigma_i^2 \qquad \text{a.s.}$$

We therefore see that the exponentially stable system $\dot{x}(t) = bx(t)$, when $b < 0$, can not be destabilized by noise if you restrict the stochastic perturbation in the linear form of $\sigma_1 x(t)\, dw_1(t) + ... + \sigma_m x(t)\, dw_m(t)$.

The above proofs can be adopted to deal with non-linear systems. We shall only state the theorems below but leave the details of the proofs to the reader.

Theorem 4.5 Assume that $f : R^d \times R_+ \to R^d$ is a locally Lipschitz continuous function, and particularly,

$$|f(x, t)| \leq K|x| \qquad \text{for all } (x, t) \in R^d \times R_+, \tag{4.9}$$

where $K > 0$. Let w(t) be a real-valued Brownian motion and σ be a constant such that $\sigma^2 > 2K$. Then the stochastic differential equation

$$dx(t) = f(x(t), t)\, dt + \sigma x(t)\, dw(t)$$

is almost surely exponentially stable and its almost sure Lyapunov exponent is not greater than $-(\sigma^2/2 - K)$. We therefore conclude that any non-linear system $\dot{x}(t) = f(x(t), t)$ can be stabilized by Brownian motion provided (4.9) is satisfied.

Theorem 4.6 Let f be the same function as in Theorem 4.5.

(i) Let $d \geq 3$ and $(w_1(t), ..., w_d(t))$ be a d-dimensional Brownian motion. Let σ be a constant such that $\sigma^2 > 6K$. Then for the stochastic differential equation

$$dx(t) = f(x(t), t)\,dt + \sigma \begin{bmatrix} x_2(t)dw_1(t) \\ \vdots \\ x_d(t)dw_{d-1}(t) \\ x_1(t)dw_d(t) \end{bmatrix} \qquad \text{on } t \geq t_o$$

with initial data $x(t_o) = \xi \neq 0$ a.s., we have

$$\liminf_{t\to\infty} \frac{1}{t} \log|x(t)| \geq \sigma^2/6 - K \qquad \text{a.s.}$$

(ii) Let d be an even number, say $d = 2m$. Let $w(t)$ be a real-valued Brownian motion and σ be a constant such that $\sigma^2 > 2K$. Then for the stochastic differential equation

$$dx(t) = f(x(t), t)\,dt + \sigma \begin{bmatrix} x_2(t) \\ -x_1(t) \\ \vdots \\ x_{2m}(t) \\ -x_{2m-1}(t) \end{bmatrix} dw(t) \qquad \text{on } t \geq t_o$$

with initial data $x(t_o) = \xi \neq 0$ a.s., we have

$$\liminf_{t\to\infty} \frac{1}{t} \log|x(t)| \geq \sigma^2/2 - K \qquad \text{a.s.}$$

(iii) From (i) and (ii) we conclude that any non-linear system $\dot{x}(t) = f(x(t), t)$ can be destabilized by Brownian motion provided the dimension of the state space is greater than 1 and (4.9) is satisfied.

4.5. LYAPUNOV EXPONENTS OF STOCHASTIC FLOWS

Throughout this section we let $\varphi_{s,t}(x)$, $0 \leq s \leq t < \infty$, $x \in R^d$ be a forward Brownian flow satisfying hypotheses (H2-7.1)-(H2-7.3) (see Section 2.7). In view of Theorem 2-7.1, this stochastic flow can be represented by the stochastic differential equation

$$\varphi_{s,t}(x) = x + \int_s^t F(\varphi_{s,r}(x), dr), \tag{5.1}$$

where F(x, t) is a C–Brownian motion with the local characteristic (a(x, y, t), b(x, t), t) in which

$$b(x, t) = (b_1, \ldots, b_d)^T = \lim_{h\to 0+} \frac{1}{h} E(\varphi_{t,t+h}(x) - x), \tag{5.2}$$

$$a(x, y, t) = (a_{ij})_{d\times d} = \lim_{h\to 0+} \frac{1}{h} E\{(\varphi_{t,t+h}(x) - x)(\varphi_{t,t+h}(y) - y)^T\}. \tag{5.3}$$

The operators L and H associated with equation (5.1) or the stochastic flow have the following forms:

$$LV(x, t) = \frac{\partial}{\partial t} V(x, t) + \sum_{i=1}^{d} b_i(x, t) \frac{\partial}{\partial x_i} V(x, t)$$

$$+ \frac{1}{2} \sum_{i,j=1}^{d} a_{ij}(x, x, t) \frac{\partial^2}{\partial x_i \partial x_j} V(x, t)$$

and

$$HV(x, t) = \sum_{i,j=1}^{d} a_{ij}(x, x, t) \frac{\partial}{\partial x_i} V(x, t) \frac{\partial}{\partial x_j} V(x, t).$$

Employing the theorems in Section 2 we can easily obtain the following results which give the estimates for the Lyapunov exponents of the stochastic flow.

Theorem 5.1 Let $V \in C^{2,1}((R^d - \{0\}) \times R_+; R)$ and $p > 0$, $\rho \geq 0$, $\lambda < \rho/2$. Assume the flow has the property that $\varphi_{s,t}(x) \neq 0$ a.s. whenever $x \neq 0$. Assume furthermore that for all $x \neq 0$ and $t \geq 0$,

(i) $|x|^p \leq V(x, t)$,

(ii) $LV(x, t) \leq \lambda V(x, t)$,

(iii) $HV(x, t) \geq \rho\, (V(x, t))^2$.

Then the flow satisfies

$$\limsup_{t\to\infty} \frac{1}{t} \log|\varphi_{s,t}(x)| \leq -\frac{1}{p}(\rho/2 - \lambda) \quad \text{a.s.}$$

for any $s \geq 0$ and $x \neq 0$.

Theorem 5.2 Let $V \in C^{2,1}((R^d - \{0\}) \times R_+; R)$ and $p > 0$, $\rho > 0$, $\lambda > \rho/2$. Assume the flow has the property that $\varphi_{s,t}(x) \neq 0$ a.s. whenever $x \neq 0$. Assume furthermore that for all $x \neq 0$ and $t \geq 0$,

(i) $|x|^p \geq V(x, t)$,

(ii) $LV(x, t) \geq \lambda V(x, t)$,

(iii) $HV(x, t) \leq \rho\,(V(x, t))^2$.

Then the flow satisfies

$$\liminf_{t\to\infty} \frac{1}{t} \log|\varphi_{s,t}(x)| \geq \frac{1}{p}(\lambda - \rho/2) \quad \text{a.s.}$$

for any $s \geq 0$ and $x \neq 0$.

Theorem 5.3 Let $V \in C^{2,1}(R^d \times R_+; R)$ and let p, λ, μ be positive constants. Assume that for all $x \in R^d$ and $t \geq 0$,

(i) $|x|^p \leq V(x, t)$,

(ii) $LV(x, t) \leq -\lambda V(x, t) + \mu\, e^{-\lambda t}$,

(iii) $HV(x, t) \leq \mu\, e^{-\lambda t}\, V(x, t)$.

Then the flow satisfies

$$\limsup_{t\to\infty} \frac{1}{t} \log|\varphi_{s,t}(x)| \leq -\lambda/p \quad \text{a.s.}$$

for all s and x.

Theorem 5.4 Let $V \in C^{2,1}(R^d \times R_+; R)$ and let $\psi_1(t)$, $\psi_2(t)$ be two nonnegative predictable processes. Let p, λ, η, q be positive constants and $0 \leq \alpha < 1$. Assume that for all $x \in R^d$ and $t \geq 0$,

(i) $e^{\lambda t}|x|^p \leq V(x, t)$,

(ii) $LV(x, t) + \eta e^{-qt} HV(x, t) \leq \psi_1(t) + \psi_2(t)\,(V(x, t))^\alpha$,

(iii) $$\limsup_{t\to\infty} \frac{1}{t} \log\Big(\int_0^t \psi_1(s)\, ds \Big) \leq q,$$

$$\limsup_{t\to\infty} \frac{1}{t} \log\Big(\int_0^t \psi_2(s)\, ds \Big) \leq (1 - \alpha)q.$$

Then the flow satisfies

$$\limsup_{t\to\infty} \frac{1}{t} \log|\varphi_{s,t}(x)| \le -(\lambda - q)/p \qquad \text{a.s.}$$

for all s and x.

We shall now give a few examples to illustrate the theory. We emphasize that all the forward Brownian flows used in these examples satisfy the hypotheses (H2–7.1)–(H2–7.3) and the local characteristics $b(x, t)$, $a(x, y, t)$ associated with the flows are defined by (5.2) and (5.3) respectively.

Example 5.1 Let $\varphi_{s,t}(x)$, $0 \le s \le t < \infty$, $x \in R$ be a real-valued forward Brownian flow with the local characteristics

$$b(x, t) = -(2 - \sin t \cos x)\, x, \qquad a(x, y, t) = e^{-2\gamma t} |xy|^{1/2},$$

where $0 < \gamma < 1/2$. Let $1/2 < p < 1$ and introduce a Lyapunov function

$$V(x, t) = e^{pt} |x|^p, \qquad (x, t) \in R \times R_+.$$

Compute

$$LV(x, t) = pe^{pt} |x|^p - e^{pt} p|x|^{p-1} \operatorname{sign}(x) (2 - \sin t \cos x)\, x$$

$$+ \frac{1}{2} e^{pt} p(p - 1)|x|^{p-2} e^{-2\gamma t} |x|$$

$$\le \frac{1}{2} p(p - 1)\, e^{(p-2\gamma)t} |x|^{p-1} \le 0$$

and

$$HV(x, t) = e^{2pt} p^2 |x|^{2(p-1)} e^{-2\gamma t} |x| = p^2 e^{2(p-\gamma)t} |x|^{2p-1}.$$

Thus

$$LV(x, t) + e^{-(1-2\gamma)pt} HV(x, t) \le e^{-(1-2\gamma)pt} p^2 e^{2(p-\gamma)t} |x|^{2p-1}$$

$$\le e^{(1-2\gamma)(1-p)t} (V(x, t))^{(2p-1)/p}.$$

Note $0 < (2p - 1)/p < 1$ and $(1-(2p - 1)/p)(1 - 2\gamma)p = (1 - p)(1 - 2\gamma)$. So by Theorem 5.4 we see that the flow satisfies

$$\limsup_{t\to\infty} \frac{1}{t} \log|\varphi_{s,t}(x)| \le -(p - (1 - 2\gamma)p)/p = -2\gamma \qquad \text{a.s.}$$

Example 5.2 Let $\varphi_{s,t}(x)$, $0 \le s \le t < \infty$, $x \in R^2$ be a forward Brownian flow in R^2 with the local characteristics

$$b(x, t) = \begin{bmatrix} x_1 - 2x_2 \\ 3x_1 - 4x_2 \end{bmatrix},$$

$$a(x, y, t) = t\, e^{-2t} \begin{bmatrix} \sin x_1 \sin y_1, & \sin x_1 \cos(y_1 - y_2) \\ \cos(x_1 - x_2) \sin y_1, & \cos(x_1 - x_2)\cos(y_1 - y_2) \end{bmatrix}.$$

Introduce a Lyapunov function

$$V(x, t) = \frac{1}{3}(\, 13\, x_1^2 - 14\, x_1 x_2 + 4\, x_2^2\,).$$

It is easy to show that

(i) $V(x, t) \ge |x|^2$,

(ii) $LV(x, t) \le -V(x, t) + \mu e^{-t}$,

(iii) $HV(x, t) \le \mu e^{-t}\, V(x, t)$,

where μ is a positive constant. Therefore, by Theorem 5.3, the flow satisfies

$$\limsup_{t \to \infty} \frac{1}{t} \log|\varphi_{s,t}(x)| \le -1/2 \quad \text{a.s.}$$

4.6. STOCHASTIC OSCILLATORS

A stochastic oscillator is described mathematically as the solution of an appropriate ordinary differential equation, which is driven by an external disturbance of white noise. In this section we shall illustrate how to employ the theory established in Sections 4.2 and 4.3 to investigate the stability of stochastic oscillators through a number of examples.

Example 6.1 Let us start with a scalar linear stochastic oscillator:

$$\ddot{y}(t) + a\dot{y}(t) + by(t) = (c\dot{y}(t) + ey(t))\, \dot{w}(t), \tag{6.1}$$

where $a > 0$, $b > 0$, $\dot{w}(t)$ is a white noise, i.e. $w(t)$ is a Brownian motion, and c, e are constants which represent the intensity of a stochastic disturbance. Introduce a R^2-vector $x = (x_1, x_2)^T = (y, \dot{y})^T$. Then the corresponding Itô's equation is

$$
\begin{aligned}
dx_1(t) &= x_2(t)\,dt \\
dx_2(t) &= (-bx_1(t) - ax_2(t))\,dt + (cx_2(t) + ex_1(t))\,dw(t).
\end{aligned}
\tag{6.1$'$}
$$

For the Lyapunov function, we try a quadratic function

$$V(x, t) = \alpha x_1^2 + \beta x_1 x_2 + x_2^2.$$

We then compute

$$LV(x, t) = -(\beta b - e^2)x_1^2 - (2a - \beta - c^2)x_2^2 + (2\alpha - \beta a - 2b + 2ce)x_1 x_2.$$

In order to convert $LV(x, t)$ to be negative definite (we here mean $LV(x, t) \le -\varepsilon|x|^2$ for some $\varepsilon > 0$), we set

$$2\alpha - \beta a - 2b + 2ce = 0, \text{ that is } \alpha = (\beta a + 2b - 2ce)/2.$$

Then V and LV become

$$V(x, t) = \frac{1}{2}(\beta a + 2b - 2ce)x_1^2 + \beta x_1 x_2 + x_2^2$$

and

$$LV(x, t) = -(\beta b - e^2)x_1^2 - (2a - \beta - c^2)x_2^2.$$

For LV to be negative definite we must have $\beta b - e^2 > 0$ and $2a - \beta - c^2 > 0$, that is

$$e^2/b < \beta < 2a - c^2. \tag{6.2}$$

For V to be positive definite (we mean $V(x, t) \ge \varepsilon|x|^2$ for some $\varepsilon > 0$), we must have

$$2(\beta a + 2b - 2ce) > \beta^2,$$

this is equivalent to

$$a - \sqrt{a^2 + 4(b - ce)} < \beta < a + \sqrt{a^2 + 4(b - ce)}. \tag{6.3}$$

Combining (6.2) and (6.3) we see that V is positive definite and LV is negative definite provided

$$\max\left(e^2/b,\ a - \sqrt{a^2 + 4(b - ce)}\right) < \beta < \min\left(2a - c^2,\ a + \sqrt{a^2 + 4(b - ce)}\right).$$

We therefore conclude, by Corollary 3.4, if

$$\max\left(e^2/b,\ a - \sqrt{a^2 + 4(b - ce)}\right) < \min\left(2a - c^2,\ a + \sqrt{a^2 + 4(b - ce)}\right) \quad (6.4)$$

then the stochastic oscillator is exponentially stable in mean square, and hence by Theorem 3.1 it is also almost surely exponentially stable. A stronger but possibly more convenient condition than (6.4) is

$$ce \le b, \quad e^2 + bc^2 < 2ab. \quad (6.5)$$

Condition (6.5) gives a quite clear estimate for the intensity of the external stochastic disturbance in the sense that the stochastic disturbance can be tolerated by the stable deterministic oscillator $\ddot{y}(t) + a\dot{y}(t) + by(t) = 0$ without losing the property of stability.

Example 6.2 Our second example is from satellite dynamics. Sagirow (1970) derived the equation

$$\ddot{y}(t) + B(1+A\dot{w}(t))\dot{y}(t) + (1 + A\dot{w}(t))y(t) - C\sin(2y(t)) = 0 \quad (6.6)$$

in the study of the influence of a rapidly fluctuating density of the atmosphere of the earth on the motion of a satellite in a circular orbit, where B, C are positive constants, $\dot{w}(t)$ is a scalar white noise and A is a constant which represents the intensity of the disturbance. Again we introduce $x = (x_1, x_2)^T = (y, \dot{y})^T$. Then the corresponding Itô's equation is

$$dx_1(t) = x_2(t)\,dt$$
$$dx_2(t) = (-x_1(t) + C\sin(2x_1(t)) - Bx_2(t))\,dt - A(x_1(t) + Bx_2(t))\,dw(t).$$

For the Lyapunov function, we try an expression consisting of a quadratic form and integral of the nonlinear component:

$$V(x, t) = ax_1^2 + bx_1x_2 + x_2^2 + e\int_0^{x_1} \sin(2y)\,dy$$
$$= ax_1^2 + bx_1x_2 + x_2^2 + e(\sin x_1)^2.$$

This yields

$$LV(x, t) = -(b - A^2)x_1^2 + bCx_1\sin(2x_1) - (2B - b - A^2B^2)x_2^2$$

$$+ (2a - bB - 2 + 2A^2B)x_1x_2 + (e + 2C)x_2\sin(2x_1).$$

Setting $2a - bB - 2 + 2A^2B = 0$ and $e + 2C = 0$ we obtain

$$V(x, t) = \frac{1}{2}(bB + 2 - 2A^2B)x_1^2 + bx_1x_2 + x_2^2 - 2C(\sin x_1)^2$$

and

$$LV(x, t) = -(b - A^2)x_1^2 + bCx_1\sin(2x_1) - (2B - b - A^2B^2)x_2^2.$$

Note

$$V(x, t) \geq \frac{1}{2}(bB + 2 - 2A^2B - 4C)x_1^2 + bx_1x_2 + x_2^2.$$

So V is positive definite if

$$2(bB + 2 - 2A^2B - 4C) \geq b^2$$

or equivalently

$$B - \sqrt{B^2 + 4 - 8C - 4BA^2} < b < B + \sqrt{B^2 + 4 - 8C - 4BA^2}. \tag{6.7}$$

Also note

$$LV(x, t) \leq -(b - A^2 - 2bC)x_1^2 - (2B - b - A^2B^2)x_2^2,$$

which is negative definite provided $b - A^2 - 2bC > 0$ and $2B - b - A^2B^2 > 0$, i.e.

$$2C < 1 \quad \text{and} \quad A^2/(1 - 2C) < b < 2B - A^2B^2. \tag{6.8}$$

We therefore conclude, by Corollary 3.4, if $C < 1/2$ and

$$\max\left(A^2/(1 - 2C),\ B - \sqrt{B^2 + 4 - 8C - 4BA^2} \right)$$
$$< \min\left(2B - A^2B^2,\ B + \sqrt{B^2 + 4 - 8C - 4BA^2} \right) \tag{6.9}$$

then equation (6.6) is exponentially stable in mean square, and hence by Theorem 3.1 it is also almost surely exponentially stable. Moreover, if $A^2 < (1 - 2C)/B$, then (6.9) reduces to $A^2/(1 - 2C) < 2B - A^2B^2$, that is, $A^2 < 2B(1 - 2C)/(1+B^2(1 - 2C))$. In other words (6.9) may be replaced by a stronger condition

$$A^2 < \min\Big((1 - 2C)/B,\ 2B(1 - 2C)/(1 + B^2(1 - 2C)) \Big), \tag{6.10}$$

which gives an explicitly estimate for the intensity of the stochastic disturbance that can be tolerated by the stable deterministic oscillator $\ddot{y}(t) + B\dot{y}(t) + y(t) - C\sin(2y(t)) = 0$ without losing the property of stability.

Example 6.3 We now consider a d-dimensional nonlinear stochastic oscillator

$$\ddot{x} + b(x, \dot{x}) + \nabla G(x) = \sum_{k=1}^{\infty} \sigma_k(x, \dot{x}, t)\, \dot{w}_k(t) + e(t) \tag{6.11}$$

motivated by the classical theory of nonlinear oscillations, where $-b(x, \dot{x})$ is a dissipative force, $-\nabla G(x) = (G_{x_1}, \dots, G_{x_d})^T$ is a restoring force, $e(t)$ is an external driving force, $\{w_k(t),\ k \geq 1\}$ is a sequence of independent Brownian motions and $\{\sigma_k(x, \dot{x}, t),\ k \geq 1\}$ represents the intensity of a stochastic disturbance. Introduce $y = \dot{x}$ and then the corresponding 2d-dimensional stochastic differential equation is

$$dx = y\, dt$$

$$dy = \Big[-b(x, y) - \nabla G(x) + e(t) \Big] dt + \sum_{k=1}^{\infty} \sigma_k(x, y, t)\, dw_k(t). \tag{6.11$'$}$$

We assume:

(i) $b(x, y)$ is locally Lipschitzian; $G(x)$ is a C^2-function such that $\nabla G(x)$ is locally Lipschitzian; $\sigma_k(x, y, t)$, $k \geq 1$ are all continuous in $(x, y, t) \in R^{2d} \times R_+$ and also locally Lipschitzian in (x, y) at each fixed $t \geq 0$; and $e(t)$, $t \geq 0$ is an d-dimensional adapted continuous process.

(ii) $G(x) > 0$ except $G(0) = 0$, $\quad y^T b(x, y) \geq \lambda\, U(x, y)$,

(iii) $|e(t)|^2 + \sum_{k=1}^{\infty} |\sigma_k(x, y, t)|^2 \leq \mu e^{-\lambda t}$

for some constants $\lambda > 0$ and $\mu > 0$, where U is the energy of the system defined by

$$U(x, y) = \frac{1}{2} |y|^2 + G(x).$$

We try to use the energy function as the Lyapunov function (Note we only require U is once differentiable in x in this special case). Let $\varepsilon \in (0, \lambda)$ be arbitrary. We then compute

$$LU(x, y) = y^T(-b(x, y) - \nabla G(x) + e(t)) + y^T\nabla G(x) + \frac{1}{2}\sum_{k=1}^{\infty} |\sigma_k(x, y, t)|^2$$

$$\leq -y^T b(x, y) + |e(t)|\,|y| + \frac{1}{2}\mu e^{-\lambda t}$$

$$\leq -\lambda U(x, y) + \frac{1}{2}\varepsilon|y|^2 + \frac{1}{2\varepsilon}|e(t)|^2 + \frac{1}{2}\mu e^{-\lambda t}$$

$$\leq -(\lambda - \varepsilon)\, U(x, y) + \frac{1}{2}(1 + \varepsilon^{-1})\mu e^{-\lambda t}$$

and

$$HU(x, y) = |y|^2 \sum_{k=1}^{\infty} |\sigma_k(x, y, t)|^2 \leq 2\mu e^{-\lambda t}\, U(x, y).$$

Hence, by Corollary 2.9 (with a careful reading of the proof of Theorem 2.4) we see that the energy of the system satisfies

$$\limsup_{t\to\infty} \frac{1}{t} \log U(x(t), y(t)) \leq -(\lambda - \varepsilon) \quad \text{a.s.}$$

Since ε is arbitrary, we obtain that

$$\limsup_{t\to\infty} \frac{1}{t} \log U(x(t), y(t)) \leq -\lambda \quad \text{a.s.}$$

Moreover, if we replace condition (iii) by

$$\text{(iii)}' \quad |e(t)|^2 \leq \mu e^{-\lambda t}, \quad \sum_{k=1}^{\infty} |\sigma_k(x, y, t)|^2 \leq \gamma|y|^2,$$

where $0 < \gamma < \lambda$, then

$$LU(x, y) = -\lambda U(x, y) + \frac{1}{2}\gamma|y|^2 \leq -(\lambda - \gamma)U(x, y).$$

By Itô's formula,

$$E\Big[e^{(\lambda-\gamma)t}\, U(x(t), y(t)) \Big] = e^{(\lambda-\gamma)t_o}\, EU(x(t_o), y(t_o))$$

$$+ \int_{t_o}^{t} e^{(\lambda-\gamma)s} \Big[(\lambda - \gamma)\, U(x(s), y(s)) + LU(x(s), y(s)) \Big] ds$$

$$\leq e^{(\lambda-\gamma)t_o}\, EU(x(t_o), y(t_o)).$$

That is

$$EU(x(t), y(t)) \le EU(x(t_o), y(t_o))\, e^{-(\lambda-\gamma)(t-t_o)} \quad \text{if } t \ge t_o.$$

Hence

$$\limsup_{t\to\infty} \frac{1}{t} \log\Big[EU(x(t), y(t)) \Big] \le -(\lambda-\gamma).$$

CHAPTER 5

Almost Sure Exponential Stability of Stochastic Differential Delay Equations

5.1. INTRODUCTION

Stochastic differential equations provide a mathematical model for sophisticated dynamical systems in physical, biological, medical and social sciences. However in many circumstances, the future state of a system depends on not only the present state but also its history. Stochastic differential delay equations or stochastic differential functional equations give a mathematical formulation for such systems. Many problems in the sciences lead to the study of almost sure exponential stability or moment exponential stability of such systems. Although there exist a certain number of papers on the moment exponential stability of stochastic differential delay equations, little was known on the almost sure exponential stability until Mohammed (1990) obtained the first result on the Lyapunov exponents for linear stochastic differential delay equations and Mao (1991a) studied the almost sure exponential stability for a class of non-linear stochastic differential delay equations.

In this chapter we shall discuss the almost sure exponential stability of stochastic differential delay equations driven by non-linear integrators. In order to make the theory more understandable, we shall begin with the stochastic equations with constant delay in Section 5.2, which perhaps are the simplest ones comparing with other delay equations discussed in later sections. We shall next deal with the stochastic equations with variable delay in Section 5.3 and, more generally, stochastic differential functional equations in Section 5.4. Finally we shall give a number of examples to illustrate the theory in Section 5.5.

5.2. EQUATIONS WITH CONSTANT DELAY

In this section we let $F(x, y, t)$, $x, y \in R^d$ be a d-dimensional continuous C-semimartingale such that it has the decomposition $F(x, y, t) = M(x, y, t) + B(x, y, t)$ and the local characteristic $(a(x, y, \bar{x}, \bar{y}, t), b(x, y, t), A(t))$ belonging to the class $B^{0,\delta}$. Let $\tau > 0$ and $t_o \geq 0$. Let $\xi(.)$ be a $C([-\tau, 0]; R^d)$-valued $\mathcal{F}_{t_o}$-measurable random variable. Consider a d-dimensional stochastic differential delay equation driven by $F(x, y, t)$ of the form

$$x(t) = \xi(0) + \int_{t_o}^{t} F(x(s), x(s-\tau), ds) \qquad \text{on } t \geq t_o \tag{2.1}$$

with initial data $x(t_o + s) = \xi(s)$, $-\tau \leq s \leq 0$. Assume the equation has a unique global solution which is denoted by $x(t; t_o, \xi)$.

Definition 2.1 Equation (2.1) is said to be *almost surely exponentially stable* if there exists a $\lambda > 0$ such that for each pair of t_o and ξ there is a finite random variable β such that

$$|x(t; t_o, \xi)| \leq \beta e^{-\lambda(t-t_o)} \qquad \text{a.s. for all } t \geq t_o,$$

or equivalently if

$$\limsup_{t\to\infty} \frac{1}{t} \log|x(t; t_o, \xi)| \leq -\lambda \qquad \text{a.s.}$$

The left hand side of this last inequality is called the *Lyapunov exponent* of the solution.

Although the definition here is made for equation (2.1), it is applicable to any other types of delay equations in this book obviously.

Theorem 2.1 Let c_1 - c_3 be positive constants. Assume

(i) $2x^T b(x, y, t) \leq -c_1|x|^2 + c_2|y|^2$,

(ii) trace $a(x, y, x, y, t) \leq c_3 e^{-c_1 t}$,

(iii) $A(t) \equiv t$

for all $x, y \in R^d$ and $t \geq 0$. If $c_1 > c_2$, then there exists a $\lambda > 0$ such that

$$\limsup_{t\to\infty} \frac{1}{t} \log|x(t; t_o, \xi)| \leq -\lambda \quad \text{a.s.} \tag{2.2}$$

for all t_o and ξ. In other words, equation (2.1) is almost surely exponentially stable.

Proof. We divide the whole proof into two steps.

Step 1. Fix t_o and ξ arbitrarily and write $x(t; t_o, \xi) = x(t)$ simply. Since $c_1 > c_2$ we can find a $\lambda \in (0, c_1)$ such that

$$\frac{c_2 e^{\lambda\tau}}{c_1 - \lambda} < 1. \tag{2.3}$$

We claim that for almost all $\omega \in \Omega$ there exists an integer $k_o = k_o(\omega)$ and a positive number $C = C(\omega)$ such that for every integer $k \geq k_o$

$$\int_{t_o}^{T} e^{\lambda t} |x(t)|^2 dt \leq Ck \qquad \text{if } t_o \leq T \leq t_o + k. \tag{2.4}$$

In fact, by Itô's formula and the assumptions,

$$\begin{aligned} e^{c_1 t} |x(t)|^2 &= e^{c_1 t_o} |x(t_o)|^2 + Z(t) \\ &+ \int_{t_o}^{t} e^{c_1 s} \Big(c_1|x(s)|^2 + 2x(s)^T b(x(s), x(s-\tau), s) \\ &\qquad + \text{trace } a(x(s), x(s-\tau), x(s), x(s-\tau), s) \Big) ds \\ &\leq e^{c_1 t_o} |\xi(0)|^2 + Z(t) + c_3 t + c_2 \int_{t_o}^{t} e^{c_1 s} |x(s-\tau)|^2 ds \end{aligned} \tag{2.5}$$

for all $t \geq t_o$, where

$$Z(t) = 2 \int_{t_o}^{t} e^{c_1 s} x(s)^T M(x(s), x(s-\tau), ds)$$

which is a continuous martingale vanishing at $t = t_o$. Let $k = 1, 2, \ldots$ and, by (2.3), one can choose $\varepsilon > 0$ so small that

$$\frac{c_2 e^{\lambda\tau}}{c_1 - \lambda} + \frac{2\varepsilon c_3}{c_1 - \lambda} < 1. \tag{2.6}$$

In view of Lemma 2-6.2 we can deduce that for almost all $\omega \in \Omega$ there exists an integer $k_o = k_o(\omega)$ such that

$$Z(t) \le \frac{1}{2}\varepsilon \langle Z(t)\rangle + 2\varepsilon^{-1}\log k \qquad \text{if } t_o \le t \le t_o + k.$$

But

$$\langle Z(t)\rangle = 4 \int_{t_o}^{t} e^{2c_1 s}\, x(s)^T a(x(s), x(s-\tau), x(s), x(s-\tau), s)\, x(s)\, ds.$$

Note

$$x^T a x = \sum_{i,j=1}^{d} x_i a_{ij} x_j = \sum_{i,j=1}^{d} |x_i| (a_{ii} a_{jj})^{1/2} |x_j|$$

$$\le \frac{1}{2} \sum_{i,j=1}^{d} \left[|x_i|^2 a_{jj} + |x_j|^2 a_{ii} \right] = |x|^2 \operatorname{trace} a,$$

where Theorem 1-3.3 has been used. Hence by Condition (ii) we see

$$\langle Z(t)\rangle \le 4c_3 \int_{t_o}^{t} e^{c_1 s} |x(s)|^2\, ds.$$

Therefore

$$Z(t) \le 2\varepsilon c_3 \int_{t_o}^{t} e^{c_1 s} |x(s)|^2\, ds + 2\varepsilon^{-1}\log k$$

for all $t_o \le t \le t_o + k$, $k \ge k_o$ almost surely. Substituting this into (2.5) gives

$$e^{c_1 t} |x(t)|^2 \le e^{c_1 t_o} |\xi(0)|^2 + c_3 t + 2\varepsilon^{-1}\log k$$

$$+ c_2 \int_{t_o}^{t} e^{c_1 s} |x(s-\tau)|^2\, ds + 2\varepsilon c_3 \int_{t_o}^{t} e^{c_1 s} |x(s)|^2\, ds$$

for all $t_o \le t \le t_o + k$, $k \ge k_o$ almost surely. Consequently

$$|x(t)|^2 \le \left[e^{c_1 t_o} |\xi(0)|^2 + c_3(t_o + k) + 2\varepsilon^{-1}\log k \right] e^{-c_1 t}$$

$$+ c_2 e^{-c_1 t} \int_{t_o}^{t} e^{c_1 s} |x(s-\tau)|^2 \, ds + 2\varepsilon c_3 e^{-c_1 t} \int_{t_o}^{t} e^{c_1 s} |x(s)|^2 \, ds$$

for all $t_o \le t \le t_o + k$, $k \ge k_o$ almost surely. Therefore, for almost all $\omega \in \Omega$, if $t_o \le T \le t_o + k$, $k \ge k_o$

$$\int_{t_o}^{T} e^{\lambda t} |x(t)|^2 \, dt$$

$$\le \left[e^{c_1 t_o} |\xi(0)|^2 + c_3(t_o + k) + 2\varepsilon^{-1} \log k \right] \int_{t_o}^{T} e^{-(c_1 - \lambda)t} \, dt + J_1 + J_2$$

$$\le \frac{1}{c_1 - \lambda} \left[e^{c_1 t_o} |\xi(0)|^2 + c_3(t_o + k) + 2\varepsilon^{-1} \log k \right] + J_1 + J_2, \tag{2.7}$$

where

$$J_1 = c_2 \int_{t_o}^{T} e^{-(c_1-\lambda)t} \int_{t_o}^{t} e^{c_1 s} |x(s-\tau)|^2 \, ds \, dt,$$

$$J_2 = 2\varepsilon c_3 \int_{t_o}^{T} e^{-(c_1-\lambda)t} \int_{t_o}^{t} e^{c_1 s} |x(s)|^2 \, ds \, dt.$$

But

$$J_1 = c_2 \int_{t_o}^{T} e^{c_1 s} |x(s-\tau)|^2 \int_{s}^{T} e^{-(c_1-\lambda)t} \, dt \, ds$$

$$\le \frac{1}{c_1 - \lambda} c_2 \int_{t_o}^{T} e^{\lambda s} |x(s-\tau)|^2 \, ds$$

$$\le \frac{1}{c_1 - \lambda} c_2 \left(\int_{t_o}^{t_o+\tau} e^{\lambda s} |x(s-\tau)|^2 \, ds + e^{\lambda\tau} \int_{t_o+\tau}^{T \vee (t_o+\tau)} e^{\lambda(s-\tau)} |x(s-\tau)|^2 \, ds \right)$$

$$\le \frac{1}{c_1 - \lambda} c_2 \int_{-\tau}^{0} e^{\lambda(t_o+\tau+s)} |\xi(s)|^2 \, ds + \frac{1}{c_1 - \lambda} c_2 e^{\lambda\tau} \int_{t_o}^{T} e^{\lambda s} |x(s)|^2 \, ds.$$

Similarly

$$J_2 \le \frac{1}{c_1 - \lambda} \, 2\varepsilon c_3 \int_{t_o}^{T} e^{\lambda s} |x(s)|^2 \, ds.$$

Hence we obtain

$$\int_{t_o}^{T} e^{\lambda t} |x(t)|^2 \, dt$$

$$\le \frac{1}{c_1 - \lambda} \left[e^{c_1 t_o} |\xi(0)|^2 + c_3(t_o + k) + 2\varepsilon^{-1} \log k + c_2 \int_{-\tau}^{0} e^{\lambda(t_o+\tau+s)} |\xi(s)|^2 \, ds \right]$$

$$+ \frac{1}{c_1 - \lambda} (c_2 e^{\lambda\tau} + 2\varepsilon c_3) \int_{t_o}^{T} e^{\lambda s} |x(s)|^2 \, ds. \tag{2.8}$$

Recalling (2.6) we get the required (2.4).

Step 2. By Itô's formula and the assumptions we can derive that

$$e^{\lambda t} |x(t)|^2 = e^{\lambda t_o} |x(t_o)|^2 + N(t)$$

$$+ \int_{t_o}^{t} e^{\lambda s} \Big(\lambda |x(s)|^2 + 2x(s)^T b(x(s), x(s-\tau), s)$$

$$+ \operatorname{trace} a(x(s), x(s-\tau), x(s), x(s-\tau), s) \Big) \, ds$$

$$\le e^{\lambda t_o} |\xi(0)|^2 + N(t) + c_3/(c_1 - \lambda) + c_2 \int_{t_o}^{t} e^{\lambda s} |x(s-\tau)|^2 \, ds \tag{2.9}$$

for all $t \ge 0$, where

$$N(t) = 2 \int_{t_o}^{t} e^{\lambda s} x(s)^T M(x(s), x(s-\tau), ds).$$

Again by Lemma 2-6.2 we can show that for almost all $\omega \in \Omega$ there exists an integer $k_1 = k_1(\omega)$ such that

$$N(t) \le \frac{1}{2} \langle N(t) \rangle + 2\log k \le 2c_3 \int_{t_o}^{t} e^{\lambda s} |x(s)|^2 \, ds + 2\log k$$

for all $t_o \le t \le t_o + k$, $k \ge k_1$. Substituting this into (2.9) we obtain

$$e^{\lambda t}|x(t)|^2 \le e^{\lambda t_o}|\xi(0)|^2 + 2\log k + c_3/(c_1-\lambda)$$
$$+ c_2\int_{t_o}^{t} e^{\lambda s}|x(s-\tau)|^2\,ds + 2c_3\int_{t_o}^{t} e^{\lambda s}|x(s)|^2\,ds$$
$$\le e^{\lambda t_o}|\xi(0)|^2 + 2\log k + c_3/(c_1-\lambda)$$
$$+ c_2\int_{-\tau}^{0} e^{\lambda(t_o+\tau+s)}|\xi(s)|^2\,ds + (2c_3 + c_2e^{\lambda\tau})\int_{t_o}^{t} e^{\lambda s}|x(s)|^2\,ds$$

for all $t_o \le t \le t_o + k$, $k \ge k_1$ almost surely. This, together with (2.4), implies

$$e^{\lambda t}|x(t)|^2 \le e^{\lambda t_o}|\xi(0)|^2 + 2\log k + c_3/(c_1-\lambda)$$
$$+ c_2\int_{-\tau}^{0} e^{\lambda(t_o+\tau+s)}|\xi(s)|^2\,ds + (2c_3 + c_2e^{\lambda\tau})Ck$$

for all $t_o \le t \le t_o + k$, $k \ge k_o \vee k_1$ almost surely. In particular, for almost all $\omega \in \Omega$, if $t_o + k - 1 \le t \le t_o + k$ and $k \ge k_o \vee k_1$,

$$\frac{1}{t}\log\left(e^{\lambda t}|x(t)|^2\right)$$
$$\le \frac{1}{t_o+k-1}\log\Big(e^{\lambda t_o}|\xi(0)|^2 + 2\log k + c_3/(c_1-\lambda)$$
$$+ c_2\int_{-\tau}^{0} e^{\lambda(t_o+\tau+s)}|\xi(s)|^2\,ds + (2c_3 + c_2e^{\lambda\tau})Ck\Big)$$

and hence

$$\limsup_{t\to\infty}\frac{1}{t}\log\left(e^{\lambda t}|x(t)|^2\right) \le 0 \qquad \text{a.s.}$$

But

$$\limsup_{t\to\infty}\frac{1}{t}\log\left(e^{\lambda t}|x(t)|^2\right) = \lambda + 2\limsup_{t\to\infty}\frac{1}{t}\log|x(t)|.$$

So

$$\limsup_{t\to\infty}\frac{1}{t}\log|x(t)| \le -\lambda/2 \qquad \text{a.s.}$$

as required. The proof is complete.

In the same way one can prove the following more general result.

Theorem 2.2 Let c_1–c_3 be three positive constants and Q be a positive definite d × d matrix. Assume for all $x, y \in R^d$ and $t \geq 0$:

(i) $x^T(Q + Q^T)b(x, y, t) \leq -c_1x^TQx + c_2y^TQy$,

(ii) trace $a(x, y, x, y, t) \leq c_3e^{-c_1t}$,

(iii) $A(t) \equiv t$.

If $c_1 > c_2$, then the delay equation (2.1) is almost surely exponentially stable.

We shall now use this theorem to establish a useful result.

Corollary 2.3 Let c_1 – c_4 be four positive constants and let Q be a positive definite d × d matrix. Assume for all $x, y \in R^d$ and $t \geq 0$:

(i) $|x|^2 \leq x^TQx$,

(ii) $x^T(Q + Q^T)b(x, 0, t) \leq -c_1x^TQx$,

(iii) $|b(x, 0, t) - b(x, y, t)| \leq c_2|y|$,

(iv) trace $a(x, y, x, y, t) \leq c_3e^{-c_4t}$,

(v) $A(t) \equiv t$.

If

$$(c_1/2) \wedge c_4 > c_2|Q| \quad \text{or} \quad (c_1 - c_4)\, c_4 > (c_2|Q|)^2,$$

then the delay equation (2.1) is almost surely exponentially stable.

Proof. Let $\varepsilon \in (0, c_1)$. Then

$$\begin{aligned} x^T(Q + Q^T)b(x, y, t) &\leq -c_1x^TQx + 2c_2|Q|\,|x|\,|y| \\ &\leq -c_1x^TQx + \varepsilon|x|^2 + \varepsilon^{-1}(c_2|Q|)^2\,|y|^2 \\ &\leq -(c_1 - \varepsilon)x^TQx + \varepsilon^{-1}(c_2|Q|)^2\,y^TQy. \end{aligned} \tag{2.10}$$

In the case $(c_1/2) \wedge c_4 > c_2|Q|$, we choose $\varepsilon = c_2|Q|$ and then (2.10) yields

$$\begin{aligned} x^T(Q + Q^T)b(x, y, t) &\leq -(c_1 - c_2|Q|)x^TQx + c_2|Q|\,y^TQy \\ &\leq -((c_1 - c_2|Q|)\wedge c_4)\,x^TQx + c_2|Q|\,y^TQy. \end{aligned}$$

But obviously

$$\text{trace } a(x, y, x, y, t) \le c_3 e^{-((c_1 - c_2|Q|)\wedge c_4)t}.$$

So by Theorem 2.2, equation (2.1) is almost surely exponentially stable if

$$(c_1 - c_2|Q|)\wedge c_4 > c_2|Q|$$

which is equivalent to $(c_1/2) \wedge c_4 > c_2|Q|$ as assumed. On the other hand, in the case $(c_1 - c_4)c_4 > (c_2|Q|)^2$, we choose $\varepsilon = c_1 - c_4$ and then (2.10) gives

$$x^T(Q + Q^T)b(x, y, t) \le -c_4 x^T Q x + ((c_2|Q|)^2/(c_1 - c_4))\, y^T Q y.$$

Hence by Theorem 2.2 again, equation (2.1) is almost surely exponentially stable. The proof is complete.

Before we generalize the above results, let us make an application. Consider a d-dimensional Itô delay equation

$$dx(t) = f(x(t), x(t-\tau), t)\, dt + \sum_{k=1}^{\infty} g_k(x(t), x(t-\tau), t)\, dw_k(t) \quad \text{on } t \ge t_o \quad (2.11)$$

with initial data $x(t_o+s) = \xi(s)$, $-\tau \le s \le 0$, where $\{w_k(t)\colon k \ge 1\}$ is a sequence of independent Brownian motions, and $f(x, y, t)$, $g_k(x, y, t)$ are all d-dimensional adapted processes with parameters $x, y \in R^d$. Assume the equation has a unique solution that is still denoted by $x(t; t_o, \xi)$. The following corollaries follow from Theorem 2.2 and Corollary 2.3 respectively.

Corollary 2.4 Let $c_1 - c_3$ be three positive constants and Q be a positive definite $d \times d$ matrix. Assume for all $x, y \in R^d$ and $t \ge 0$:

(i) $x^T(Q + Q^T)f(x, y, t) \le -c_1 x^T Q x + c_2 y^T Q y$,

(ii) $\sum_{k=1}^{\infty} |g_k(x, y, t)|^2 \le c_3 e^{-c_1 t}$.

If $c_1 > c_2$, then equation (2.11) is almost surely exponentially stable.

Corollary 2.5 Let $c_1 - c_4$ be four positive constants and let Q be a positive definite $d \times d$ matrix. Assume for all $x, y \in R^d$ and $t \ge 0$:

(i) $|x|^2 \le x^T Q x$,

(ii) $x^T(Q + Q^T)f(x, 0, t) \le -c_1 x^T Q x$,

(iii) $|f(x, 0, t) - f(x, y, t)| \le c_2|y|$,

(iv) $\sum_{k=1}^{\infty} |g_k(x, y, t)|^2 \le c_3 e^{-c_4 t}$.

If

$$(c_1/2) \wedge c_4 > c_2|Q| \quad \text{or} \quad (c_1 - c_4)\, c_4 > (c_2|Q|)^2,$$

then the delay equation (2.11) is almost surely exponentially stable.

We shall now modify the above theory to deal with a little more general stochastic differential delay equation. Let $\rho : R_+ \to R$ be a continuously differentiable function such that

$$\rho(t) \le t \quad \text{and} \quad \dot{\rho}(t) \ge 1 \quad \text{for all } t \ge 0. \tag{2.12}$$

It is easy to show that

$$t + \rho(0) \le \rho(t) \quad \text{and} \quad \rho^{-1}(t) \le t - \rho(0) \quad \text{for all } t \ge 0, \tag{2.13}$$

where $\rho^{-1}(.)$ is the inverse function of $\rho(.)$. Set $\tau = -\rho(0)$ which is assumed to be positive (otherwise $\rho(t) \equiv t$ that is not what we are interested in here). Consider a stochastic differential delay equation

$$x(t) = \xi(0) + \int_{t_o}^{t} F(x(s), x(\rho(s)), ds) \quad \text{on } t \ge t_o \tag{2.14}$$

with initial data $x(t_o + s) = \xi(s)$, $-\tau \le s \le 0$. This is a more general delay equation than equation (2.1) since if we let $\rho(t) = t - \tau$ then equation (2.14) reduces to equation (2.1).

Theorem 2.6 Let (2.12) hold. Let $c_1 - c_3$ be positive constants and let Q be a positive definite $d \times d$ matrix. Assume

(i) $x^T(Q + Q^T)b(x, y, t) \le -c_1 x^T Q x + c_2 y^T Q y$,

(ii) $\text{trace } a(x, y, x, y, t) \le c_3 e^{-c_1 t}$,

(iii) $A(t) \equiv t$

for all $x, y \in R^d$ and $t \ge 0$. If $c_1 > c_2$, then the delay equation (2.14) is almost surely exponentially stable.

Proof. We shall only prove the theorem in the case Q = the identity matrix, but the general case can be done in the same way. We use the same notations as the proof of Theorem 2.1. In this case, (2.5) becomes

$$e^{c_1 t} |x(t)|^2 \le e^{c_1 t_o} |\xi(0)|^2 + Z(t) + c_3 t + c_2 \int_{t_o}^{t} e^{c_1 s} |x(\rho(s))|^2 \, ds, \tag{2.15}$$

where

$$Z(t) = 2 \int_{t_o}^{t} e^{c_1 s} x(s)^T M(x(s), x(\rho(s)), ds).$$

It is not difficult to see that the almost sure estimate for Z(t) in the proof of Theorem 2.1 is still valid. So we can derive from (2.15) that for almost all $\omega \in \Omega$, if $t_o \le T \le t_o + k$, $k \ge k_o$

$$\int_{t_o}^{T} e^{\lambda t} |x(t)|^2 \, dt$$

$$\le \frac{1}{c_1 - \lambda} \left[e^{c_1 t_o} |\xi(0)|^2 + c_3(t_o + k) + 2\varepsilon^{-1} \log k \right] + J_1 + J_2, \tag{2.16}$$

where J_2 is the same as before but

$$J_1 = c_2 \int_{t_o}^{T} e^{-(c_1 - \lambda)t} \int_{t_o}^{t} e^{c_1 s} |x(\rho(s))|^2 \, ds \, dt.$$

We now compute

$$J_1 = c_2 \int_{t_o}^{T} e^{c_1 s} |x(\rho(s))|^2 \int_{s}^{T} e^{-(c_1 - \lambda)t} \, dt \, ds$$

$$\le \frac{1}{c_1 - \lambda} c_2 \int_{t_o}^{T} e^{\lambda s} |x(\rho(s))|^2 \, ds.$$

But by changing variable $u = \rho(s)$ and making use of (2.12) and (2.13) we derive

$$\int_{t_o}^{T} e^{\lambda s} |x(\rho(s))|^2 \, ds \le \int_{t_o - \tau}^{T} e^{\lambda(u+\tau)} |x(u)|^2 \, du$$

$$\leq \int_{-\tau}^{0} e^{\lambda(t_o+\tau+u)} |\xi(u)|^2 \, du + e^{\lambda\tau} \int_{t_o}^{T} e^{\lambda u} |x(u)|^2 \, du. \tag{2.17}$$

Hence

$$J_1 \leq \frac{1}{c_1 - \lambda} c_2 \int_{-\tau}^{0} e^{\lambda(t_o+\tau+u)} |\xi(u)|^2 \, du + \frac{1}{c_1 - \lambda} c_2 e^{\lambda\tau} \int_{t_o}^{T} e^{\lambda u} |x(u)|^2 \, du$$

which is the same estimate as before. Therefore we can obtain (2.8) from (2.16) and then (2.4) follows. Moreover, (2.7) becomes

$$e^{\lambda t} |x(t)|^2 \leq e^{\lambda t_o} |\xi(0)|^2 + N(t) + c_3/(c_1 - \lambda) + c_2 \int_{t_o}^{t} e^{\lambda s} |x(\rho(s))|^2 \, ds, \tag{2.18}$$

where

$$N(t) = 2 \int_{t_o}^{t} e^{\lambda s} x(s)^T M(x(s), x(\rho(s)), ds).$$

Again the almost sure estimate for N(t) before is still valid and therefore we can get from (2.18) that

$$e^{\lambda t} |x(t)|^2 \leq e^{\lambda t_o} |\xi(0)|^2 + 2\log k + c_3/(c_1 - \lambda)$$

$$+ c_2 \int_{t_o}^{t} e^{\lambda s} |x(\rho(s))|^2 \, ds + 2 \int_{t_o}^{t} e^{\lambda s} |x(s)|^2 \, ds$$

$$\leq e^{\lambda t_o} |\xi(0)|^2 + 2\log k + c_3/(c_1 - \lambda)$$

$$+ c_2 \int_{-\tau}^{0} e^{\lambda(t_o+\tau+s)} |\xi(s)|^2 \, ds + (2 + c_2 e^{\lambda\tau}) \int_{t_o}^{t} e^{\lambda s} |x(s)|^2 \, ds$$

for all $t_o \leq t \leq t_o + k$, $k \geq k_1$ almost surely, where (2.17) has been used as well. This, together with (2.4), implies

$$e^{\lambda t} |x(t)|^2 \leq e^{\lambda t_o} |\xi(0)|^2 + 2\log k + c_3/(c_1 - \lambda)$$

$$+ c_2 \int_{-\tau}^{0} e^{\lambda(t_o+\tau+s)} |\xi(s)|^2 \, ds + (2 + c_2 e^{\lambda\tau}) Ck$$

for all $t_o \leq t \leq t_o + k$, $k \geq k_o \vee k_1$ almost surely. The remainder of the proof is the same as before. The proof is complete.

Obviously we also have the following corollary.

Corollary 2.7 Let (2.12) hold. Then under the same assumptions of Corollary 2.3 the delay equation (2.14) is almost surely exponentially stable.

Now we replace $t - \tau$ by $\rho(t)$ in equation (2.9). Then the equation becomes

$$dx(t) = f(x(t), x(\rho(t)), t)\, dt + \sum_{k=1}^{\infty} g_k(x(t), x(\rho(t)), t)\, dw_k(t). \tag{2.19}$$

For this equation the following result is immediate.

Corollary 2.8 Let (2.12) hold. Then under the same assumptions of Corollary 2.4 or Corollary 2.5, the delay equation (2.19) is almost surely exponentially stable.

5.3. EQUATIONS WITH VARIABLE DELAY

In this section we still let $F(x, y, t)$, $x, y \in R^d$ be a d-dimensional continuous C-semimartingale such that it has the decomposition $F(x, y, t) = M(x, y, t) + B(x, y, t)$ and the local characteristic $(a(x, y, \bar{x}, \bar{y}, t), b(x, y, t), A(t))$ belonging to the class $B^{0,\delta}$. Also let $\tau > 0$ and $t_o \geq 0$ and $\xi(.)$ be a $C([-\tau, 0]; R^d)$-valued $\mathcal{F}_{t_o}$-measurable random variable. Let $\delta(.) : R_+ \to [-\tau, 0]$. Particularly, for the sake of convenience we shall let

$$A(t) = 0 \qquad \text{on } -\tau \leq t \leq 0$$

in the sequel of this book unless specified. Consider a d-dimensional stochastic differential delay equation driven by $F(x, y, t)$ of the form

$$x(t) = \xi(0) + \int_{t_o}^{t} F(x(s), x(s-\delta(s)), ds) \qquad \text{on } t \geq t_o \tag{3.1}$$

with initial data $x(t_o+s) = \xi(s)$, $-\tau \leq s \leq 0$. Assume the equation has a unique global solution which is denoted by $x(t; t_o, \xi)$.

Theorem 3.1 Let $c_1 - c_5$ be positive constants. Assume for all $x, y \in R^d$ and $t \geq 0$:

(i) $2x^T b(x, y, t) \leq -c_1|x|^2 + c_2|y|^2$,

(ii) $\text{trace}\, a(x, y, x, y, t) \leq c_3 e^{-c_1 A(t)}$,

(iii) $A(t+\tau) - A(t) \le c_4\tau$ and $\liminf_{t\to\infty} \frac{A(t)}{t} \ge c_5$.

If

$$c_1c_5 > c_2c_4e^{c_1c_4\tau}, \tag{3.2}$$

then

$$\limsup_{t\to\infty} \frac{1}{t} \log|x(t; t_o, \xi)| \le -\frac{1}{2}\left(c_1c_5 - c_2c_4e^{c_1c_4\tau} \right) \quad \text{a.s.} \tag{3.3}$$

That is the delay equation (3.1) is almost surely exponentially stable. Moreover, a necessary condition for (3.2) to be satisfied is $c_2 < c_5/(\tau e c_4^2)$.

In order to prove the theorem, we need to prepare a useful inequality.

Lemma 3.2 Let $T > \tau$ and $c \ge 0$. Let $u(t)$ be a continuous nonnegative function defined on $[-\tau, T]$. Let $\mu_1(t)$ and $\mu_2(t)$ be both continuous nonnegative nondecreasing functions defined on $[0, T]$. Assume

$$u(t) \le c + \int_0^t u(s)\, d\mu_1(s) + \int_0^t u(s-\delta(s))\, d\mu_2(s), \quad 0 \le t \le T. \tag{3.4}$$

Then

$$u(t) \le \left(c + \mu_2(\tau)\left[\sup_{-\tau\le r\le 0} u(r) \right]\right) \exp(\mu_1(t) + \mu_2(t)), \quad 0 \le t \le T. \tag{3.5}$$

Proof. Define

$$z(t) = c + \int_0^t u(s)\, d\mu_1(s) + \int_0^t u(s-\delta(s))\, d\mu_2(s) \quad \text{for } 0 \le t \le T,$$

which is nondecreasing. By (3.4),

$$u(t) \le z(t) \quad \text{for } 0 \le t \le T. \tag{3.6}$$

Moreover

$$u(s-\delta(s)) \le \left[\sup_{-\tau\le r\le 0} u(r) \right] + z(s) \quad \text{if } 0 \le s \le \tau$$

and

$$u(s-\delta(s)) \le z(s) \quad \text{if } \tau \le s \le T.$$

Therefore

$$z(t) \le c + \int_0^t z(s)\, d\mu_1(s) + \int_0^\tau \Big[\sup_{-\tau\le r\le 0} u(r) \Big] d\mu_2(s) + \int_0^t z(s)\, d\mu_2(s)$$

$$\le c + \mu_2(\tau)\Big[\sup_{-\tau\le r\le 0} u(r) \Big] + \int_0^t z(s)\, d(\mu_1(s) + \mu_2(s))$$

for $0 \le t \le T$. In view of Theorem 1-7.1

$$z(t) \le \Big(c + \mu_2(\tau)\Big[\sup_{-\tau\le r\le 0} u(r) \Big]\Big) \exp(\mu_1(t) + \mu_2(t)), \quad 0 \le t \le T$$

and the required inequality (3.4) follows from (3.5). The proof is complete.

Proof of Theorem 3.1. Fix t_o and ξ arbitrarily and write $x(t; t_o, \xi) = x(t)$ simply. By Itô's formula and conditions (i) and (ii),

$$e^{c_1 A(t)} |x(t)|^2 = e^{c_1 A(t_o)} |x(t_o)|^2 + N(t)$$

$$+ \int_{t_o}^t e^{c_1 A(s)} \Big(c_1 |x(s)|^2 + 2x(s)^T b(x(s), x(s-\delta(s)), s)$$

$$+ \text{trace } a(x(s), x(s-\delta(s)), x(s), x(s-\delta(s)), s) \Big) dA(s)$$

$$\le e^{c_1 A(t_o)} |x(t_o)|^2 + N(t) + \int_{t_o}^t e^{c_1 A(s)} \Big(c_2 |x(s-\delta(s))|^2 + c_3 e^{-c_1 A(s)} \Big) dA(s),$$

where

$$N(t) = 2 \int_{t_o}^t e^{c_1 A(s)} x(s)^T M(x(s), x(s-\delta(s)), ds).$$

Note from condition (iii) that

$$A(s) \le A(s) - A(s-\tau) + A(s-\delta(s)) \le c_4\tau + A(s-\delta(s)).$$

Therefore

$$e^{c_1 A(t)} |x(t)|^2 \le e^{c_1 A(t_o)} |x(t_o)|^2 + c_3(A(t) - A(t_o)) + N(t)$$

$$+ c_2 e^{c_1 c_4 \tau} \int_{t_o}^t e^{c_1 A(s-\delta(s))} |x(s-\delta(s))|^2 \, dA(s), \tag{3.7}$$

Let k = 1, 2, ... and assign $\theta > 1$ arbitrarily. Applying Lemma 2–6.2 we can derive that for almost all $\omega \in \Omega$ there exists a random integer $k_o = k_o(\omega)$ such that if $k \geq k_o$

$$N(t) \leq \frac{1}{2}\theta^{-k} \langle N(t) \rangle + \theta^{k+1} \log k$$

for $t_o \leq t \leq k_o + \theta^k$. But

$$\begin{aligned}
&\langle N(t) \rangle \\
&\leq 4 \int_{t_o}^{t} e^{2c_1A(s)} |x(s)|^2 \,\mathrm{trace}\, a(x(s), x(s-\delta(s)), x(s), x(s-\delta(s)), s)\, dA(s) \\
&\leq 4c_3 \int_{t_o}^{t} e^{c_1A(s)} |x(s)|^2 \, dA(s).
\end{aligned}$$

So

$$N(t) \leq 2c_3\theta^{-k} \int_{t_o}^{t} e^{c_1A(s)} |x(s)|^2 \, dA(s) + \theta^{k+1} \log k$$

for all $t_o \leq t \leq t_o + \theta^k$, $k \geq k_o$ almost surely. Substituting this into (3.7) yields

$$\begin{aligned}
e^{c_1A(t)} |x(t)|^2 &\leq e^{c_1A(t_o)} |x(t_o)|^2 + c_3(A(t_o+\theta^k) - A(t_o)) + \theta^{k+1} \log k \\
&+ 2c_3\theta^{-k} \int_{t_o}^{t} e^{c_1A(s)} |x(s)|^2 \, dA(s) \\
&+ c_2 e^{c_1c_4\tau} \int_{t_o}^{t} e^{c_1A(s-\delta(s))} |x(s-\delta(s))|^2 \, dA(s)
\end{aligned}$$

for all $t_o \leq t \leq t_o + \theta^k$, $k \geq k_o$ almost surely. By Lemma 3.2,

$$\begin{aligned}
e^{c_1A(t)} |x(t)|^2 \leq &\Big(C + c_3(A(t_o+\theta^k) - A(t_o)) + \theta^{k+1} \log k \Big) \\
&\times \exp\Big(\Big[2c_3\theta^{-k} + c_2 e^{c_1c_4\tau} \Big] (A(t_o+\theta^k) - A(t_o)) \Big)
\end{aligned}$$

for all $t_o \leq t \leq t_o + \theta^k$, $k \geq k_o$ almost surely, where

$$C = e^{c_1A(t_o)} \Big(|x(t_o)|^2 + c_2c_4\tau e^{c_1c_4\tau} \|\xi\|_\tau^2 \Big).$$

Note from condition (iii) that

$$A(t_o+\theta^k) - A(t_o) \le c_4\tau(1 + \theta^k/\tau) \le c_4(\tau + \theta^k).$$

Then

$$e^{c_1A(t)}\,|x(t)|^2 \le \Big(C + c_3c_4(\tau + \theta^k) + \theta^{k+1}\log k \Big)$$
$$\times \exp\Big(\Big[2c_3\theta^{-k} + c_2e^{c_1c_4\tau} \Big] c_4(\tau + \theta^k) \Big)$$

for all $t_o \le t \le t_o + \theta^k$, $k \ge k_o$ almost surely. Consequently, for almost all $\omega \in \Omega$, if $t_o + \theta^{k-1} \le t \le t_o + \theta^k$, $k \ge k_o$,

$$\frac{1}{t}\log\Big[e^{c_1A(t)}\,|x(t)|^2 \Big] \le \Big\{ \log\Big(C + c_3c_4(\tau + \theta^k) + \theta^{k+1}\log k \Big)$$
$$+ \Big[2c_3\theta^{-k} + c_2e^{c_1c_4\tau} \Big] c_4(\tau + \theta^k) \Big\} (t_o + \theta^{k-1})^{-1}.$$

Thus

$$\limsup_{t\to\infty} \frac{1}{t}\log\Big[e^{c_1A(t)}\,|x(t)|^2 \Big] \le c_2c_4\theta e^{c_1c_4\tau} \qquad \text{a.s.}$$

Letting $\theta \to 1$ we obtain

$$\limsup_{t\to\infty} \frac{1}{t}\log\Big[e^{c_1A(t)}\,|x(t)|^2 \Big] \le c_2c_4e^{c_1c_4\tau} \qquad \text{a.s.}$$

Finally

$$\limsup_{t\to\infty} \frac{1}{t}\log|x(t)| = \frac{1}{2}\limsup_{t\to\infty} \frac{1}{t}\log\Big[e^{-c_1A(t)}\,e^{c_1A(t)}\,|x(t)|^2 \Big]$$

$$\le \frac{1}{2}\Big(-c_1 \liminf_{t\to\infty}\frac{A(t)}{t} + c_2c_4e^{c_1c_4\tau} \Big)$$

$$\le -\frac{1}{2}\Big(c_1c_5 - c_2c_4e^{c_1c_4\tau} \Big) \qquad \text{a.s.}$$

as required. Finally, (3.2) is equivalent to

$$c_1c_5e^{-c_1c_4\tau}/c_4 > c_2.$$

Note

$$\max_{c_1>0}\, c_1e^{-c_1c_4\tau} = \frac{1}{\tau e c_4}.$$

So for (2.2) to be satisfied we must have

$$c_2 < \frac{c_5}{\tau e c_4^2}$$

as required. The proof is complete.

Similarly we can prove a more general result but leave the details of proof to the reader.

Theorem 3.3 The conclusion of Theorem 3.1 holds if condition (i) is replaced by

$$x^T(Q + Q^T)b(x, y, t) \le -c_1 x^T Q x + c_2 y^T Q y$$

for some positive definite $d \times d$ matrix Q.

From this theorem one can obtain the following corollary in the same way as the proof of Corollary 2.3.

Corollary 3.4 Let $c_1 - c_6$ be positive constants and let Q be a positive definite $d \times d$ matrix. Assume for all $x, y \in R^d$ and $t \ge 0$:

(i) $|x|^2 \le x^T Q x$,

(ii) $x^T(Q + Q^T)b(x, 0, t) \le -c_1 x^T Q x$,

(iii) $|b(x, 0, t) - b(x, y, t)| \le c_2|y|$,

(iv) $\operatorname{trace} a(x, y, x, y, t) \le c_3 e^{-c_6 A(t)}$,

(iv) $A(t+\tau) - A(t) \le c_4\tau$ and $\liminf_{t\to\infty} \frac{A(t)}{t} \ge c_5$.

If

$$((c_1 - c_2|Q|)\wedge c_6)c_5 > c_2 c_4 e^{((c_1 - c_2|Q|)\wedge c_6)c_4\tau},$$

or

$$(c_1 - c_6)c_6 c_5 > c_4(c_2|Q|)^2 e^{c_6 c_4 \tau},$$

then the delay equation (3.1) is almost surely exponentially stable.

As an application, let us consider a d-dimensional Itô delay equation

$$dx(t) = f(x(t), x(t-\delta(t)), t)\,dt + \sum_{k=1}^{\infty} g_k(x(t), x(t-\delta(t)), t)\,dw_k(t) \tag{3.8}$$

on $t \geq t_o$ with initial data $x(t_o + s) = \xi(s)$, $-\tau \leq s \leq 0$, where $\{w_k(t): k \geq 1\}$ is a sequence of independent Brownian motions, and $f(x, y, t)$, $g_k(x, y, t)$ are all d-dimensional adapted processes with parameters $x, y \in R^d$. Assume the equation has a unique solution which is still denoted by $x(t; t_o, \xi)$. The following corollaries follow from Theorem 3.3 and Corollary 3.4 respectively.

Corollary 3.5 Let c_1–c_3 be three positive constants and Q be a positive definite $d \times d$ matrix. Assume for all $x, y \in R^d$ and $t \geq 0$:

(i) $x^T(Q + Q^T)f(x, y, t) \leq -c_1 x^T Q x + c_2 y^T Q y$,

(ii) $\sum_{k=1}^{\infty} |g_k(x, y, t)|^2 \leq c_3 e^{-c_1 t}$.

If $c_1 > c_2 e^{c_1 \tau}$, then

$$\limsup_{t\to\infty} \frac{1}{t} \log|x(t; t_o, \xi)| \leq -\frac{1}{2}\left(c_1 - c_2 e^{c_1 \tau} \right) \quad \text{a.s.} \tag{3.9}$$

That is the delay equation (3.8) is almost surely exponentially stable.

Corollary 3.6 Let c_1 - c_4 be four positive constants and Q be a positive definite $d \times d$ matrix. Assume for all $x, y \in R^d$ and $t \geq 0$:

(i) $|x|^2 \leq x^T Q x$,

(ii) $2x^T(Q + Q^T)f(x, 0, t) \leq -c_1 x^T Q x$,

(iii) $|f(x, 0, t) - f(x, y, t)| \leq c_2|y|$,

(iv) $\sum_{k=1}^{\infty} |g_k(x, y, t)|^2 \leq c_3 e^{-c_4 t}$.

If

$$(c_1 - c_2|Q|) \wedge c_4 > c_2|Q| \, e^{((c_1 - c_2|Q|)\wedge c_4)\tau}$$

or

$$(c_1 - c_4)c_4 > (c_2|Q|)^2 \, e^{c_4 \tau},$$

then the delay equation (3.8) is almost surely exponentially stable.

5.4. STOCHASTIC DIFFERENTIAL FUNCTIONAL EQUATIONS

In this section we still let $F(x, y, t)$, $x \in R^d$ and $y \in R^m$, be a d-dimensional continuous C–semimartingale such that it has the decomposition $F(x, y, t) = M(x, y, t) + B(x, y, t)$ and the local characteristic $(a(x, y, \bar{x}, \bar{y}, t), b(x, y, t), A(t))$ belonging to the class $B^{0,\delta}$. Also let $\tau > 0$ and $t_o \geq 0$ and $\xi(.)$ be a $C([-\tau, 0]; R^d)$–valued $\mathcal{F}_{t_o}$–measurable random variable. Let $D : C([-\tau, 0]; R^d) \to R^m$ be an operator. Consider a d–dimensional stochastic differential functional equation

$$x(t) = \xi(0) + \int_{t_o}^{t} F(x(s), D(\hat{x}(s)), ds) \qquad \text{on } t \geq t_o \tag{4.1}$$

with initial data $x(t_o + s) = \xi(s)$, $-\tau \leq s \leq 0$, where $\hat{x}(t) = \{x(t + s): -\tau \leq s \leq 0\}$. Assume the equation has a unique global solution which is denoted by $x(t; t_o, \xi)$.

Theorem 4.1 Let $c_1 - c_5$ be positive constants and $Q = (q_{ij})_{d\times d}$ be a positive definite matrix. Assume for all $x \in R^d$, $y \in R^m$, $t \geq 0$ and $u \in C([-\tau, 0]; R^d)$:

(i) $u(0)^T (Q + Q^T)\, b(u(0), D(u), t) \leq -c_1 u(0)^T Q u(0) + c_2 \|u^T Q u\|_\tau$,

(ii) $\text{trace } a(x, y, x, y, t) \leq c_3 e^{-c_1 A(t)}$,

(iii) $A(t+\tau) - A(t) \leq c_4 \tau$ and $\liminf\limits_{t\to\infty} \dfrac{A(t)}{t} \geq c_5$,

where and in the sequel we use notation $u^T Q u = \{u(s)^T Q u(s): -\tau \leq s \leq 0\}$. If

$$c_1 c_5 > c_2 c_4 e^{c_1 c_4 \tau}, \tag{4.2}$$

then

$$\limsup_{t\to\infty} \frac{1}{t} \log|x(t; t_o, \xi)| \leq -\frac{1}{2}\left(c_1 c_5 - c_2 c_4 e^{c_1 c_4 \tau} \right) \qquad \text{a.s.} \tag{4.3}$$

That is the delay equation (4.1) is almost surely exponentially stable. Moreover, a necessary condition for (4.2) to be satisfied is $c_2 < c_5/(\tau e c_4{}^2)$.

Proof. Without losing any generality, we can assume

$$|x|^2 \leq x^T Q x \qquad \text{for all } x \in R^d. \tag{4.4}$$

Otherwise we can find a $\alpha > 0$ large enough so that $|x|^2 \leq \alpha x^T Q x$ since Q is positive definite and then replace Q by αQ. Fix t_o and ξ arbitrarily and write $x(t; t_o, \xi) = x(t)$ simply. By Itô's formula,

$$e^{c_1 A(t)}\, x(t)^T Q x(t) = e^{c_1 A(t_o)}\, x(t_o)^T Q x(t_o) + N(t)$$

$$+ \int_{t_o}^{t} e^{c_1 A(s)} \Big(c_1 x(s)^T Q x(s) + x(s)^T (Q + Q^T)\, b(x(s), D(\hat{x}(s)), s)$$

$$+ \frac{1}{2} \sum_{i,j=1}^{d} (q_{ij} + q_{ji})\, a_{ij}(x(s), D(\hat{x}(s)), x(s), D(\hat{x}(s)), s) \Big)\, dA(s),$$

where

$$N(t) = 2 \int_{t_o}^{t} e^{c_1 A(s)}\, x(s)^T (Q + Q^T)\, M(x(s), D(\hat{x}(s)), ds).$$

Note from condition (i)

$$x(s)^T (Q + Q^T)\, b(x(s), D(\hat{x}(s)), s) \le -c_1 x(s)^T Q x(s) + c_2 \|\hat{x}(s)^T Q \hat{x}(s)\|_\tau,$$

and from condition (ii)

$$\frac{1}{2} \sum_{i,j=1}^{d} (q_{ij} + q_{ji})\, a_{ij} = \sum_{i,j=1}^{d} q_{ij} a_{ij} \le |Q| \sum_{i,j=1}^{d} |a_{ij}| \le |Q| \sum_{i,j=1}^{d} (a_{ii} a_{ij})^{1/2}$$

$$\le \frac{1}{2} |Q| \sum_{i,j=1}^{d} (a_{ii} + a_{jj}) = d|Q| \text{ trace } a \le dc_3 |Q|\, e^{-c_1 A(s)}.$$

Hence

$$e^{c_1 A(t)}\, x(t)^T Q x(t) \le |Q|\, |\xi(0)|^2\, e^{c_1 A(t_o)} + dc_3 |Q|\, (A(t) - A(t_o)) + N(t)$$

$$+ c_2 \int_{t_o}^{t} e^{c_1 A(s)}\, \|\hat{x}(s)^T Q \hat{x}(s)\|_\tau\, dA(s). \tag{4.5}$$

Let $k = 1, 2, \ldots$. Applying Lemma 2-6.2 we can derive that for almost all $\omega \in \Omega$ there exists a random integer $k_o = k_o(\omega)$ such that if $k \ge k_o$

$$N(t) \le \frac{1}{2k} \langle N(t) \rangle + 2k \log k, \qquad t_o \le t \le k_o + k.$$

However

$$\langle N(t) \rangle$$

$$= \int_{t_o}^{t} e^{2c_1 A(s)}\, x^T (Q + Q^T)\, a(x, D(\hat{x}), x, D(\hat{x}), s)\, (Q + Q^T)\, x\, dA(s)$$

$$\leq 4|Q|^2 \int_{t_o}^{t} e^{2c_1A(s)} |x(s)|^2 \operatorname{trace} a(x, D(\hat{x}), x, D(\hat{x}), s)\, dA(s)$$

$$\leq 4c_3|Q|^2 \int_{t_o}^{t} e^{c_1A(s)} \|\hat{x}(s)^T Q\hat{x}(s)\|_\tau \, dA(s).$$

So

$$N(t) \leq (2c_3|Q|^2/k) \int_{t_o}^{t} e^{c_1A(s)} \|\hat{x}(s)^T Q\hat{x}(s)\|_\tau \, dA(s) + 2k \log k$$

for all $t_o \leq t \leq t_o+k$, $k \geq k_o$ almost surely. Substituting this into (4.5) and making use of condition (iii) we obtain

$$e^{c_1A(t)} x(t)^T Qx(t) \leq |Q|\, |\xi(0)|^2 e^{c_1A(t_o)} + dc_3c_4|Q|\, (k+\tau) + 2k \log k$$

$$+ (c_2 + 2c_3|Q|^2/k) \int_{t_o}^{t} e^{c_1A(s)} \|\hat{x}(s)^T Q\hat{x}(s)\|_\tau \, dA(s) \tag{4.6}$$

for all $t_o \leq t \leq t_o + k$, $k \geq k_o$ almost surely. Note, if $t_o \leq t \leq t_o + \tau$,

$$e^{c_1A(t)} \|\hat{x}(t)^T Q\hat{x}(t)\|_\tau \leq e^{c_1A(t)} \Big(|Q|\|\xi\|_\tau^2 + \sup_{t_o\leq s\leq t} \big[x(s)^T Qx(s) \big] \Big)$$

$$\leq |Q|\, \|\xi\|_\tau^2\, e^{c_1A(t_o+\tau)} + e^{c_1(A(t)-A(t_o))} \sup_{t_o\leq s\leq t} \big[e^{c_1A(s)} x(s)^T Qx(s) \big]$$

$$\leq |Q|\, \|\xi\|_\tau^2\, e^{c_1A(t_o)+c_1c_4\tau} + e^{c_1c_4\tau} \sup_{t_o\leq s\leq t} \big[e^{c_1A(s)} x(s)^T Qx(s) \big],$$

and if $t \geq t_o + \tau$,

$$e^{c_1A(t)} \|\hat{x}(t)^T Q\hat{x}(t)\|_\tau \leq e^{c_1c_4\tau} \sup_{t-\tau\leq s\leq t} \big[e^{c_1A(s)} x(s)^T Qx(s) \big].$$

Thus if $t \geq t_o$,

$$e^{c_1A(t)} \|\hat{x}(t)^T Q\hat{x}(t)\|_\tau$$

$$\leq |Q|\, \|\xi\|_\tau^2\, e^{c_1A(t_o)+c_1c_4\tau} + e^{c_1c_4\tau} \sup_{t_o\vee(t-\tau)\leq s\leq t} \big[e^{c_1A(s)} x(s)^T Qx(s) \big].$$

Using (4.6) we see that

$$e^{c_1 A(t)} \|\hat{x}(t)^T Q \hat{x}(t)\|_\tau \le C + e^{c_1 c_4 \tau}\Big[d c_3 c_4 |Q| (k+\tau) + 2k \log k \Big]$$

$$+ (c_2 + 2c_3|Q|^2/k) e^{c_1 c_4 \tau} \int_{t_o}^{t} e^{c_1 A(s)} \|\hat{x}(s)^T Q \hat{x}(s)\|_\tau \, dA(s) \tag{4.7}$$

for all $t_o \le t \le t_o + k$, $k \ge k_o$ almost surely, where

$$C = |Q|\, e^{c_1 A(t_o) + c_1 c_4 \tau}\Big[\|\xi\|_\tau^2 + |\xi(0)|^2\Big].$$

Applying Theorem 1-7.1 yields

$$e^{c_1 A(t)} \|\hat{x}(t)^T Q \hat{x}(t)\|_\tau \le \Big(C + e^{c_1 c_4 \tau}\Big[d c_3 c_4 |Q| (k+\tau) + 2k \log k \Big]\Big)$$

$$\times \exp\Big[c_4 (k+\tau)(c_2 + 2c_3|Q|^2/k) e^{c_1 c_4 \tau}\Big] \tag{4.8}$$

for all $t_o \le t \le t_o + k$, $k \ge k_o$ almost surely. This implies immediately that

$$\limsup_{t\to\infty} \frac{1}{t} \log\Big[e^{c_1 A(t)} \|\hat{x}(t)^T Q \hat{x}(t)\|_\tau \Big] \le c_2 c_4 e^{c_1 c_4 \tau} \qquad \text{a.s.}$$

But

$$\limsup_{t\to\infty} \frac{1}{t} \log\Big[e^{c_1 A(t)} \|\hat{x}(t)^T Q \hat{x}(t)\|_\tau \Big]$$

$$\ge \limsup_{t\to\infty} \frac{1}{t}\Big[c_1 A(t) + 2\log|x(t)| \Big]$$

$$\ge c_1 \liminf_{t\to\infty} \frac{A(t)}{t} + 2\limsup_{t\to\infty} \frac{1}{t} \log|x(t)|$$

$$\ge c_1 c_5 + 2\limsup_{t\to\infty} \frac{1}{t} \log|x(t)| \qquad \text{a.s.}$$

Thus

$$\limsup_{t\to\infty} \frac{1}{t} \log|x(t)| \le -\frac{1}{2}\Big(c_1 c_5 - c_2 c_4 e^{c_1 c_4 \tau}\Big) \qquad \text{a.s.}$$

as required. Finally, we recall that it has been showed in the proof of Theorem 3.1 that a necessary condition for (4.2) to be satisfied is $c_2 < c_5/(\tau e c_4^2)$. The proof is complete.

We now use this theorem to establish a useful corollary.

Corollary 4.2 Let c_1 - c_7 be positive constants and $Q = (q_{ij})_{d\times d}$ be a positive definite matrix. Assume for all $x \in R^d$, $y \in R^m$, $t \ge 0$ and $u \in C([-\tau, 0]; R^d)$:

(i) $|x|^2 \le x^T Q x$,

(ii) $x^T (Q + Q^T)\, b(x, 0, t) \le -c_1 x^T Q x$,

(iii) $|b(x, 0, t) - b(x, y, t)| \le c_2 |y|$,

(iv) $\operatorname{trace} a(x, y, x, y, t) \le c_3 e^{-c_6 A(t)}$,

(v) $|D(u)| \le c_7 \|u\|_\tau$,

(vi) $A(t+\tau) - A(t) \le c_4 \tau$ and $\liminf_{t\to\infty} \frac{A(t)}{t} \ge c_5$,

Then equation (4.1) is almost surely exponentially stable if one of the following further conditions holds:

(a) $(c_1 \wedge c_6) c_5 > 2 c_2 c_4 c_7 |Q|\, e^{(c_1 \wedge c_6) c_4 \tau}$,

(b) $((c_1 - c_2 c_7 |Q|) \wedge c_6) c_5 \ge c_2 c_4 c_7 |Q| \exp\left[\, ((c_1 - c_2 c_7 |Q|) \wedge c_6) c_4 \tau \,\right]$,

(c) $c_5 c_6 (c_1 - c_6) \ge c_4 (c_2 c_7 |Q|)^2\, e^{c_4 c_6 \tau}$.

Proof. Case (a). For any $u \in C([-\tau, 0], R^d)$,

$$
\begin{aligned}
&u(0)^T (Q + Q^T)\, b(u(0), D(u), t) \\
&= u(0)^T (Q + Q^T)\, b(u(0), 0, t) \\
&+ u(0)^T (Q + Q^T)(b(u(0), D(u), t) - b(u(0), 0, t)) \\
&\le -c_1 u(0)^T Q u(0) + 2 c_2 c_7 |Q|\, |u(0)|\, \|u\|_\tau \\
&\le -c_1 u(0)^T Q u(0) + 2 c_2 c_7 |Q|\, \|u^T Q u\|_\tau.
\end{aligned} \tag{4.9}
$$

So the conclusion follows from Theorem 4.1.

Case (b). Note for any $\varepsilon > 0$,

$$
\begin{aligned}
&u(0)^T (Q + Q^T)\, b(u(0), D(u), t) \\
&\le -c_1 u(0)^T Q u(0) + \varepsilon |u(0)|^2 + (c_2 c_7 |Q|\, \|u\|_\tau)^2/\varepsilon \\
&\le -(c_1 - \varepsilon) u(0)^T Q u(0) + ((c_2 c_7 |Q|)^2/\varepsilon)\, \|u^T Q u\|_\tau.
\end{aligned} \tag{4.10}
$$

Let $\varepsilon = c_2 c_7 |Q|$. Hence

$$
\begin{aligned}
&u(0)^T (Q + Q^T)\, b(u(0), D(u), t) \\
&\le -(c_1 - c_2 c_7 |Q|) u(0)^T Q u(0) + c_2 c_7 |Q|\, \|u^T Q u\|_\tau
\end{aligned} \tag{4.11}
$$

and the conclusion follows from Theorem 4.1.

Case (c). Letting $\varepsilon = c_1 - c_6$ in (4.10) gives

$$u(0)^T (Q + Q^T)\, b(u(0), D(u), t)$$

$$\le -c_6\, u(0)^T Q u(0) + ((c_2 c_7 |Q|)^2/(c_1 - c_6))\, \|u^T Q u\|_\tau$$

and then the conclusion follows from Theorem 4.1 again. The proof is complete.

We now prove a new theorem that in fact is a generalization of the results in Section 4.2.

Theorem 4.3 Let $c_1 - c_3$ be positive constants and $Q = (q_{ij})_{d\times d}$ be a positive definite matrix. Let $\beta : [-\tau, 0] \to R_+$ be a Borel measurable function such that

$$\int_{-\tau}^{0} \beta(r)\, dr = 1. \tag{4.12}$$

Assume for all $x \in R^d$, $y \in R^m$, $t \ge 0$ and $u \in C([-\tau, 0]; R^d)$:

(i) $u(0)^T(Q + Q^T) b(u(0), D(u), t) \le -c_1 u(0)^T Q u(0) + c_2 \int_{-\tau}^{0} \beta(r) u(r)^T Q u(r)\, dr$

(ii) trace $a(x, y, x, y, t) \le c_3 e^{-c_1 t}$,

(iii) $A(t) \equiv t$.

If $c_1 > c_2$, then there exists a $\lambda > 0$ such that

$$\limsup_{t\to\infty} \frac{1}{t} \log|x(t; t_o, \xi)| \le -\lambda \qquad \text{a.s.}$$

That is equation (4.1) is almost surely exponentially stable.

Before the proof of the theorem it should be pointed out that condition (4.12) is not essential and it is just a normalization. As a matter of fact, once condition (i) is satisfied we can replace $\beta(r)$ and c_2 by $\beta(r)/\|\beta\|$ and $c_2\|\beta\|$ respectively, where

$$\|\beta\| = \int_{-\tau}^{0} \beta(r)\, dr.$$

Obviously this normalization makes (4.12) satisfied but the condition $c_1 > c_2$ should become $c_1 > c_2\|\beta\|$. Let us now begin to prove the theorem.

Proof. Similarly we can assume, without losing any generality, that

$$|x|^2 \le x^T Q x \quad \text{for all } x \in R^d. \tag{4.13}$$

Fix t_o and ξ arbitrarily and write $x(t; t_o, \xi) = x(t)$ simply. Since $c_1 > c_2$ we can find a $\lambda \in (0, c_1)$ such that

$$\frac{c_2 e^{\lambda\tau}}{c_1 - \lambda} < 1. \tag{4.14}$$

We first claim that for almost all $\omega \in \Omega$ there exists an integer $k_o = k_o(\omega)$ and a positive number $C = C(\omega)$ such that for every integer $k \geq k_o$

$$\int_0^T e^{\lambda t} |x(t)|^2 \, dt \leq Ck \qquad \text{if } 0 \leq T \leq k. \tag{4.15}$$

In fact, by Itô's formula and the hypotheses we can derive that

$$e^{c_1 t} x(t)^T Q x(t) \leq |Q| \, |\xi(0)|^2 e^{c_1 t_o} + dc_3 |Q| \, (t - t_o) + Z(t)$$

$$+ c_2 \int_{t_o}^{t} e^{c_1 s} \int_{-\tau}^{0} \beta(r) \, x(s+r)^T Q x(s+r) \, dr \, ds, \tag{4.16}$$

where

$$Z(t) = 2 \int_{t_o}^{t} e^{c_1 s} x(s)^T (Q + Q^T) \, M(x(s), D(\hat{x}(s)), ds).$$

By (4.14), one can choose $\varepsilon > 0$ so small that

$$\frac{1}{c_1 - \lambda} \left[c_2 e^{\lambda\tau} + 2\varepsilon c_3 |Q|^2 \right] < 1. \tag{4.17}$$

Let $k = 1, 2, \ldots$. In view of Lemma 2-6.2 we can deduce that for almost all $\omega \in \Omega$ there exists an integer $k_o = k_o(\omega)$ such that

$$Z(t) \leq \frac{1}{2} \varepsilon \langle Z(t) \rangle + 2\varepsilon^{-1} \log k \qquad \text{if } t_o \leq t \leq t_o + k.$$

However

$$\langle Z(t) \rangle = \int_{t_o}^{t} e^{2c_1 s} x(s)^T (Q + Q^T) \, a(x, D(\hat{x}), x, D(\hat{x}), s) \, (Q + Q^T) \, x(s) \, ds$$

$$\leq 4|Q|^2 \int_{t_o}^{t} e^{2c_1 s} |x(s)|^2 \, \mathrm{trace}\, a(x, D(\hat{x}), x, D(\hat{x}), s) \, ds$$

$$\le 4c_3|Q|^2 \int_{t_o}^{t} e^{c_1 s}\, x(s)^T Q x(s)\, ds.$$

So

$$Z(t) \le 2\varepsilon c_3 |Q|^2 \int_{t_o}^{t} e^{c_1 s}\, x(s)^T Q x(s)\, ds + 2\varepsilon^{-1} \log k$$

for all $t_o \le t \le t_o + k$, $k \ge k_o$ almost surely. Substituting this into (4.16) gives

$$e^{c_1 t}\, x(t)^T Q x(t) \le |Q|\, |\xi(0)|^2\, e^{c_1 t_o} + dkc_3|Q| + 2\varepsilon^{-1} \log k$$

$$+ c_2 \int_{t_o}^{t} e^{c_1 s} \int_{-\tau}^{0} \beta(r)\, x(s+r)^T Q x(s+r)\, dr\, ds$$

$$+ 2\varepsilon c_3 |Q|^2 \int_{t_o}^{t} e^{c_1 s}\, x(s)^T Q x(s)\, ds$$

for all $t_o \le t \le t_o + k$, $k \ge k_o$ almost surely. We then can easily derive that for almost all $\omega \in \Omega$, if $t_o \le T \le t_o + k$, $k \ge k_o$,

$$\int_{t_o}^{T} e^{\lambda t}\, x(t)^T Q x(t)\, dt$$

$$\le \Big[\, |Q|\, |\xi(0)|^2\, e^{c_1 t_o} + dkc_3|Q| + 2\varepsilon^{-1} \log k \,\Big] \int_{t_o}^{T} e^{-(c_1-\lambda)t}\, dt + J_1 + J_2$$

$$\le \frac{1}{c_1-\lambda}\Big[\, |Q|\, |\xi(0)|^2\, e^{c_1 t_o} + dkc_3|Q| + 2\varepsilon^{-1} \log k \,\Big] + J_1 + J_2, \qquad (4.18)$$

where

$$J_1 = c_2 \int_{t_o}^{T} e^{-(c_1-\lambda)t} \int_{t_o}^{t} e^{c_1 s} \int_{-\tau}^{0} \beta(r)\, x(s+r)^T Q x(s+r)\, dr\, ds\, dt,$$

$$J_2 = 2\varepsilon c_3 |Q|^2 \int_{t_o}^{T} e^{-(c_1-\lambda)t} \int_{t_o}^{t} e^{c_1 s}\, x(s)^T Q x(s)\, ds\, dt.$$

We now compute

$$J_2 = 2\varepsilon c_3|Q|^2 \int_{t_o}^{T} e^{c_1 s}\, x(s)^T Q x(s) \int_{s}^{T} e^{-(c_1-\lambda)t}\, dt\, ds$$

$$\le \frac{1}{c_1 - \lambda}\, 2\varepsilon c_3|Q|^2 \int_{t_o}^{T} e^{\lambda s}\, x(s)^T Q x(s)\, ds.$$

Also

$$J_1 = c_2 \int_{t_o}^{T} e^{c_1 s} \int_{-\tau}^{0} \beta(r)\, x(s+r)^T Q x(s+r)\, dr \int_{s}^{T} e^{-(c_1-\lambda)t}\, dt\, ds$$

$$\le \frac{1}{c_1 - \lambda}\, c_2 \int_{t_o}^{T} e^{\lambda s} \int_{-\tau}^{0} \beta(r)\, x(s+r)^T Q x(s+r)\, dr\, ds.$$

But

$$\int_{t_o}^{T} e^{\lambda s} \int_{-\tau}^{0} \beta(r)\, x(s+r)^T Q x(s+r)\, dr\, ds$$

$$= \int_{t_o}^{T} e^{\lambda s} \int_{s-\tau}^{s} \beta(r-s)\, x(r)^T Q x(r)\, dr\, ds$$

$$\le \int_{t_o-\tau}^{T} x(r)^T Q x(r) \int_{r}^{r+\tau} \beta(r-s)\, e^{\lambda s}\, ds\, dr$$

$$\le \int_{t_o-\tau}^{T} e^{\lambda(r+\tau)}\, x(r)^T Q x(r) \int_{r}^{r+\tau} \beta(r-s)\, ds\, dr$$

$$= \int_{t_o-\tau}^{t_o} e^{\lambda(r+\tau)}\, x(r)^T Q x(r)\, dr + \int_{t_o}^{T} e^{\lambda(r+\tau)}\, x(r)^T Q x(r)\, dr$$

$$\le |Q| \int_{-\tau}^{0} e^{\lambda(t_o+\tau+s)}\, |\xi(s)|^2\, ds + e^{\lambda\tau} \int_{t_o}^{T} e^{\lambda r}\, x(r)^T Q x(r)\, dr. \tag{4.19}$$

So we have

$$J_1 \le \frac{1}{c_1 - \lambda} c_2 \left(|Q| \int_{-\tau}^{0} e^{\lambda(t_o+\tau+s)} |\xi(s)|^2 ds + e^{\lambda\tau} \int_{t_o}^{T} e^{\lambda r} x(r)^T Q x(r)\, dr \right).$$

We therefore see from (4.18) that

$$\int_{t_o}^{T} e^{\lambda t} x(t)^T Q x(t)\, dt$$

$$\le \frac{1}{c_1 - \lambda} \left[|Q|\, |\xi(0)|^2 e^{c_1 t_o} + dkc_3|Q| + 2\varepsilon^{-1} \log k \right]$$

$$+ \frac{1}{c_1 - \lambda} c_2 |Q| \int_{-\tau}^{0} e^{\lambda(t_o+\tau+s)} |\xi(s)|^2 ds$$

$$+ \frac{1}{c_1 - \lambda} \left[c_2 e^{\lambda\tau} + 2\varepsilon c_3 |Q|^2 \right] \int_{t_o}^{T} e^{\lambda s} x(s)^T Q x(s)\, ds.$$

Recalling (4.17) we obtain (4.15).

Now by Itô's formula and the hypotheses we can similarly derive that

$$e^{\lambda t} x(t)^T Q x(t) \le |Q|\, |\xi(0)|^2 e^{\lambda t_o} + dc_3|Q|/(c_1 - \lambda) + N(t)$$

$$+ c_2 \int_{t_o}^{t} e^{\lambda s} \int_{-\tau}^{0} \beta(r)\, x(s+r)^T Q x(s+r)\, dr\, ds, \tag{4.20}$$

where

$$N(t) = 2 \int_{t_o}^{t} e^{\lambda s} x(s)^T (Q+Q^T) M(x(s), D(\hat{x}(s)), ds).$$

Again by Lemma 2-6.2 we can show that for almost all $\omega \in \Omega$ there exists an integer $k_1 = k_1(\omega)$ such that

$$N(t) \le \frac{1}{2} \langle N(t) \rangle + 2\log k \le 2c_3|Q|^2 \int_{t_o}^{t} e^{\lambda s} x(s)^T Q x(s) ds + 2\log k$$

for all $t_o \le t \le t_o + k$, $k \ge k_1$. Substituting this into (4.20) and making use of (4.19) yield

$$e^{\lambda t} x(t)^T Q x(t) \le |Q|\, |\xi(0)|^2 e^{\lambda t_o} + dc_3|Q|/(c_1 - \lambda) + 2\log k$$

$$+ c_2|Q| \int_{-\tau}^{0} e^{\lambda(t_o+\tau+s)} |\xi(s)|^2 \, ds + \left[c_2 e^{\lambda\tau} + 2c_3|Q|^2 \right] \int_{t_o}^{t} e^{\lambda s} x(s)^T Q x(s) ds.$$

for all $t_o \le t \le t_o + k$, $k \ge k_1$ almost surely. Using (4.13) and (4.15) we see

$$e^{\lambda t} |x(t)|^2 \le |Q| \, |\xi(0)|^2 e^{\lambda t_o} + dc_3|Q|/(c_1 - \lambda) + 2\log k$$

$$+ c_2|Q| \int_{-\tau}^{0} e^{\lambda(t_o+\tau+s)} |\xi(s)|^2 \, ds + \left[c_2 e^{\lambda\tau} + 2c_3|Q|^2 \right] Ck$$

for all $t_o \le t \le t_o + k$, $k \ge k_o \vee k_1$ almost surely. The remainder of the proof is the same as that of Theorem 2.1. The proof is complete.

Corollary 4.4 Let $c_1 - c_3$ be positive constants and $Q = (q_{ij})_{d\times d}$ be a positive definite matrix. Let $\beta : [-\tau, 0] \to R_+$ be a Borel measurable function such that

$$\int_{-\tau}^{0} \beta(r) \, dr = 1.$$

Assume for all $x \in R^d$, $y \in R^m$, $t \ge 0$ and $u \in C([-\tau, 0]; R^d)$:

(i) $|x|^2 \le x^T Q x$,

(ii) $x^T (Q + Q^T) \, b(x, 0, t) \le -c_1 x^T Q x$,

(iii) $|b(x, 0, t) - b(x, y, t)| \le c_2|y|$,

(iv) trace $a(x, y, x, y, t) \le c_3 e^{-c_4 t}$,

(v) $|D(u)| \le c_5 \left(\int_{-\tau}^{0} \beta(r) \, |u(r)|^2 \, dr \right)^{1/2}$,

(iv) $A(t) \equiv t$.

If

$$(c_1/2) \wedge c_4 > c_2 c_5 |Q| \quad \text{or} \quad c_4(c_1 - c_4) > (c_2 c_5 |Q|)^2$$

then equation (4.1) is almost surely exponentially stable.

Proof. Let $\varepsilon > 0$. For any $u \in C([-\tau, 0], R^d)$ and $t \ge 0$,

$$u(0)^T (Q + Q^T) \, b(u(0), D(u), t)$$

$$\le -c_1 u(0)^T Q u(0) + 2c_2 |Q|\, |u(0)|\, |D(u)|$$
$$\le -c_1 u(0)^T Q u(0) + \varepsilon |u(0)|^2 + (c_2 |Q|\, |D(u)|)^2/\varepsilon$$
$$\le -(c_1 - \varepsilon) u(0)^T Q u(0) + ((c_2 c_5 |Q|)^2/\varepsilon) \int_{-\tau}^{0} \beta(r)\, u(r)^T Q u(r)\, dr.$$

We first choose $\varepsilon = c_2 c_5 |Q|$. Then

$$u(0)^T (Q + Q^T)\, b(u(0), D(u), t)$$
$$\le -(c_1 - c_2 c_5 |Q|) u(0)^T Q u(0) + c_2 c_5 |Q| \int_{-\tau}^{0} \beta(r)\, u(r)^T Q u(r)\, dr.$$

So by Theorem 4.3, equation (4.1) is almost surely exponentially stable if

$$(c_1 - c_2 c_5 |Q|) \wedge c_4 > c_2 c_5 |Q|,$$

but this is the same as $(c_1/2) \wedge c_4 > c_2 c_5 |Q|$ as assumed. We next choose $\varepsilon = c_1 - c_4$. Then

$$u(0)^T (Q + Q^T)\, b(u(0), D(u), t)$$
$$\le -c_4 u(0)^T Q u(0) + ((c_2 c_5 |Q|)^2/(c_1 - c_4)) \int_{-\tau}^{0} \beta(r)\, u(r)^T Q u(r)\, dr.$$

So by Theorem 4.3 again, equation (4.1) is almost surely exponentially stable if

$$c_4 > (c_2 c_5 |Q|)^2/(c_1 - c_4)$$

as assumed. The proof is complete.

As an application, let us consider a d-dimensional Itô stochastic differential functional equation

$$dx(t) = f(x(t), \hat{x}(t), t)\, dt + \sum_{k=1}^{\infty} g_k(\hat{x}(t), t)\, dw_k(t) \quad \text{on } t \ge t_o \tag{4.21}$$

with initial data $x(t_o + s) = \xi(s)$, $-\tau \le s \le 0$, where

$$f : R^d \times C([-\tau, 0]; R^d) \times R_+ \to R^d \quad \text{and} \quad g_k : C([-\tau, 0]; R^d) \times R_+ \to R^d,$$

and $\{w_k(t): k \ge 1\}$ is a sequence of independent Brownian motions. Assume that the equation has a unique solution which is still denoted by $x(t; t_o, \xi)$. For this equation we immediately have the following results.

Corollary 4.5 Let $c_1 - c_3$ be positive constants and Q be a positive definite $d \times d$ matrix. Assume for all $u \in C([-\tau, 0]; R^d)$ and $t \geq 0$:

(i) $u(0)^T (Q + Q^T) f(u(0), u, t) \leq -c_1 u(0)^T Q u(0) + c_2 \|u^T Q u\|_\tau$,

(ii) $\sum_{k=1}^{\infty} |g_k(u, t)|^2 \leq c_3 e^{-c_1 t}$.

If $c_1 > c_2 e^{c_1 \tau}$, then equation (4.21) is almost surely exponentially stable.

Corollary 4.6 Let $c_1 - c_4$ be positive constants and Q be a positive definite $d \times d$ matrix. Assume for all $x \in R^d$, $t \geq 0$ and $u \in C([-\tau, 0]; R^d)$:

(i) $|x|^2 \leq x^T Q x$,

(ii) $x^T (Q+Q^T) f(x, 0, t) \leq -c_1 x^T Q x$,

(iii) $|f(x, u, t) - f(x, 0, t)| \leq c_2 \|u\|_\tau$,

(iv) $\sum_{k=1}^{\infty} |g_k(u, t)|^2 \leq c_3 e^{-c_4 t}$,

Then equation (4.21) is almost surely exponentially stable if one of the following further conditions holds:

(a) $c_1 \wedge c_4 > 2c_2 |Q| \, e^{(c_1 \wedge c_4)\tau}$,

(b) $(c_1 - c_2|Q|) \wedge c_4 \geq c_2 |Q| \exp\big[((c_1 - c_2|Q|) \wedge c_4)\tau \big]$,

(c) $c_4(c_1 - c_4) \geq (c_2|Q|)^2 \, e^{c_4 \tau}$.

Corollary 4.7 Let $c_1 - c_3$ be positive constants and Q be a positive definite $d \times d$ matrix. Let $\beta : [-\tau, 0] \to R_+$ be a Borel measurable function such that

$$\int_{-\tau}^{0} \beta(r)\, dr = 1.$$

Assume for all $u \in C([-\tau, 0]; R^d)$ and $t \geq 0$:

(i) $u(0)^T (Q + Q^T) f(u(0), D(u), t)$

$$\leq -c_1 u(0)^T Q u(0) + c_2 \int_{-\tau}^{0} \beta(r)\, u(r)^T Q u(r)\, dr$$

(ii) $\sum_{k=1}^{\infty} |g_k(u, t)|^2 \le c_3 e^{-c_1 t}$.

If $c_1 > c_2$, then equation (4.21) is almost surely exponentially stable.

Corollary 4.8 Let c_1 - c_3 be positive constants and Q be a positive definite $d \times d$ matrix. Let $\beta : [-\tau, 0] \to R_+$ be a Borel measurable function such that

$$\int_{-\tau}^{0} \beta(r)\, dr = 1.$$

Assume for all $x \in R^d$, $t \ge 0$ and $u \in C([-\tau, 0]; R^d)$:

(i) $|x|^2 \le x^T Q x$,

(ii) $x^T (Q + Q^T) f(x, 0, t) \le -c_1 x^T Q x$,

(iii) $|f(x, 0, t) - f(x, u, t)| \le c_2 \left(\int_{-\tau}^{0} \beta(r)\, |u(r)|^2\, dr \right)^{1/2}$,

(iv) $\sum_{k=1}^{\infty} |g_k(u, t)|^2 \le c_3 e^{-c_4 t}$,

If

$$(c_1/2) \wedge c_4 > c_2|Q| \quad \text{or} \quad c_4(c_1 - c_4) > (c_2|Q|)^2$$

then equation (4.21) is almost surely exponentially stable.

5.5. EXAMPLES

In this section we shall give a number of examples to illustrate our theory.

Example 5.1 Let us first consider a stochastic oscillator with delay

$$\begin{aligned} &\ddot{z}(t) + 5\dot{z}(t) + 6z(t) + \theta_1 z(t-\tau) \cos\dot{z}(t) \\ &+ \theta_2 \dot{z}(t-\tau) e^{-|z(t)|} = p(t)\, e^{-t}\, \dot{w}(t) \end{aligned} \tag{5.1}$$

on $t \ge 0$ with a suitable initial data, where θ_1 and θ_2 are small parameters, p(t) is a polynomial of t and $\dot{w}(t)$ is a white noise, i.e. w(t) is a scale Brownian motion. We claim that if

$$|\theta| := \sqrt{\theta_1^2 + \theta_2^2} < \sqrt{2/15}\ , \tag{5.2}$$

then the stochastic oscillator (5.1) is almost surely exponentially stable, that is there exists a $\lambda > 0$ such that

$$\limsup_{t\to\infty} \frac{1}{2t} \log(|z(t)|^2 + |\dot{z}(t)|^2) \le -\lambda \qquad \text{a.s.}$$

In order to show this, we introduce a new variable $x = (x_1, x_2)^T = (z, \dot{z})^T$ and write equation (5.1) as an Itô equation

$$dx(t) = f(x(t), x(t-\tau))\, dt + \sigma(t)\, dw(t), \tag{5.3}$$

where

$$f(x, y) = (x_2,\ -6x_1 - 5x_2 - \theta_1 y_1 \cos x_2 - \theta_2 y_2 e^{-|x_1|}\,)^T$$

and

$$\sigma(t) = (0, p(t)\, e^{-t})^T.$$

Note

$$f(x, 0) = Kx = \begin{bmatrix} 0 & 1 \\ -6 & -5 \end{bmatrix} x.$$

It is easy to find

$$H = \begin{bmatrix} 1 & 1 \\ -2 & -3 \end{bmatrix} \quad \text{and then} \quad H^{-1} = \begin{bmatrix} 3 & 1 \\ -2 & -1 \end{bmatrix}$$

such that

$$H^{-1}KH = \begin{bmatrix} -2 & 0 \\ 0 & -3 \end{bmatrix} := \bar{H}.$$

Set

$$Q = (H^{-1})^T H^{-1} = \begin{bmatrix} 13 & 5 \\ 5 & 2 \end{bmatrix}.$$

Obviously Q is positive definite. In fact

$$|x|^2/15 \le x^T Q x \qquad \text{for any } x \in R^2.$$

We now compute

$$x^T(Q + Q^T)f(x, y) \le 2x^TQKx + 2x^TQ(f(x, y) - f(x, 0)).$$

But

$$2x^TQKx = 2x^T(H^{-1})^T H^{-1} Kx = 2x^T(H^{-1})^T \bar{K}H^{-1} x$$
$$\le -4x^T(H^{-1})^T H^{-1} x = -4x^TQx.$$

Moreover,

$$2x^T(H^{-1})^T H^{-1} (f(x, y)-f(x, 0))$$

$$\le 2|x^T(H^{-1})^T|^2 + \frac{1}{2} |H^{-1} (f(x, y) - f(x, 0))|^2$$

$$= 2x^T(H^{-1})^TH^{-1}x + (\theta_1 y_1 \cos x_2 + \theta_2 y_2 e^{-|x_1|})^2$$
$$\le 2x^TQx + |\theta|^2 |y|^2 \le 2x^TQx + 15|\theta|^2 y^TQy.$$

So

$$x^T(Q + Q^T)f(x, y) \le -2x^TQx + 15|\theta|^2 y^TQy.$$

On the other hand, for arbitrary $\varepsilon > 0$ one can find a sufficiently large C so that

$$|\sigma(t)|^2 = (p(t)\, e^{-t})^2 \le Ce^{-(2-\varepsilon)t}.$$

Thus by Corollary 2.4, equation (5.3), i.e. the stochastic oscillator (5.1) is almost surely exponential stable if $15|\theta|^2 < 2-\varepsilon$, but this is the same as (5.2) since ε is arbitrary.

Another method to handle this example is to make a linear transformation and then apply our result. Define a new stochastic process $\bar{x}(t) = H^{-1}x(t)$. Then $\bar{x}(t)$ satisfies the following delay equation

$$d\bar{x}(t) = g(\bar{x}(t), \bar{x}(t-\tau))\, dt + H^{-1}\sigma(t)\, dw(t), \tag{5.4}$$

where

$$g(\bar{x}, \bar{y}) = H^{-1} f(H\bar{x}, H\bar{y}).$$

Obviously the almost sure exponential stability of equation (5.1) is now equivalent to that of equation (5.4). We compute

$$2\bar{x}^Tg(\bar{x}, 0) = 2\bar{x}^TH^{-1} f(H\bar{x}, 0) = 2\bar{x}^TH^{-1}KH\bar{x} \le -4|\bar{x}|^2$$

and, with $y = H\bar{y}$ and $x = H\bar{x}$,

$$|g(\bar{x}, \bar{y}) - g(\bar{x}, 0)| \le |H^{-1}(f(H\bar{x}, H\bar{y}) - f(H\bar{x}, 0))|$$

$$\le \sqrt{2}\ |\theta_1 y_1 \cos x_2 + \theta_2 y_2 e^{-|x_1|}| \le \sqrt{2}\ |\theta|\ |y|$$

$$= \sqrt{2}\ |\theta|\ |H\bar{y}| \le \sqrt{30}\ |\theta|\ |\bar{y}|.$$

Hence by Corollary 2.5 with Q = identity matrix we can derive that equation (5.4), hence equation (5.1) is almost surely exponentially stable if

$$2 > \sqrt{30}\ |\theta|, \quad \text{i.e.} \quad |\theta| < \sqrt{2/15}$$

the same as before.

Example 5.2 Let $\rho : R_+ \to R$ be a continuously differentiable function such that

$$\rho(t) \le t \quad \text{and} \quad \dot{\rho}(t) \ge 1 \quad \text{for all } t \ge 0.$$

If $t-\tau$ is replaced by $\rho(t)$ then equation (5.1) becomes

$$\ddot{z}(t) + 5\dot{z}(t) + 6z(t) + \theta_1 z(\rho(t)) \cos\dot{z}(t) + \theta_2 \dot{z}(\rho(t)) e^{-|z(t)|} = p(t)\, e^{-t}\, \dot{w}(t). \quad (5.5)$$

By Corollary 2.8 and the arguments in Example 5.1, we conclude that this stochastic oscillator is almost surely exponentially stable if $\theta_1^2 + \theta_2^2 < 2/15$.

Example 5.3 Let $\delta(.)$ be a mapping from R_+ to $[-\tau, 0]$. Replacing τ in equation (5.1) by $\delta(t)$ we arrive at a stochastic oscillator with variable delay

$$\ddot{z}(t) + 5\dot{z}(t) + 6z(t) + \theta_1 z(t-\delta(t)) \cos\dot{z}(t)$$
$$+ \theta_2 \dot{z}(t-\delta(t)) e^{-|z(t)|} = p(t)\, e^{-t}\, \dot{w}(t). \quad (5.6)$$

In the same way as Example 5.1 but using Corollary 3.5 or 3.6 we can derive that this equation is almost surely exponentially stable if

$$2 > 15|\theta|^2 e^{2\tau}, \quad \text{i.e.} \quad \theta_1^2 + \theta_2^2 < 2e^{-2\tau}/15.$$

Example 5.4 Let us consider a stochastic oscillator of the form

$$\ddot{z}(t) + 5\dot{z}(t) + 6z(t) + \theta_1 \cos\dot{z}(t) \int_{-\tau}^{0} z(t+r)dr$$

$$+ \theta_2 e^{-|z(t)|} \int_{-\tau}^{0} \dot{z}(t+r)dr = p(t)\, e^{-t}\, \dot{w}(t) \quad (5.7)$$

on $t \geq 0$ with a suitable initial data, where θ_1, θ_2, $p(t)$ and $\dot{w}(t)$ are all the same as before. Again introduce a new variable $x = (x_1, x_2)^T = (z, \dot{z})^T$ and write the equation as a stochastic differential functional equation

$$dx(t) = f(x(t), \hat{x}(t))\,dt + \sigma(t)\,dw(t), \tag{5.8}$$

where

$$f(x, u) = \Big(x_2, -6x_1 - 5x_2 - \theta_1 \cos x_2 \int_{-\tau}^{0} u_1(r)\,dr - \theta_2 e^{-|x_1|} \int_{-\tau}^{0} u_2(r)\,dr \Big)^T$$

and

$$\sigma(t) = (0, p(t)\,e^{-t})^T$$

for $x \in R^2$, $u \in C([-\tau, 0]; R^2)$ and $t \geq 0$. Define a new process $\bar{x}(t) = H^{-1}x(t)$, where H is the same as defined in Example 5.1. Then $\bar{x}(t)$ satisfies the following stochastic differential functional equation

$$d\bar{x}(t) = g(\bar{x}(t), \hat{\bar{x}}(t))\,dt + H^{-1}\sigma(t)\,dw(t), \tag{5.9}$$

where $\hat{\bar{x}}(t) = \{\bar{x}(t+s): -\tau \leq s \leq 0\}$ and

$$g(\bar{x}, \bar{u}) = H^{-1} f(H\bar{x}, H\bar{u}) \qquad \text{for } \bar{x} \in R^2,\ \bar{u} \in C([-\tau, 0]; R^2)$$

in which we use the notation $H\bar{u} = \{H\bar{u}(s): -\tau \leq s \leq 0\}$. Obviously the almost sure exponential stability of equation (5.7) is now equivalent to that of equation (5.9). We compute

$$2\bar{x}^T g(\bar{x}, 0) = 2\bar{x}^T H^{-1} f(H\bar{x}, 0) = 2\bar{x}^T H^{-1} A H\bar{x} \leq -4|\bar{x}|^2$$

and, setting $u = H\bar{u}$ and $|\theta|$ the same as defined in Example 5.1,

$$|g(\bar{x}, \bar{u}) - g(\bar{x}, 0)| \leq |\, H^{-1}(f(H\bar{x}, H\bar{u}) - f(H\bar{x}, 0))\,|$$

$$\leq \sqrt{2}\,|\theta| \Big(\Big[\int_{-\tau}^{0} u_1(r)\,dr \Big]^2 + \Big[\int_{-\tau}^{0} u_2(r)\,dr \Big]^2 \Big)^{1/2}$$

$$\leq \sqrt{2}\,|\theta| \Big(\tau \int_{-\tau}^{0} |u_1(r)|^2\,dr + \tau \int_{-\tau}^{0} |u_2(r)|^2\,dr \Big)^{1/2}$$

$$\le \sqrt{2}\,|\theta| \Big(\tau \int_{-\tau}^{0} |u(r)|^2\,dr \Big)^{1/2} = \sqrt{2}\,|\theta| \Big(\tau \int_{-\tau}^{0} |H\bar{u}(r)|^2\,dr \Big)^{1/2}$$

$$\le \sqrt{2}\,|\theta| \Big(\tau \int_{-\tau}^{0} |H|^2\,|\bar{u}(r)|^2\,dr \Big)^{1/2} = \sqrt{2}\,\tau|\theta|\,|H| \Big(\frac{1}{\tau} \int_{-\tau}^{0} |\bar{u}(r)|^2\,dr \Big)^{1/2}$$

$$\le \sqrt{30}\,\tau|\theta| \Big(\frac{1}{\tau} \int_{-\tau}^{0} |\bar{u}(r)|^2\,dr \Big)^{1/2}.$$

Hence, by Corollary 4.8 with Q = identity matrix and $\beta(r) = 1/\tau$, we can derive that equation (5.6) is almost surely exponentially stable if

$$2 > \sqrt{30}\,\tau|\theta|, \qquad \text{i.e.} \qquad \theta_1^2 + \theta_2^2 < 2/15\tau^2.$$

Example 5.5 Consider a semilinear stochastic differential functional equation

$$dx(t) = Kx(t)\,dt + f(\hat{x}(t), t)\,dt + \sum_{k=1}^{\infty} g_k(\hat{x}(t), t)\,dw_k(t) \quad \text{on } t \ge t_o \tag{5.10}$$

with initial data $x(t_o + s) = \xi(s)$, $-\tau \le s \le 0$, where $\{w_k(t): k \ge 1\}$ is a sequence of independent Brownian motions, and f, g_k are all mappings from $C([-\tau, 0]; R^d) \times R_+$ to R^d, and K is a d × d matrix. Assume

(i) K has eigenvalues $-\lambda_1 < -\lambda_2 < \ldots < -\lambda_d < 0$;

(ii) $|f(u, t)| \le c_1 \|u\|_\tau$,

(iii) $\sum_{k=1}^{\infty} |g_k(u, t)|^2 \le c_2 e^{c_3 t}$

for all $u \in C([-\tau, 0]; R^d)$ and $t \ge 0$, where $c_1 - c_3$ are all positive constants. Note that there exists an invertible d × d matrix H such that

$$H^{-1}KH = \text{diag.}(-\lambda_1, -\lambda_2, \ldots, -\lambda_d) := \bar{K}. \tag{5.11}$$

Define a new stochastic process $\bar{x}(t) = H^{-1}x(t)$. Then $\bar{x}(t)$ satisfies the following stochastic differential functional equation

$$d\bar{x}(t) = \bar{K}\bar{x}(t)\,dt + H^{-1}f(H\hat{\bar{x}}(t), t)\,dt + \sum_{k=1}^{\infty} H^{-1}g_k(H\hat{\bar{x}}(t), t)\,dw_k(t). \tag{5.12}$$

Obviously the almost sure exponential stability of equation (5.10) is equivalent to that of equation (5.12). It is easy to see

$$2\bar{x}^T\bar{K}\bar{x} \le -2\lambda_d|\bar{x}|^2,$$

$$|H^{-1}f(H\bar{u}, t)| \le c_1|H^{-1}|\,|H|\,\|\bar{u}\|_\tau$$

for all $\bar{x} \in R^d$ and $\bar{u} \in C([-\tau, 0]; R^d)$. Hence, by Corollary 4.6, we conclude that equation (5.12), hence equation (5.10) is almost surely exponentially stable if one of the following condition holds:

(a) $(2\lambda_d \wedge c_3) > 2c_1|H^{-1}|\,|H|\, e^{(2\lambda_d \wedge c_3)\tau}$,

(b) $(2\lambda_d - c_1|H^{-1}|\,|H|) \wedge c_3 \ge c_1|H^{-1}|\,|H| \exp\Big[\,((2\lambda_d - c_1|H^{-1}|\,|H|) \wedge c_3)\tau\,\Big]$,

(c) $c_3(2\lambda_d - c_3) \ge (c_1|H^{-1}|\,|H|)^2\, e^{c_3\tau}$.

We now replace condition (ii) by

$$\text{(ii)}' \quad |f(u, t)| \le c_2\Big(\int_{-\tau}^{0} \beta(r)\, |u(r)|^2\, dr\Big)^{1/2},$$

where $\beta : [-\tau, 0] \to R_+$ is a Borel measurable function such that

$$\int_{-\tau}^{0} \beta(r)\, dr = 1.$$

Then

$$|H^{-1}f(H\bar{u}, t)| \le c_2|H^{-1}|\Big(\int_{-\tau}^{0} \beta(r)\, |H\bar{u}(r)|^2\, dr\Big)^{1/2}$$

$$\le c_2|H^{-1}|\,|H|\Big(\int_{-\tau}^{0} \beta(r)\, |\bar{u}(r)|^2\, dr\Big)^{1/2}.$$

So by Corollary 4.8, equation (5.10) is almost surely exponentially stable if

$$\lambda_d \wedge c_3 > c_2|H^{-1}|\,|H| \quad \text{or} \quad c_3(2\lambda_d - c_3) > (c_2|H^{-1}|\,|H|)^2.$$

Example 5.6 Let $F(x, y, t)$, $x, y \in R^2$ be a 2-dimensional continuous C-semimartingale with the local characteristic $(a(x, y, \bar{x}, \bar{y}, t), b(x, y, t), A(t))$. Let $\tau > 0$ and $\delta(.) : R_+ \to [-\tau, 0]$. Consider a 2-dimensional stochastic differential delay equation driven by $F(x, y, t)$ of the form

$$x(t) = \xi(0) + \int_{t_o}^{t} F(x(s), x(s-\delta(s)), ds) \qquad \text{on } t \ge t_o \tag{5.13}$$

with initial data $x(t_o + s) = \xi(s)$, $-\tau \le s \le 0$. We assume

(i) $b(x, y, t) = (-x_1 - x_2^2 \sin t + \theta y_2 \cos x_1, -x_2 + x_1 x_2 \sin t + \theta y_1 \sin x_2)^T$;

(ii) $\operatorname{trace} a(x, y, \bar{x}, \bar{y}, t) \le p(t)\, e^{-t}$;

(iii) $A(t+\tau) - A(t) \le 2\tau$ and

$$\liminf_{t\to\infty} \frac{A(t)}{t} \ge 1,$$

where $\theta > 0$ and $p(t)$ is a polynomial of t. Note

$$2x^T b(x, 0, t) = 2x_1(-x_1 - x_2^2 \sin t) + 2x_2(-x_2 + x_1 x_2 \sin t) = -2|x|^2,$$

$$|b(x, 0, t) - b(x, y, t)| \le \theta |y|.$$

Hence, we can apply Corollary 3.4 to derive that the delay equation (5.13) is almost surely exponentially stable if one of the followings is satisfied:

(a) $(2 - \theta) \wedge 1 > 2\theta e^{((2-\theta)\wedge 1)2\tau}$,

(b) $1 > 2\theta^2 e^{2\tau}$.

Case (b) is equivalent to

$$\theta < e^{-\tau}/\sqrt{2}. \tag{5.14}$$

In case (a) we must have $\theta < 1$, otherwise if $\theta \ge 1$, (a) implies $\theta < 2$ and

$$1 \ge (2 - \theta) > 2\theta e^{(2-\theta)2\tau} \ge 2\theta \ge 2$$

a contradiction. Since $\theta < 1$, (a) gives

$$1 > 2\theta e^{2\tau}, \qquad \text{i.e. } \theta < e^{-\tau}/2,$$

but this is worse than (5.14). We therefore summarize that equation (5.13) is almost surely exponentially stable if (5.14) is satisfied.

CHAPTER 6

Moment Exponential Stability of Stochastic Differential Delay Equations

6.1. INTRODUCTION

Unlike the almost sure exponential stability of stochastic differential delay equations which had almost not been studied until 1990, the moment exponential stability of stochastic differential delay equation of Itô's type has been extensively studied by many mathematicians. For example we mention Curtain and Pritchard (1978), Ladde (1974, 1981), Mao (1991b, f), Mizel and Trutzer (1984), Mohammed (1984) among others. However, so far little has been done about the moment exponential stability for stochastic differential delay equations driven by non-linear integrators.

In this chapter we shall systematically discuss the pth moment exponential stability of stochastic differential delay or functional equations driven by non-linear integrators. We first study the moment exponential stability of stochastic differential delay equations in Sections 6.2 and 6.3, where the delay is a constant or a variable respectively. We shall next deal with the pth moment exponential stability of stochastic differential functional equations in Section 6.4. Furthermore, we shall study the pth moment exponential stability of stochastic differential functional equations with respect to stochastic perturbation in Section 6.5. Finally we shall give a number of examples to illustrate the theory in Section 6.6.

6.2. Pth MOMENT EXPONENTIAL STABILITY OF DELAY EQUATIONS

In this section we let $F(x, y, t)$, $x, y \in R^d$ be a d-dimensional continuous C-semimartingale such that it has the decomposition $F(x, y, t) = M(x, y, t) + B(x, y, t)$ and the local characteristic $(a(x, y, \bar{x}, \bar{y}, t), b(x, y, t), A(t))$ belonging to the class $B^{0,\delta}$. Throughout this chapter we let $p > 0$ unless specified. Let $\tau > 0$ and $t_o \geq 0$. Let $\xi(.)$ be a $C([-\tau, 0]; R^d)$-valued $\mathcal{F}_{t_o}$-measurable random variable such that $E(\|\xi\|_\tau)^p < \infty$. Consider a d-dimensional stochastic differential delay equation driven by $F(x, y, t)$ of the form

$$x(t) = \xi(0) + \int_{t_o}^{t} F(x(s), x(s-\tau), ds) \qquad \text{on } t \geq t_o \tag{2.1}$$

with initial data $x(t_o + s) = \xi(s)$, $-\tau \leq s \leq 0$. Assume the equation has a unique global solution which is denoted by $x(t; t_o, \xi)$.

Definition 2.1 Equation (2.1) is said to be *pth moment exponentially stable* if there exists a pair of constants $\lambda > 0$ and $C > 0$ such that for all t_o and ξ

$$E|x(t; t_o, \xi)|^p \leq CE\|\xi\|_\tau^p e^{-\lambda(t-t_o)}, \qquad t \geq t_o. \tag{2.2}$$

When $p = 2$, it is usually called the *exponential stability in mean square*. Moreover, (2.2) implies

$$\limsup_{t\to\infty} \frac{1}{t} \log\Big(E|x(t; t_o, \xi)|^p \Big) \leq -\lambda \qquad \text{a.s.} \tag{2.3}$$

This left hand side term is called the *pth moment Lyapunov exponent* of the solution.

Although the definition here is made for equation (2.1), similar ones for other types of delay equations can be made obviously. For the convenience, we introduce an operator L acting on $C^2(R^d; R)$-valued functions $V(x)$ as follows:

$$(LV)(x, y, t) = \sum_{i=1}^{d} b_i(x, y, t) \frac{\partial}{\partial x_i} V(x)$$

$$+\frac{1}{2}\sum_{i,j=1}^{d} a_{ij}(x, y, x, y, t)\frac{\partial^2}{\partial x_i \partial x_j}V(x), \tag{2.4}$$

where $b_i(x, y, t)$ is the ith component of $b(x, y, t)$ and $a_{ij}(x, y, x, y, t)$ the ijth component of $a(x, y, x, y, t)$. Note LV is defined on $(x, y, t) \in R^d \times R^d \times R_+$.

Theorem 2.1 Let c_1–c_4 be positive constants and $V(x) \in C^2(R^d; R)$. Assume

(i) $c_1|x|^p \le V(x) \le c_2|x|^p$,

(ii) $(LV)(x, y, t) \le -c_3V(x) + c_4V(y)$,

(iii) $A(t) \equiv t$

for all $x, y \in R^d$ and $t \ge 0$. If $c_3 > c_4$, then equation (2.1) is pth moment exponentially stable.

Proof. Since $c_3 > c_4$ we can find a $\lambda \in (0, c_3)$ such that

$$\frac{c_4 e^{\lambda\tau}}{c_3 - \lambda} < 1. \tag{2.5}$$

We first claim that there is a constant β such that

$$\int_{t_o}^{\infty} e^{\lambda t} EV(x(t; t_o, \xi))\, dt \le \beta e^{\lambda t_o} E\|\xi\|_\tau^p \tag{2.6}$$

for all t_o and ξ. To show this we fix t_o and ξ arbitrarily and write $x(t; t_o, \xi) = x(t)$ simply. By Itô's formula and the assumptions,

$$e^{c_3 t} EV(x(t))$$

$$= e^{c_3 t_o} EV(x(t_o)) + E\int_{t_o}^{t} e^{c_3 s}\Big(c_3V(x(s)) + (LV)(x(s), x(s-\tau), s)\Big)\, ds$$

$$\le c_2 e^{c_3 t_o} E\|\xi\|_\tau^p + c_4 \int_{t_o}^{t} e^{c_3 s} EV(x(s-\tau))\, ds$$

for all $t \ge t_o$. Consequently

$$EV(x(t)) \le c_2 e^{c_3 t_o} E\|\xi\|_\tau^p\, e^{-c_3 t} + c_4 e^{-c_3 t}\int_{t_o}^{t} e^{c_3 s} EV(x(s-\tau))\, ds.$$

Therefore, for any $T \geq t_o$

$$\int_{t_o}^{T} e^{\lambda t} EV(x(t))\,dt$$

$$\leq c_2 e^{c_3 t_o} E\|\xi\|_\tau^p \int_{t_o}^{T} e^{-(c_3-\lambda)t}\,dt + c_4 \int_{t_o}^{T} e^{-(c_3-\lambda)t} \int_{t_o}^{t} e^{c_3 s} EV(x(s-\tau))\,ds\,dt$$

$$\leq \frac{1}{c_3 - \lambda} c_2 e^{\lambda t_o} E\|\xi\|_\tau^p + c_4 \int_{t_o}^{T} e^{c_3 s} EV(x(s-\tau)) \int_{s}^{T} e^{-(c_3-\lambda)t}\,dt\,ds$$

$$\leq \frac{1}{c_3 - \lambda} c_2 e^{\lambda t_o} E\|\xi\|_\tau^p + \frac{1}{c_3 - \lambda} c_4 \int_{t_o}^{T} e^{\lambda s} EV(x(s-\tau))ds. \tag{2.7}$$

But

$$\int_{t_o}^{T} e^{\lambda s} EV(x(s-\tau))ds$$

$$= \int_{t_o}^{t_o+\tau} e^{\lambda s} EV(x(s-\tau))\,ds + e^{\lambda\tau} \int_{t_o+\tau}^{T\vee(t_o+\tau)} e^{\lambda(s-\tau)} EV(x(s-\tau))\,ds$$

$$\leq \tau c_2\, e^{\lambda(t_o+\tau)} E\|\xi\|_\tau^p + e^{\lambda\tau} \int_{t_o}^{T} e^{\lambda s} EV(x(s))ds. \tag{2.8}$$

Hence we obtain

$$\int_{t_o}^{T} e^{\lambda t} EV(x(t))\,dt$$

$$\leq \frac{1}{c_3 - \lambda} c_2(1 + \tau e^{\lambda\tau})\, e^{\lambda t_o} E\|\xi\|_\tau^p + \frac{c_4\, e^{\lambda\tau}}{c_3 - \lambda} \int_{t_o}^{T} e^{\lambda s} EV(x(s))ds. \tag{2.9}$$

So

$$\left[1 - \frac{c_4\, e^{\lambda\tau}}{c_3 - \lambda} \right] \int_{t_o}^{T} e^{\lambda t} EV(x(t))\,dt \leq \frac{1}{c_3 - \lambda} c_2(1 + \tau e^{\lambda\tau})\, e^{\lambda t_o} E\|\xi\|_\tau^p .$$

Recalling (2.6) we get

$$\int_{t_o}^{T} e^{\lambda t} EV(x(t))\, dt \le \beta e^{\lambda t_o} E\|\xi\|_\tau^p,$$

where

$$\beta = c_2(1 + \tau e^{\lambda\tau})/(c_3 - \lambda - c_4 e^{\lambda\tau})$$

and so (2.6) follows by letting $T \to \infty$. We can now apply Itô's formula, the assumptions, (2.8) and (2.6) to derive that

$$e^{\lambda t} EV(x(t))$$

$$= e^{\lambda t_o} EV(x(t_o)) + E \int_{t_o}^{t} e^{\lambda s} \Big(\lambda V(x(s)) + (LV)(x(s), x(s-\tau), s) \Big) ds$$

$$\le c_2 e^{\lambda t_o} E\|\xi\|_\tau^p + c_4 \int_{t_o}^{t} e^{\lambda s} EV(x(s-\tau))\, ds$$

$$\le c_2 e^{\lambda t_o} E\|\xi\|_\tau^p + c_4 \Big(\tau c_2 e^{\lambda(t_o+\tau)} E\|\xi\|_\tau^p + e^{\lambda\tau} \int_{t_o}^{t} e^{\lambda s} EV(x(s))ds \Big)$$

$$\le c_2 e^{\lambda t_o} E\|\xi\|_\tau^p + c_4 \Big(\tau c_2 e^{\lambda(t_o+\tau)} E\|\xi\|_\tau^p + e^{\lambda\tau} \beta e^{\lambda t_o} E\|\xi\|_\tau^p \Big)$$

$$\le (c_2 + c_4(\tau c_2 + \beta)e^{\lambda\tau}) e^{\lambda t_o} E\|\xi\|_\tau^p. \tag{2.10}$$

But by condition (i),

$$EV(x(t)) \ge c_1 E|x(t)|^p.$$

It then follows

$$E|x(t)|^p \le \Big[(c_2 + c_4(\tau c_2 + \beta)e^{\lambda\tau})/c_1 \Big] E\|\xi\|_\tau^p e^{\lambda(t-t_o)}.$$

In other words, equation (2.1) is pth moment exponentially stable. The proof is complete.

Corollary 2.2 Let $p = 2$. Let $c_1 - c_4$ be positive constants. Assume for all $x, y \in R^d$

and $t \geq 0$:

(i) $2x^T b(x, 0, t) \leq -c_1|x|^2$,

(ii) $|b(x, y, t) - b(x, 0, t)| \leq c_2|y|$,

(iii) $\text{trace } a(x, y, x, y, t) \leq c_3|x|^2 + c_4|y|^2$,

(iv) $A(t) \equiv t$.

If $c_1 > 2c_2 + c_3 + c_4$, then equation (2.1) is exponentially stable in mean square.

Proof. Let $V(x) = |x|^2$. Note

$$(LV)(x, y, t) = 2x^T b(x, y, t) + \text{trace } a(x, y, x, y, t)$$

$$\leq -(c_1 - c_3)|x|^2 + 2c_2|x|\,|y| + c_4|y|^2 \leq -(c_1 - c_2 - c_3)|x|^2 + (c_2 + c_4)|y|^2.$$

Then the conclusion follows from Theorem 2.1. The proof is complete.

As an application, let us consider a d-dimensional Itô delay equation

$$dx(t) = f(x(t), x(t-\tau), t)\, dt + \sum_{k=1}^{\infty} g_k(x(t), x(t-\tau), t)\, dw_k(t) \quad \text{on } t \geq t_o \tag{2.11}$$

with initial data $x(t_o + s) = \xi(s)$, $-\tau \leq s \leq 0$, where $\{w_k(t): k \geq 1\}$ is a sequence of independent Brownian motions, and $f(x, y, t)$, $g_k(x, y, t)$ are all R^d-valued functions defined on $R^d \times R^d \times R_+$. Assume the equation has a unique solution that is still denoted by $x(t; t_o, \xi)$. Note the corresponding operator L takes the form

$$(LV)(x, y, t) = \sum_{i=1}^{d} f_i(x, y, t) \frac{\partial}{\partial x_i} V(x)$$

$$+ \frac{1}{2} \sum_{i,j=1}^{d} \sum_{k=1}^{\infty} g_{ki}(x, y, t)\, g_{kj}(x, y, t) \frac{\partial^2}{\partial x_i \partial x_j} V(x), \tag{2.12}$$

where f_i and g_{ki} are the ith components of f and g_k respectively. For this stochastic delay equation, we have the following corollaries that follow from Theorem 2.1 and Corollary 2.2 respectively.

Corollary 2.3 Let $c_1 - c_4$ be positive constants and $V(x) \in C^2(R^d; R)$. Assume that for all $x, y \in R^d$ and $t \geq 0$:

(i) $c_1|x|^p \leq V(x) \leq c_2|x|^p$,

(ii) $(LV)(x, y, t) \leq -c_3 V(x) + c_4 V(y)$.

If $c_3 > c_4$, then equation (2.11) is pth moment exponentially stable.

Corollary 2.4 Let $p = 2$. Let $c_1 - c_4$ be positive constants. Assume for all $x, y \in R^d$ and $t \geq 0$:

(i) $2x^T f(x, 0, t) \leq -c_1|x|^2$,

(ii) $|f(x, y, t) - f(x, 0, t)| \leq c_2|y|$,

(iii) $\sum_{k=1}^{\infty} |g_k(x, y, t)|^2 \leq c_3|x|^2 + c_4|y|^2$.

If $c_1 > 2c_2 + c_3 + c_4$, then equation (2.11) is exponentially stable in mean square.

6.3. GENERALIZATIONS

In this section we shall generalize the results in the previous section. We shall not only allow the time lag to be a variable rather than a constant but also remove the restriction of $A(t) \equiv t$.

Let $F(x, y, t)$ be the same as in the previous section. In the first half of this section, we let $\rho : R_+ \to R$ be a continuously differentiable function such that

$$\rho(t) \leq t \quad \text{and} \quad \dot{\rho}(t) \geq 1 \quad \text{for all } t \geq 0. \tag{3.1}$$

It is easy to see that

$$t + \rho(0) \leq \rho(t) \quad \text{and} \quad \rho^{-1}(t) \leq t - \rho(0) \quad \text{for all } t \geq 0, \tag{3.2}$$

where $\rho^{-1}(.)$ is the inverse function of $\rho(.)$. Set $\tau = -\rho(0)$ which is assumed to be positive. Consider a stochastic differential delay equation

$$x(t) = \xi(0) + \int_{t_o}^{t} F(x(s), x(\rho(s)), ds) \quad \text{on } t \geq t_o \tag{3.3}$$

with initial data $x(t_o+s) = \xi(s)$, $-\tau \leq s \leq 0$, where ξ is a $C([-\tau, 0]; R^d)$-valued $\mathcal{F}_{t_o}$-measurable random variable such that $E(\|\xi\|_\tau)^p < \infty$. Assume the equation has a unique solution which is denoted by $x(t; t_o, \xi)$. The corresponding operator L is the same as defined by (2.4).

Theorem 3.1 Let (3.1) hold. Let $c_1 - c_4$ be positive constants and $V(x) \in C^2(R^d; R)$. Assume

(i) $c_1|x|^p \le V(x) \le c_2|x|^p$,

(ii) $(LV)(x, y, t) \le -c_3V(x) + c_4V(y)$,

(iii) $A(t) \equiv t$

for all $x, y \in R^d$ and $t \ge 0$. If $c_3 > c_4$, then equation (3.3) is pth moment exponentially stable.

Proof. This theorem can be proved in the same way as Theorem 2.1. In fact, everything in the proof of Theorem 2.1 is valid except $s-\tau$ is replaced by $\rho(s)$ and (2.8) needs a modification as follows: By changing variable $u = \rho(s)$ and making use of (3.1) and (3.2) we deduce

$$\int_{t_o}^{T} e^{\lambda s} EV(x(\rho(s))ds \le \int_{t_o-\tau}^{T} e^{\lambda(u+\tau)} EV(x(u))\, du$$

$$\le c_2 \int_{t_o-\tau}^{t_o} e^{\lambda(u+\tau)} E|\xi(u)|^p\, du + e^{\lambda\tau} \int_{t_o}^{T} e^{\lambda u} EV(x(u))\, du$$

$$\le \tau c_2\, e^{\lambda(t_o+\tau)} E\|\xi\|_\tau^p + e^{\lambda\tau} \int_{t_o}^{T} e^{\lambda s} EV(x(s))\, ds,$$

the same estimate as before. The proof is complete.

Corollary 3.2 Let (3.1) hold. Then under the same conditions of Corollary 2.2, equation (3.3) is exponentially stable in mean square.

It is rather easy to see this corollary holds. Now replacing $t-\tau$ in equation (2.11) by $\rho(t)$, we arrive at an Itô delay equation

$$dx(t) = f(x(t), x(\rho(t)), t)\, dt + \sum_{k=1}^{\infty} g_k(x(t), x(\rho(t)), t)\, dw_k(t) \quad \text{on } t \ge t_o \qquad (3.4)$$

with initial data $x(t_o + s) = \xi(s)$, $-\tau \le s \le 0$. For this equation we have:

Corollary 3.3 Let (3.1) hold. (i) Under the same conditions of Corollary 2.3, equation (3.4) is pth moment exponentially stable. (ii) Under the same conditions of Corollary 2.4, equation (3.4) is exponentially stable in mean square.

In the sequel of this section, let $\tau > 0$ and let $\delta(.)$ be any Borel measurable function from R_+ to $[-\tau, 0]$. Consider a d-dimensional stochastic differential delay equation driven by $F(x, y, t)$ of the form

$$x(t) = \xi(0) + \int_{t_o}^{t} F(x(s), x(s-\delta(s)), ds) \qquad \text{on } t \geq t_o \tag{3.5}$$

with initial data $x(t_o + s) = \xi(s)$, $-\tau \leq s \leq 0$, where ξ is the same as before. Assume the equation has a unique global solution which is denoted by $x(t; t_o, \xi)$. The corresponding operator L is still defined as (2.4).

Theorem 3.4 Let $c_1 - c_4$ be positive constants and $V(x) \in C^2(R^d; R)$. Assume

(i) $c_1|x|^p \leq V(x) \leq c_2|x|^p$,

(ii) $(LV)(x, y, t) \leq -c_3V(x) + c_4V(y)$,

(iii) $A(t) = t$

for all $x, y \in R^d$ and $t \geq 0$. Then

$$E|x(t; t_o, \xi)|^p \leq CE\|\xi\|_\tau^p \exp\Big[-(c_3 - c_4e^{c_3\tau})(t - t_o)\Big] \qquad \text{on } t \geq t_o \tag{3.6}$$

for all t_o and ξ, where $C = c_2(1 + c_4e^{c_3\tau})/c_1$. Therefore, equation (3.5) is pth moment exponentially stable if one of the following further conditions holds:

(a) $c_3e^{-c_3\tau} > c_4$,

(b) $c_3 \geq 1/\tau$ and $c_4 < 1/\tau e$.

Proof. Fix t_o and ξ arbitrarily and write $x(t; t_o, \xi) = x(t)$ simply. By Itô's formula and the hypotheses,

$$e^{c_3t}\, EV(x(t))$$

$$= e^{c_3t_o}\, EV(x(t_o)) + E\int_{t_o}^{t} e^{c_3s}\Big(c_3V(x(s)) + (LV)(x(s), x(s-\delta(s)), s)\Big)\, ds$$

$$\leq c_2e^{c_3t_o}\, E\|\xi\|_\tau^p + c_4\int_{t_o}^{t} e^{c_3s}\, EV(x(s-\delta(s)))\, ds$$

$$\leq c_2e^{c_3t_o}\, E\|\xi\|_\tau^p + c_4\, e^{c_3\tau}\int_{t_o}^{t} e^{c_3(s-\delta(s))}\, EV(x(s-\delta(s)))\, ds$$

for all $t \geq t_o$. Applying Lemma 5-3.2 we derive that

$$e^{c_3 t}\,EV(x(t))$$
$$\le \Big(c_2 e^{c_3 t_o} E\|\xi\|_\tau^p + c_4\, e^{c_3\tau}\Big[\sup_{t_o-\tau\le r\le t_o} e^{c_3 r}\, EV(x(r)) \Big]\Big) \exp\Big[c_4 e^{c_3\tau}(t - t_o) \Big]$$
$$\le c_2 e^{c_3 t_o} E\|\xi\|_\tau^p (1 + c_4 e^{c_3\tau}) \exp\Big[c_4 e^{c_3\tau}(t - t_o) \Big].$$

Hence by hypothesis (i),

$$c_1 E|x(t)|^p \le c_2(1 + c_4 e^{c_3\tau})\, E\|\xi\|_\tau^p \,\exp\Big[-(c_3 - c_4 e^{c_3\tau})\,(t - t_o) \Big],$$

which is the required (3.6). Consequently, if (a) holds, equation (3.5) is pth moment exponentially stable. In the case of (b), note from hypothesis (ii) that

$$(LV)(x, y, t) \le -\frac{1}{\tau} V(x) + c_4 V(y).$$

Hence, by (3.6), we see that

$$E|x(t; t_o, \xi)|^p \le C E\|\xi\|_\tau^p \,\exp\Big[-(\frac{1}{\tau} - c_4 e)\,(t - t_o) \Big], \quad t \ge t_o$$

where $C = c_2(1 + c_4 e)/c_1$. So equation (3.5) is also pth moment exponentially stable since $c_4 < 1/\tau e$. The proof is complete.

So far we have not removed the restriction of $A(t) \equiv t$ yet and it is time to do so. Of course we need strengthen other conditions. For convenience let us introduce another operator H acting on $C^2(R^d; R)$-valued functions V(x) defined as:

$$(HV)(x, y, t) = \sum_{i,j=1}^{d} a_{ij}(x, y, x, y, t)\, \frac{\partial}{\partial x_i} V(x)\, \frac{\partial}{\partial x_j} V(x). \tag{3.7}$$

Note HV is defined on $(x, y, t) \in R^d \times R^d \times R_+$.

Theorem 3.5 Let $c_1 - c_7$ be positive constants and $V(x) \in C^2(R^d; R)$. Assume

(i) $c_1|x|^p \le V(x) \le c_2|x|^p$,

(ii) $(LV)(x, y, t) \le -c_3 V(x) + c_4 V(y)$,

(iii) $c_5\tau \le A(t+\tau) - A(t) \le c_6\tau$,

(iv) $(HV)(x, y, t) \le c_7 V(x) V(y)$

for all $x, y \in R^d$ and $t \ge 0$. If

$$c_3 c_5 > 2c_6(c_4 + 4c_7)e^{c_3 c_6 \tau}, \tag{3.8}$$

then equation (3.5) is pth moment exponentially stable.

Proof. Fix t_o and ξ arbitrarily and write $x(t; t_o, \xi) = x(t)$ simply. By Itô's formula and the assumptions,

$$e^{c_3(A(t)-A(t_o))} V(x(t)) = V(x(t_o))$$

$$+ \int_{t_o}^{t} e^{c_3(A(s)-A(t_o))} \Big(c_3 V(x(s)) + (LV)(x(s), x(s-\delta(s)), s) \Big) dA(s) + N(t)$$

$$\le c_2 E\|\xi\|_\tau^p + c_4 \int_{t_o}^{t} e^{c_3(A(s)-A(t_o))} V(x(s-\delta(s)))\, dA(s) + N(t)$$

for all $t \ge t_o$, where

$$N(t) = \sum_{i=1}^{d} \int_{t_o}^{t} e^{c_3(A(s)-A(t_o))} \frac{\partial}{\partial x_i} V(x(s))\, M(x(s), x(s-\delta(s)), ds),$$

which is a continuous martingale vanishing at $t = t_o$. Hence for any bounded stopping time $\sigma \ge t_o$,

$$E\Big(\sup_{t_o \le t \le \sigma} e^{c_3(A(t)-A(t_o))} V(x(t)) \Big) \le c_2 E\|\xi\|_\tau^p$$

$$+ c_4 E \int_{t_o}^{\sigma} e^{c_3(A(s)-A(t_o))} V(x(s-\delta(s)))\, dA(s) + E\Big(\sup_{t_o \le t \le \sigma} N(t) \Big). \tag{3.9}$$

But by Theorem 1-2.6 and the hypothesis

$$E\Big(\sup_{t_o \le t \le \sigma} N(t) \Big)$$

$$\le 2\sqrt{2}\, E\Big(\int_{t_o}^{\sigma} e^{2c_3(A(s)-A(t_o))} (HV)(x(s), x(s-\delta(s)), s)\, dA(s) \Big)^{1/2}$$

$$\le 2\sqrt{2}\, E\Big(c_7 \int_{t_o}^{\sigma} e^{2c_3(A(s)-A(t_o))} V(x(s)\, V(x(s-\delta(s)))\, dA(s) \Big)^{1/2}$$

$$\le 2\sqrt{2}\, E\Big(c_7 \Big[\sup_{t_o \le t \le \sigma} e^{c_3(A(t)-A(t_o))} V(x(t)) \Big]$$

$$\times \int_{t_o}^{\sigma} e^{c_3(A(s)-A(t_o))} V(x(s-\delta(s)))\, dA(s) \Big)^{1/2}$$

$$\leq \frac{1}{2} E\Big(\sup_{t_o \leq t \leq \sigma} e^{c_3(A(t)-A(t_o))} V(x(t)) \Big)$$

$$+ 4c_7 E \int_{t_o}^{\sigma} e^{c_3(A(s)-A(t_o))} V(x(s-\delta(s)))\, dA(s).$$

So we have

$$E\Big(\sup_{t_o \leq t \leq \sigma} e^{c_3(A(t)-A(t_o))} V(x(t)) \Big) \leq 2c_2 E\|\xi\|_\tau^p + 2(c_4 + 4c_7)\, J,$$

where

$$J = E \int_{t_o}^{\sigma} e^{c_3(A(s)-A(t_o))} V(x(s-\delta(s)))\, dA(s)$$

$$\leq E \int_{t_o}^{t_o+\tau} e^{c_3c_6\tau} V(x(s-\delta(s)))\, \mathbb{1}_{[t_o-\tau,\, t_o]}(s-\delta(s))\, dA(s)$$

$$+ E \int_{t_o}^{\sigma} e^{c_3(A(s)-A(t_o))} V(x(s-\delta(s)))\, \mathbb{1}_{[t_o,\, \infty)}(s-\delta(s))\, dA(s)$$

$$\leq c_2c_6\tau e^{c_3c_6\tau} E\|\xi\|_\tau^p$$

$$+ e^{c_3c_6\tau} E \int_{t_o}^{\sigma} e^{c_3(A(s-\delta(s))-A(t_o))} V(x(s-\delta(s)))\, \mathbb{1}_{[t_o,\, \infty)}(s-\delta(s))\, dA(s)$$

$$\leq c_2c_6\tau e^{c_3c_6\tau} E\|\xi\|_\tau^p$$

$$+ e^{c_3c_6\tau} E \int_{t_o}^{\sigma} \Big(\sup_{t_o \leq r \leq s} e^{c_3(A(r)-A(t_o))} V(x(r)) \Big)\, dA(s), \tag{3.10}$$

where condition (iii) has been used. Therefore

$$E\Big(\sup_{t_o\le t\le\sigma} e^{c_3(A(t)-A(t_o))}\,V(x(t))\Big)$$

$$\le \alpha E\|\xi\|_\tau^p + \beta E\int_{t_o}^{\sigma}\Big(\sup_{t_o\le r\le s} e^{c_3(A(r)-A(t_o))}\,V(x(r))\Big)\,dA(s), \tag{3.11}$$

where

$$\alpha = 2c_2\Big[\,1 + c_6\tau(c_4+4c_7)e^{c_3c_6\tau}\,\Big],$$

$$\beta = 2(c_4+4c_7)e^{c_3c_6\tau}.$$

We now let $T > t_o$ be arbitrary. Note from condition (iii) that

$$A(T) - A(t_o) \le c_6(T - t_o + \tau).$$

So if we restrict the stopping time $\sigma \le T$, we can apply the stochastic Gronwall inequality (i.e. Theorem 1–7.11) to (3.11) to obtain that

$$E\Big(\sup_{t_o\le t\le T} e^{c_3(A(t)-A(t_o))}\,V(x(t))\Big) \le \alpha E\|\xi\|_\tau^p \exp\Big[\,\beta c_6(T - t_o + \tau)\,\Big].$$

But it is easy to see

$$E\Big(\sup_{t_o\le t\le T} e^{c_3(A(t)-A(t_o))}\,V(x(t))\Big) \ge \exp\Big[\,c_3c_5(T - t_o - \tau)\,\Big]\,c_1E|X(T)|^p.$$

Hence

$$E|X(T)|^p \le (\alpha/c_1)\,E\|\xi\|_\tau^p\,\exp\Big[\,-(c_3c_5 - \beta c_6)(T - t_o) + (c_3c_5 + \beta c_6)\tau\,\Big].$$

Recalling assumption (3.8) and the definition of β we see $c_3c_5 - \beta c_6 > 0$. Therefore equation (3.5) is pth moment exponentially stable. The proof is complete.

Theorem 3.6 Let c_1–c_8 be positive constants and $V(x) \in C^2(R^d; R)$. Assume

(i) $c_1|x|^p \le V(x) \le c_2|x|^p$,

(ii) $(LV)(x, y, t) \le -c_3V(x) + c_4V(y)$,

(iii) $c_5\tau \le A(t+\tau) - A(t) \le c_6\tau$,

(iv) $(HV)(x, y, t) \le c_7(V(x))^2 + c_8(V(y))^2$

for all $x, y \in R^d$ and $t \ge 0$. Then equation (3.5) is pth moment exponentially stable if

$$c_3c_5 > 3c_6\Big[\,c_4e^{c_3c_6\tau} + 6(c_7 + c_8e^{2c_3c_6\tau})\,\Big]. \tag{3.12}$$

Proof. We use the same notations as in the proof of Theorem 3.5. We first estimate

$$E\Big(\sup_{t_o\le t\le\sigma} N(t)\Big)$$

$$\le 2\sqrt{2}\,E\Big(\int_{t_o}^{\sigma} e^{2c_3(A(s)-A(t_o))}\big[\,c_7(V(x(s))^2 + c_8(V(x(s-\delta(s)))^2\,\big]\,dA(s)\Big)^{1/2}$$

$$\le \frac{1}{3}E\Big(\sup_{t_o\le t\le\sigma} e^{c_3(A(t)-A(t_o))}\,V(x(t))\Big)$$

$$+ 6c_7\,E\int_{t_o}^{\sigma}\Big[\sup_{t_o\le r\le s} e^{c_3(A(r)-A(t_o))}\,V(x(r))\Big]\,dA(s) + J,$$

where

$$J = 2\sqrt{2}\,E\Big(c_8\int_{t_o}^{\sigma} e^{2c_3(A(s)-A(t_o))}\,(V(x(s-\delta(s)))^2\,dA(s)\Big)^{1/2}$$

$$\le 2\sqrt{2}\,E\Big(c_8\int_{t_o}^{t_o+\tau} e^{2c_3c_6\tau}\,(V(x(s-\delta(s)))^2\,\mathbb{1}_{[t_o-\tau,\,t_o]}(s-\delta(s))\,dA(s)\Big)^{1/2}$$

$$+ 2\sqrt{2}\,E\Big(c_8\int_{t_o}^{\sigma} e^{2c_3(A(s)-A(t_o))}\,(V(x(s-\delta(s)))^2\,\mathbb{1}_{[t_o,\,\infty)}(s-\delta(s))\,dA(s)\Big)^{1/2}$$

$$\le 2e^{c_3c_6\tau}\,(2c_6c_8\tau)^{1/2}\,E\|\xi\|_\tau^p$$

$$+ 2\sqrt{2}\,E\Big(c_8e^{2c_3c_6\tau}\int_{t_o}^{\sigma}\Big[\sup_{t_o\le r\le s} e^{2c_3(A(r)-A(t_o))}\,(V(x(r)))^2\Big]\,dA(s)\Big)^{1/2}$$

$$\le 2e^{c_3c_6\tau}\,(2c_6c_8\tau)^{1/2}\,E\|\xi\|_\tau^p + \frac{1}{3}E\Big(\sup_{t_o\le t\le\sigma} e^{c_3(A(t)-A(t_o))}\,V(x(t))\Big)$$

$$+ 6c_8e^{2c_3c_6\tau}\,E\int_{t_o}^{\sigma}\Big[\sup_{t_o\le r\le s} e^{c_3(A(r)-A(t_o))}\,V(x(r))\Big]\,dA(s).$$

So

$$E\Big(\sup_{t_o \le t \le \sigma} N(t)\Big)$$

$$\le 2e^{c_3c_6\tau}(2c_6c_8\tau)^{1/2}\, E\|\xi\|_\tau^p + \frac{2}{3}E\Big(\sup_{t_o \le t \le \sigma} e^{c_3(A(t)-A(t_o))}\, V(x(t))\Big)$$

$$+ 6(c_7 + c_8e^{2c_3c_6\tau})\, E\int_{t_o}^{\sigma} \Big[\sup_{t_o \le r \le s} e^{c_3(A(r)-A(t_o))}\, V(x(r))\Big]\, dA(s). \qquad (3.13)$$

Substituting (3.13), (3.10) into (3.9) gives (3.11) but with

$$\alpha = 3\Big[c_4 + 2e^{c_3c_6\tau}(2c_6c_8\tau)^{1/2}\Big],$$

$$\beta = 3\Big[c_4e^{c_3c_6\tau} + 6(c_7 + c_8e^{2c_3c_6\tau})\Big].$$

The remainder of the proof is the same as that of Theorem 3.5. The proof is complete.

To close this section, let us consider an Itô delay equation

$$dx(t) = f(x(t), x(t-\delta(t)), t)\, dt + \sum_{k=1}^{\infty} g_k(x(t), x(t-\delta(s)), t)\, dw_k(t) \qquad (3.14)$$

on $t \ge t_o$ with initial data $x(t_o + s) = \xi(s)$, $-\tau \le s \le 0$, where f, g_k and w_k are all the same as introduced in the precious section. The corresponding operator L is the same as defined by (2.12).

Corollary 3.7 Let c_1 - c_4 be positive constants and $V(x) \in C^2(R^d; R)$. Assume

(i) $c_1|x|^p \le V(x) \le c_2|x|^p$,

(ii) $(LV)(x, y, t) \le -c_3V(x) + c_4V(y)$

for all $x, y \in R^d$ and $t \ge 0$. Then equation (3.14) is pth moment exponentially stable if one of the following further conditions holds:

(a) $c_3e^{-c_3\tau} > c_4$,

(b) $c_3 \ge 1/\tau$ and $c_4 < 1/\tau e$.

This corollary follows from Theorem 3.4 directly. Using this corollary one can obtain a useful result immediately.

Corollary 3.8 Let $p = 2$. Let c_1, c_2 be two positive constants and Q be a positive definite $d \times d$ matrix. Assume that

$$x^T(Q + Q^T)f(x, y, t) + \sum_{k=1}^{\infty} g_k(x, y, t)^T Q\, g_k(x, y, t)$$

$$\le - c_1 x^T Q x + c_2 y^T Q y$$

for all $x, y \in R^d$ and $t \ge 0$. Then equation (3.14) is exponentially stable in mean square if one of the following further conditions holds:

(a) $c_1 e^{-c_1 \tau} > c_2$,

(b) $c_1 \ge 1/\tau$ and $c_2 < 1/\tau e$.

6.4. Pth MOMENT EXPONENTIAL STABILITY OF FUNCTIONAL EQUATIONS

In this section we shall let $F(x, y, t)$, $x \in R^d$ and $y \in R^m$, be a d-dimensional continuous C-semimartingale such that it has the decomposition $F(x, y, t) = M(x, y, t) + B(x, y, t)$ and the local characteristic $(a(x, y, \bar{x}, \bar{y}, t), b(x, y, t), A(t))$ belonging to the class $B^{0,\delta}$. Also let $\tau > 0$ and $t_o \ge 0$ and $\xi(.)$ be a $C([-\tau, 0]; R^d)$-valued $\mathcal{F}_{t_o}$-measurable random variable. Let $D : C([-\tau, 0]; R^d) \to R^m$ be an operator. Consider a d-dimensional stochastic differential functional equation

$$x(t) = \xi(0) + \int_{t_o}^{t} F(x(s), D(\hat{x}(s)), ds) \qquad \text{on } t \ge t_o \tag{4.1}$$

with initial data $x(t_o + s) = \xi(s)$, $-\tau \le s \le 0$, where $\hat{x}(t) = \{x(t + s): -\tau \le s \le 0\}$. Assume the equation has a unique global solution which is denoted by $x(t; t_o, \xi)$. Also the operators L and H related to the equation are defined as (2.4) and (3.7) respectively.

Theorem 4.1 Let $c_1 - c_4$ be positive constants and $V(x) \in C^2(R^d; R)$. Let β be a Borel measurable function from $[-\tau, 0]$ to R_+ such that

$$\int_{-\tau}^{0} \beta(r)\, dr = 1.$$

Assume for all $x \in R^d$, $u \in C([-\tau, 0]; R^d)$ and $t \ge 0$:

(i) $c_1|x|^p \le V(x) \le c_2|x|^p$,

(ii) $(LV)(x, D(u), t) \le - c_3 V(x) + c_4 \int_{-\tau}^{0} \beta(r)\, V(u(r))\, dr$,

(iii) $A(t) \equiv t$.

If $c_3 > c_4$, then equation (4.1) is pth moment exponentially stable.

Proof. This theorem can be proved in a similar way as Theorem 2.1. Fix t_o and ξ arbitrarily and write $x(t; t_o, \xi) = x(t)$ simply. By Itô's formula and the assumptions,

$$e^{c_3 t} EV(x(t))$$

$$\leq c_2 e^{c_3 t_o} E\|\xi\|_\tau^p + c_4 \int_{t_o}^{t} e^{c_3 s} \int_{-\tau}^{0} \beta(r)\, EV(x(s+r))\, dr\, ds \tag{4.2}$$

for all $t \geq t_o$. Since $c_3 > c_4$ one can find a $\lambda \in (0, c_3)$ such that

$$\frac{c_4\, e^{\lambda\tau}}{c_3 - \lambda} < 1. \tag{4.3}$$

It is easy to derive from (4.2) that for any $T \geq t_o$

$$\int_{t_o}^{T} e^{\lambda t}\, EV(x(t))\, dt$$

$$\leq \frac{1}{c_3 - \lambda}\, c_2 e^{\lambda t_o} E\|\xi\|_\tau^p + \frac{1}{c_3 - \lambda}\, c_4 \int_{t_o}^{T} e^{\lambda s} \int_{-\tau}^{0} \beta(r)\, EV(x(s+r))\, dr\, ds. \tag{4.4}$$

But by changing the order of integrations one can show (cf. (5–4.19) on page 210) that

$$\int_{t_o}^{T} e^{\lambda s} \int_{-\tau}^{0} \beta(r)\, EV(x(s+r))\, dr\, ds$$

$$\leq \tau c_2\, e^{\lambda(t_o+\tau)} E\|\xi\|_\tau^p + e^{\lambda\tau} \int_{t_o}^{T} e^{\lambda r}\, EV(x(r))dr.$$

Substituting this into (4.4) gives (2.9) and then (2.6) follows. The remainder of the proof is almost the same as that of Theorem 2.1. The proof is complete.

Corollary 4.2 Let $p = 2$. Let $c_1 - c_2$ be positive constants and let β be a Borel measurable function from $[-\tau, 0]$ to R_+ such that

$$\int_{-\tau}^{0} \beta(r)\, dr = 1.$$

Assume for all $x \in R^d$, $y \in R^m$, $u \in C([-\tau, 0]; R^d)$ and $t \geq 0$:

(i) $2x^T b(x, 0, t) \leq -c_1|x|^2$,

(ii) $|b(x, y, t) - b(x, 0, t)| \leq -c_2|y|$,

(iii) $\text{trace}\, a(x, y, x, y, t) \leq c_3|x|^2 + c_4|y|^2$,

(iv) $|D(u)|^2 \leq c_5 \int_{-\tau}^{0} \beta(r)\, |u(r)|^2\, dr$,

(v) $A(t) \equiv t$.

If $c_1 > c_2 + c_3 + (c_2 + c_4)c_5$, then equation (4.1) is exponentially stable in mean square.

Proof. Let $V(x) = |x|^2$. It is easy to show that

$$(LV)(x, D(u), t) = 2x^T b(x, D(u), t) + \text{trace}\, a(x, D(u), x, D(u), t)$$

$$\leq -(c_1 - c_2 - c_3)|x|^2 + (c_2+c_4)c_5 \int_{-\tau}^{0} \beta(r)\, |u(r)|^2\, dr.$$

So the conclusion follows from Theorem 4.1. The proof is complete.

Theorem 4.3 Let c_1–c_7 be positive constants and $V(x) \in C^2(R^d; R)$. Assume

(i) $c_1|x|^p \leq V(x) \leq c_2|x|^p$,

(ii) $(LV)(x, D(u), t) \leq -c_3V(x) + c_4\left[\sup_{-\tau\leq r\leq 0} V(u(r))\right]$,

(iii) $c_5\tau \leq A(t+\tau) - A(t) \leq c_6\tau$,

(iv) $(HV)(x, D(u), t) \leq c_7V(x)\left[\sup_{-\tau\leq r\leq 0} V(u(r))\right]$

for all $x \in R^d$, $u \in C([-\tau, 0]; R^d)$ and $t \geq 0$. If

$$c_3c_5 > 2c_6(c_4 + 4c_7)e^{c_3c_6\tau},$$

then equation (4.1) is pth moment exponentially stable.

Proof. This theorem can be proved in a similar way as Theorem 3.5 so only the outlines are given here. Again fix t_o and ξ arbitrarily and write $x(t; t_o, \xi) = x(t)$ simply. Let σ be any bounded stopping time such that $\sigma \geq t_o$. By Itô's formula and the assumptions one can show that

$$E\Big(\sup_{t_o \le t \le \sigma} e^{c_3(A(t)-A(t_o))} V(x(t))\Big) \le c_2 E\|\xi\|_\tau^p$$

$$+ c_4 E \int_{t_o}^{\sigma} e^{c_3(A(s)-A(t_o))}\Big[\sup_{-\tau \le r \le 0} V(x(s+r))\Big] dA(s) + E\Big(\sup_{t_o \le t \le \sigma} N(t)\Big),$$

where

$$N(t) = \sum_{i=1}^{d} \int_{t_o}^{t} e^{c_3(A(s)-A(t_o))} \frac{\partial}{\partial x_i} V(x(s))\; M(x(s), D(\hat{x}(s)), ds).$$

On the other hand one can also show that

$$E\Big(\sup_{t_o \le t \le \sigma} N(t)\Big)$$

$$\le \frac{1}{2} E\Big(\sup_{t_o \le t \le \sigma} e^{c_3(A(t)-A(t_o))} V(x(t))\Big)$$

$$+ 4c_7 E \int_{t_o}^{\sigma} e^{c_3(A(s)-A(t_o))}\Big[\sup_{-\tau \le r \le 0} V(x(s+r))\Big] dA(s).$$

So we have

$$E\Big(\sup_{t_o \le t \le \sigma} e^{c_3(A(t)-A(t_o))} V(x(t))\Big) \le 2c_2 E\|\xi\|_\tau^p + 2(c_4 + 4c_7)\, J,$$

where

$$J = E \int_{t_o}^{\sigma} e^{c_3(A(s)-A(t_o))}\Big[\sup_{-\tau \le r \le 0} V(x(s+r))\Big] dA(s)$$

$$\le E \int_{t_o}^{t_o+\tau} e^{c_3(A(s)-A(t_o))}\Big[\sup_{s-\tau \le r \le t_o} V(x(r))\Big] dA(s)$$

$$+ E \int_{t_o}^{\sigma} e^{c_3(A(s)-A(s-\tau))}\Big[\sup_{t_o \vee (s-\tau) \le r \le s} e^{c_3(A(s-\tau)-A(t_o))} V(x(r))\Big] dA(s)$$

$$\le c_2 c_6 \tau e^{c_3 c_6 \tau} E\|\xi\|_\tau^p$$

$$+ e^{c_3 c_6 \tau} E \int_{t_o}^{\sigma} \Big(\sup_{t_o \le r \le s} e^{c_3(A(r)-A(t_o))} V(x(r)) \Big) dA(s).$$

We therefore obtain (3.11). The remainder of the proof is the same as that of Theorem 3.5. The proof is complete.

The following theorem can be proved similarly.

Theorem 4.4 Replace condition (iv) of Theorem 4.3 by

$$\text{(iv)}' \quad (HV)(x, D(u), t) \le c_7(V(x))^2 + c_8 \big[\sup_{-\tau \le r \le 0} V(u(r)) \big]^2,$$

where c_8 is one more positive constant. If

$$c_3 c_5 > 3c_6 \big[c_4 e^{c_3 c_6 \tau} + 6(c_7 + c_8 e^{2c_3 c_6 \tau}) \big],$$

then equation (4.1) is pth moment exponentially stable.

To close this section, let us consider a d-dimensional Itô stochastic differential functional equation

$$dx(t) = f(x(t), \hat{x}(t), t)\, dt + \sum_{k=1}^{\infty} g_k(x(t), \hat{x}(t), t)\, dw_k(t) \quad \text{on } t \ge t_o \tag{4.5}$$

with initial data $x(t_o + s) = \xi(s)$, $-\tau \le s \le 0$, where $\{w_k(t)\colon k \ge 1\}$ is a sequence of independent Brownian motions, f and g_k are mappings from $R^d \times C([-\tau, 0]; R^d) \times R_+$ to R^d. Assume that the equation has a unique solution which is still denoted by $x(t; t_o, \xi)$. The corresponding operator L, acting on C^2-functions V(x), is defined as

$$(LV)(x, u, t) = \sum_{i=1}^{d} f_i(x, u, t) \frac{\partial}{\partial x_i} V(x)$$

$$+ \frac{1}{2} \sum_{i,j=1}^{d} \sum_{k=1}^{\infty} g_{ki}(x, u, t)\, g_{kj}(x, u, t) \frac{\partial^2}{\partial x_i \partial x_j} V(x),$$

where f_i and g_{ki} are the ith components of f and g_k respectively. Note LV is defined on $(x, u, t) \in R^d \times C([-\tau, 0]; R^d) \times R_+$. For this equation we immediately have the following results.

Corollary 4.5 Let $c_1 - c_4$ be positive constants and $V(x) \in C^2(R^d; R)$. Let β be a Borel measurable function from $[-\tau, 0]$ to R_+ such that

$$\int_{-\tau}^{0} \beta(r)\, dr = 1.$$

Assume for all $x \in R^d$, $u \in C([-\tau, 0]; R^d)$ and $t \geq 0$:

(i) $c_1|x|^p \leq V(x) \leq c_2|x|^p$,

(ii) $(LV)(x, u, t) \leq -c_3V(x) + c_4 \int_{-\tau}^{0} \beta(r)\, V(u(r))\, dr.$

If $c_3 > c_4$, then equation (4.5) is pth moment exponentially stable.

Corollary 4.6 Let $p = 2$. Let c_1–c_2 be positive constants and let β be a Borel measurable function from $[-\tau, 0]$ to R_+ such that

$$\int_{-\tau}^{0} \beta(r)\, dr = 1.$$

Assume for all $x \in R^d$, $u \in C([-\tau, 0]; R^d)$ and $t \geq 0$:

(i) $2x^Tf(x, 0, t) \leq -c_1|x|^2$,

(ii) $|f(x, u, t) - f(x, 0, t)| \leq c_2 \left(\int_{-\tau}^{0} \beta(r)\, |u(r)|^2\, dr \right)^{1/2}$,

(iii) $\sum_{k=1}^{\infty} |g_k(x, u, t)|^2 \leq c_3|x|^2 + c_4 \int_{-\tau}^{0} \beta(r)\, |u(r)|^2\, dr.$

If $c_1 > 2c_2 + c_3 + c_4$, then equation (4.1) is exponentially stable in mean square.

6.5. Pth MOMENT EXPONENTIAL STABILITY OF FUNCTIONAL EQUATIONS WITH PERTURBATION

In this section we still let $F(x, y, t)$, $x \in R^d$ and $y \in R^m$, be a d-dimensional continuous C-semimartingale such that it has the decomposition $F(x, y, t) = M(x, y, t) + B(x, y, t)$ and its local characteristic $(a(x, y, \bar{x}, \bar{y}, t), b(x, y, t), A(t))$ belongs to the class $B^{0,\delta}$. Also let $\tau > 0$ and $t_o \geq 0$ and $\xi(.)$ be a $C([-\tau, 0]; R^d)$-valued $\mathcal{F}_{t_o}$-measurable random variable. Let $D : C([-\tau, 0]; R^d) \to R^m$ be an operator. Furthermore, let $\tilde{F}(x, y, t)$, $x \in R^d$ and $y \in R^m$, be another d-dimensional continuous C-semimartingale such that it has the decomposition $\tilde{F}(x, y, t) = \tilde{M}(x, y, t) + \tilde{B}(x, y, t)$ and its local characteristic $(\tilde{a}(x, y, \bar{x}, \bar{y}, t), \tilde{b}(x, y, t), A(t))$ belongs to the class $B^{0,\delta}$. Consider a d-dimensional stochastic differential functional equation

$$x(t) = \xi(0) + \int_{t_o}^{t} F(x(s), D(\hat{x}(s)), ds) + \int_{t_o}^{t} \tilde{F}(x(s), D(\hat{x}(s)), ds) \tag{5.1}$$

on $t \geq t_o$, with initial data $x(t_o + s) = \xi(s)$, $-\tau \leq s \leq 0$, where $\hat{x}(t) = \{x(t + s): -\tau \leq s \leq 0\}$ as before. This equation can be regarded as a stochastically perturbed equation of the stochastic equation

$$y(t) = \xi(0) + \int_{t_o}^{t} F(y(s), D(\hat{y}(s)), ds) \tag{5.2}$$

on $t \geq t_o$, with initial data $y(t_o + s) = \xi(s)$, $-\tau \leq s \leq 0$. Assume that both of the equations have unique global solutions which are denoted by $x(t; t_o, \xi)$ and $y(t; t_o, \xi)$ respectively. Suppose equation (5.2) is pth moment exponentially stable. What we are interested in this section is to see how much of the stochastic perturbation this stable equation can tolerate without losing the property of stability, i.e. the stochastically perturbed equation (5.1) remains pth moment exponentially stable.

Theorem 5.1 Let $p \geq 2$. Assume equation (5.2) is pth moment exponentially stable, that is there exists a pair of positive constants c_1 and λ such that for all t_o and ξ

$$E|y(t; t_o, \xi)|^p \leq c_1 \Big[\sup_{-\tau \leq s \leq 0} E|\xi(s)|^p \Big] e^{-\lambda(t-t_o)}, \qquad t \geq t_o. \tag{5.3}$$

Let $c_2 - c_6$ be positive constants and let $\beta : [-\tau, 0] \rightarrow R_+$ satisfy

$$\int_{-\tau}^{0} \beta(s)\, ds = 1.$$

Assume furthermore that for all $x, \bar{x} \in R^d$, $y, \bar{y} \in R^m$, $u, v \in C([-\tau, 0]; R^d)$ and $t \geq 0$:

(i) $|b(x, y, t) - b(\bar{x}, \bar{y}, t)|^2 \vee \text{trace}\Big[a(x, y, x, y, t) - 2a(x, y, \bar{x}, \bar{y}, t)$

$$+ a(\bar{x}, \bar{y}, \bar{x}, \bar{y}, t) \Big] \leq c_2|x - \bar{x}|^2 + c_3|y - \bar{y}|^2;$$

(ii) $|\tilde{b}(x, y, t)|^2 \vee \text{trace}\, \tilde{a}(x, y, x, y, t) \leq c_4|x|^2 + c_5|y|^2$;

(iii) $D(0) = 0$ and $|D(u) - D(v)|^p \leq c_6 \int_{-\tau}^{0} \beta(s)\, |u(s) - v(s)|^p\, ds$;

(iv) $A(t) \equiv t$.

Then the perturbed system (5.1) is pth moment exponentially stable if

$$\delta 2^{p+1} K_1 K_3 \exp\left[2\delta K_1 K_2 \right] < 1, \tag{5.4}$$

where

$$\delta = \tau + \lambda^{-1} \log(c_1 2^p),$$

$$K_1 = 4^{p-1}\left[(2\delta - \tau)^{p-1}\, 2^{(p-2)/2} + C_p\,(2(2\delta - \tau))^{(p-2)/2} \right],$$

$$K_2 = c_2{}^{p/2} + c_6 c_3{}^{p/2},$$

$$K_3 = c_4{}^{p/2} + c_6 c_5{}^{p/2},$$

and C_p is the universal constant determined by Theorem 1-2.6.

Remark 5.1 Before the proof of this theorem it should be pointed out that comparing (5.3) with Definition 2.1, we use here $\sup(E|\xi(s)|^p: -\tau \le s \le 0\}$ instead of $E(\|\xi\|_\tau{}^p)$. Of course $\sup(E|\xi(s)|^p: -\tau \le s \le 0\} \le E(\|\xi\|_\tau{}^p)$ so (5.3) is slightly stronger than the definition. On the other hand, in the case of $A(t) \equiv t$, one can replace $E(\|\xi\|_\tau{}^p)$ in Definition 2.1 by $\sup(E|\xi(s)|^p: -\tau \le s \le 0\}$ and all the results in the previous sections hold still.

Proof. Fix any t_o and ξ and write $x(t; t_o, \xi) = x(t)$ and $y(t; t_o, \xi) = y(t)$ simply. For any $t_o \le t \le t_o - \tau + 2\delta$, we have

$$E|x(t) - y(t)|^p \le 4^{p-1}\left[J_1 + J_2 + J_3 + J_4 \right], \tag{5.5}$$

where

$$J_1 = E\left| \int_{t_o}^{t} \left[b(x(s), D(\hat{x}(s)), s) - b(y(s), D(\hat{y}(s)), s) \right] ds \right|^p,$$

$$J_2 = E\left| \int_{t_o}^{t} \left[M(x(s), D(\hat{x}(s)), ds) - M(y(s), D(\hat{y}(s)), ds) \right] \right|^p,$$

$$J_3 = E\left| \int_{t_o}^{t} \tilde{b}(x(s), D(\hat{x}(s)), s)\, ds \right|^p,$$

$$J_4 = E\left| \int_{t_o}^{t} \tilde{M}(x(s), D(\hat{x}(s)), ds) \right|^p.$$

Using Hölder's inequality and the assumptions we estimate

$$J_1 \le (2\delta - \tau)^{p-1} E \int_{t_o}^{t} | b(x(s), D(\hat{x}(s)), s) - b(y(s), D(\hat{y}(s)), s) |^p ds$$

$$\le (2\delta - \tau)^{p-1} E \int_{t_o}^{t} \left[c_2|x(s) - y(s)|^2 + c_3|D(\hat{x}(s)) - D(\hat{y}(s))|^2 \right]^{p/2} ds$$

$$\le (2\delta-\tau)^{p-1} 2^{(p-2)/2} E \int_{t_o}^{t} \left[c_2^{p/2}|x(s)-y(s)|^p + c_3^{p/2}|D(\hat{x}(s))-D(\hat{y}(s))|^p\right] ds.$$

But

$$E \int_{t_o}^{t} |D(\hat{x}(s)) - D(\hat{y}(s))|^p ds \le c_6 E \int_{t_o}^{t} \int_{-\tau}^{0} \beta(r) |x(s+r) - y(s+r)|^p dr\, ds$$

$$\le c_6 \int_{t_o-\tau}^{t} E|x(r) - y(r)|^p dr = c_6 \int_{t_o}^{t} E|x(r) - y(r)|^p dr.$$

So

$$J_1 \le (2\delta - \tau)^{p-1} 2^{(p-2)/2} \left[c_2^{p/2} + c_6 c_3^{p/2} \right] \int_{t_o}^{t} E|x(r) - y(r)|^p dr. \tag{5.6}$$

Also by Burkholder–Davis–Gundy's inequality, we have

$$J_2 = C_p E\Big(\int_{t_o}^{t} \mathrm{trace}\Big[a(x, D(\hat{x}), x, D(\hat{x}), s) - 2a(x, D(\hat{x}), y, D(\hat{y}), s)$$

$$+ a(y, D(\hat{y}), y, D(\hat{y}), s) \Big] ds \Big)^{p/2}.$$

Then we can derive in the same way as above that

$$J_2 = C_p (2(2\delta - \tau))^{(p-2)/2} \left[c_2^{p/2} + c_6 c_3^{p/2} \right] \int_{t_o}^{t} E|x(r) - y(r)|^p dr. \tag{5.7}$$

Similarly, we can show

$$J_3 \le (2\delta - \tau)^{p-1} 2^{(p-2)/2} \left[c_4^{p/2} + c_6 c_5^{p/2} \right] \int_{t_o-\tau}^{t} E|x(r)|^p dr \tag{5.8}$$

and

$$J_4 = C_p\,(2(2\delta - \tau))^{(p-2)/2}\left[c_4{}^{p/2} + c_6 c_5{}^{p/2}\right]\int_{t_o-\tau}^{t} E|x(r)|^p\,dr. \tag{5.9}$$

Substituting (5.6)–(5.9) into (5.5) gives

$$E|x(t) - y(t)|^p \le K_1K_2\int_{t_o}^{t} E|x(r) - y(r)|^p\,dr + K_1K_3\int_{t_o-\tau}^{t} E|x(r)|^p\,dr, \tag{5.10}$$

where K_1 - K_3 have been defined in the statement of the theorem. In view of the Gronwall inequality we see

$$E|x(t) - y(t)|^p \le K_4\int_{t_o-\tau}^{t} E|x(r)|^p\,dr, \qquad t_o \le t \le t_o + 2\delta, \tag{5.11}$$

where

$$K_4 = K_1K_3\exp\left[2\delta K_1K_2\right].$$

On the other hand,

$$E|x(t)|^p \le 2^{p-1}E|y(t)|^p + 2^{p-1}E|x(t) - y(t)|^p.$$

So if $t_o - \tau + \delta \le t \le t_o - \tau + 2\delta$, one can use (5.3) and (5.11) to deduce

$$E|x(t)|^p \le c_1 2^{p-1}\left[\sup_{-\tau\le s\le 0} E|\xi(s)|^p\right]e^{-\lambda(\delta-\tau)}$$

$$+\ \delta 2^{p-1}K_4\left[\sup_{t_o-\tau\le r\le t_o-\tau+\delta} E|x(r)|^p\right] + 2^{p-1}K_4\int_{t_o-\tau+\delta}^{t} E|x(r)|^p\,dr.$$

But by (5.4) one can find a sufficiently small $\varepsilon > 0$ such that

$$c_1 2^{p-1}e^{\lambda\tau-(\lambda-\varepsilon)\delta} \le 0.525 \quad\text{and}\quad \delta 2^{p-1}K_4 \le 0.25\,e^{-\varepsilon\delta}. \tag{5.12}$$

We see

$$E|x(t)|^p \le 0.775\,e^{-\varepsilon\delta}\left[\sup_{t_o-\tau\le r\le t_o-\tau+\delta} E|x(r)|^p\right] + 2^{p-1}K_4\int_{t_o-\tau+\delta}^{t} E|x(r)|^p\,dr$$

if $t_o - \tau + \delta \le t \le t_o - \tau + 2\delta$. Using the Gronwall inequality once again we obtain

$$\sup_{t_o-\tau+\delta\le t\le t_o-\tau+2\delta} E|x(t)|^p$$

$$\le 0.775\, e^{-\varepsilon\delta}\Big[\sup_{t_o-\tau\le r\le t_o-\tau+\delta} E|x(r)|^p\Big]\exp\Big[\delta 2^{p-1} K_4\Big]$$

$$\le e^{-\varepsilon\delta}\Big[\sup_{t_o-\tau\le r\le t_o-\tau+\delta} E|x(r)|^p\Big], \tag{5.13}$$

where (5.12) has been used once more.

Now let $k = 1, 2, \ldots$. Note the solution has the flow property

$$x(t; t_o, \xi) = x(t; t_o + (k-1)\delta, \hat{x}(t_o + (k-1)\delta; t_o, \xi))$$

for any $t \ge t_o + (k-1)\delta$. So by (5.13)

$$\sup\{E|x(t; t_o, \xi)|^p: t_o - \tau + k\delta \le t \le t_o - \tau + (k+1)\delta\}$$

$$\le e^{-\varepsilon\delta} \sup\{E|x(t; t_o, \xi)|^p: t_o - \tau + (k-1)\delta \le t \le t_o - \tau + k\delta\}$$

and then by induction

$$\sup\{E|x(t; t_o, \xi)|^p: t_o - \tau + k\delta \le t \le t_o - \tau + (k+1)\delta\}$$

$$\le e^{-\varepsilon k\delta} \sup\{E|x(t; t_o, \xi)|^p: t_o - \tau \le t \le t_o - \tau + \delta\}. \tag{5.14}$$

On the other hand, it is not difficult to show that there exists a $K_5 > 0$ such that

$$\sup\{E|x(t; t_o, \xi)|^p: t_o - \tau \le t \le t_o - \tau + \delta\} \le K_5\Big[\sup_{-\tau\le s\le 0} E|\xi(s)|^p\Big]. \tag{5.15}$$

Consequently

$$\sup\{E|x(t; t_o, \xi)|^p: t_o - \tau + k\delta \le t \le t_o - \tau + (k+1)\delta\}$$

$$\le K_5\Big[\sup_{-\tau\le s\le 0} E|\xi(s)|^p\Big] e^{-\varepsilon k\delta}.$$

Now for any $t > t_o - \tau + \delta$, find a k such that $t_o - \tau + k\delta \le t < t_o - \tau + (k+1)\delta$ and so

$$E|x(t; t_o, \xi)|^p \le K_5\Big[\sup_{-\tau\le s\le 0} E|\xi(s)|^p\Big] e^{-\varepsilon(t-t_o)},$$

but this holds for any $t_o \le t \le t_o - \tau + \delta$ as well. The proof is complete.

Note that once the coefficients λ, c_1 - c_3, c_6 related to the unperturbed system (5.2) are given, then (5.4) can be satisfied provided c_4 and c_5 small enough. In other words, Theorem 5.1 tells us that provided the intensity of the stochastic perturbation, which is measured by c_4 and c_5, is small enough, the perturbed system (5.1) remains stable. A special case is described by the following corollary.

Corollary 5.2 Assume (5.3) and hypotheses (i), (iii) and (iv) of Theorem 5.1 hold but hypothesis (ii) is replaced by

(ii) $|\tilde{b}(x, y, t)|^2 \vee \text{trace}\, \tilde{a}(x, y, x, y, t) \le \varepsilon(t)\,(|x|^2 + |y|^2)$,

where $\varepsilon(t)$ is a positive real-valued bounded function defined on R_+ such that $\varepsilon(t) \to 0$ as $t \to \infty$. Then the perturbed system (5.1) is pth moment exponentially stable.

Proof. Since $\varepsilon(t) \to 0$ as $t \to \infty$, we can, by Theorem 5.1, find a sufficiently large T such that for all $t_o \ge T$ and ξ,

$$E|x(t; t_o, \xi)|^p \le K_1 \Big[\sup_{-\tau \le s \le 0} E|\xi(s)|^p \Big] e^{-\lambda(t-t_o)} \qquad \text{on } t \ge t_o,$$

where K_1 and λ are positive constants. But it is not difficult to show that there exists a $K_2 \ge 1$ such that for all $0 \le t_o \le T$ and ξ,

$$E|x(t; t_o, \xi)|^p \le K_2 \Big[\sup_{-\tau \le s \le 0} E|\xi(s)|^p \Big] \qquad \text{on } t_o \le t \le T.$$

Hence, for any $0 \le t_o \le T$ and ξ,

$$E|x(t; t_o, \xi)|^p \le K_2\, e^{\lambda T} \Big[\sup_{-\tau \le s \le 0} E|\xi(s)|^p \Big] e^{-\lambda(t-t_o)}$$

if $t_o \le t \le T$, and

$$E|x(t; t_o, \xi)|^p \le E|x(t; T, \hat{x}(T; t_o, \xi)|^p$$

$$\le K_1 \Big[\sup_{-\tau \le s \le 0} E|x(T+s; t_o, \xi)|^p \Big] e^{-\lambda(t-T)}$$

$$\le K_1 K_2\, e^{\lambda T} \Big[\sup_{-\tau \le s \le 0} E|\xi(s)|^p \Big] e^{-\lambda(t-t_o)}$$

if $t > T$. In other words, we have already showed that for all $t_o \ge 0$ and ξ,

$$E|x(t; t_o, \xi)|^p \le K_1 K_2\, e^{\lambda T} \Big[\sup_{-\tau \le s \le 0} E|\xi(s)|^p \Big] e^{-\lambda(t-t_o)} \qquad \text{on } t \ge t_o.$$

The proof is complete.

Let us now consider a d-dimensional stochastic differential functional equation of

Itô's type

$$dx(t) = f(x(t), \hat{x}(t), t)\, dt + \sum_{k=1}^{\infty} g_k(x(t), \hat{x}(t), t)\, dw_k(t)$$

$$+ \tilde{f}(x(t), \hat{x}(t), t)\, dt + \sum_{k=1}^{\infty} \tilde{g}_k(x(t), \hat{x}(t), t)\, dw_k(t) \quad \text{on } t \geq t_o \tag{5.16}$$

with initial data $x(t_o + s) = \xi(s)$, $-\tau \leq s \leq 0$, where f, $\tilde{f}$, g_k and $\tilde{g}_k$ are all mappings from $R^d \times C([-\tau, 0]; R^d) \times R_+$ to R^d and $\{w_k(t): k \geq 1\}$ is a sequence of independent Brownian motions as before. Regard this equation as a stochastically perturbed system of the following equation

$$dy(t) = f(y(t), \hat{y}(t), t)\, dt + \sum_{k=1}^{\infty} g_k(y(t), \hat{y}(t), t)\, dw_k(t) \quad \text{on } t \geq t_o \tag{5.17}$$

with initial data $y(t_o + s) = \xi(s)$, $-\tau \leq s \leq 0$. Assume that both of equations (5.16) and (5.17) have unique solutions denoted by $x(t; t_o, \xi)$ and $y(t; t_o, \xi)$ respectively.

Corollary 5.3 Let $p \geq 2$. Assume equation (5.17) is pth moment exponentially stable, that is there exists a pair of positive constants c_1 and λ such that for all t_o and ξ

$$E|y(t; t_o, \xi)|^p \leq c_1 \Big[\sup_{-\tau \leq s \leq 0} E|\xi(s)|^p \Big] e^{-\lambda(t-t_o)}, \qquad t \geq t_o.$$

Let c_2 - c_4 be positive constants and let $\beta : [-\tau, 0] \to R_+$ satisfy

$$\int_{-\tau}^{0} \beta(s)\, ds = 1.$$

Assume furthermore that for all $x, \bar{x} \in R^d$, $u, v \in C([-\tau, 0]; R^d)$ and $t \geq 0$:

(i) $$|f(x, u, t) - f(\bar{x}, v, t)|^2 \vee \sum_{k=1}^{\infty} |g_k(x, u, t) - g_k(\bar{x}, v, t)|^2$$

$$\leq c_2|x - \bar{x}|^2 + c_3 \int_{-\tau}^{0} (\beta(s))^{2/p} |u(s) - v(s)|^2\, ds;$$

(ii) $$|\tilde{f}(x, u, t)|^2 \vee \sum_{k=1}^{\infty} |\tilde{g}_k(x, u, t)|^2 \leq c_4|x|^2 + c_5 \int_{-\tau}^{0} (\beta(s))^{2/p} |u(s)|^2\, ds.$$

Then the perturbed system (5.16) is pth moment exponentially stable if

$$\delta 2^{p+1} K_1 K_3 \exp\Big[2\delta K_1 K_2 \Big] < 1, \tag{5.4}$$

where

$$\delta = \tau + \lambda^{-1} \log(c_1 2^p),$$

$$K_1 = 4^{p-1}\Big[(2\delta - \tau)^{p-1}\, 2^{(p-2)/2} + C_p\, (2(2\delta - \tau))^{(p-2)/2} \Big],$$

$$K_2 = c_2{}^{p/2} + c_3{}^{p/2},$$

$$K_3 = c_4{}^{p/2} + c_5{}^{p/2},$$

and C_p is the universal constant determined by Theorem 1–2.6.

6.6. EXAMPLES

The aim of this section is to study several examples for illustration. In the sequel of this section we always let $\{w_k(t): k \geq 1\}$ be a sequence of independent Brownian motions.

Example 6.1 Consider a stochastic oscillator described by a stochastic delay equation

$$\ddot{z}(t) + 3\dot{z}(t) + 2z(t) + \sigma_0(z(t-\tau), \dot{z}(t-\tau), t)$$
$$= \sum_{k=1}^{\infty} \sigma_k(z(t-\tau), \dot{z}(t-\tau), t)\, \dot{w}_k(t) \tag{6.1}$$

on $t \geq 0$ with suitable initial data, where σ_k, $k \geq 0$ are all real-valued functions defined on $R^2 \times R_+$. Assume

$$|\sigma_0(y, t)|^2 \leq \varepsilon_1 |y|^2 \quad \text{and} \quad \sum_{k=1}^{\infty} |\sigma_k(y, t)|^2 \leq \varepsilon_2 |y|^2 \tag{6.2}$$

for $y \in R^2$ and $t \geq 0$, where ε_1 and ε_2 are both positive constants. Introduce a new variable $x = (x_1, x_2)^T = (z, \dot{z})^T$ and then equation (6.1) can be read as an Itô delay equation

$$dx(t) = f(x(t), x(t-\tau), t)\, dt + \sum_{k=1}^{\infty} g_k(x(t-\tau), t)\, dw_k(t), \tag{6.3}$$

where

$$f(x, y, t) = Kx + (0, \sigma_0(y, t))^T, \qquad K = \begin{bmatrix} 0 & 1 \\ -2 & -3 \end{bmatrix},$$

$$g_k(y, t) = (0, \sigma_k(y, t))^T$$

for $x, y \in R^2$. It is rather easy to obtain

$$H = \begin{bmatrix} 1 & 1 \\ -1 & -2 \end{bmatrix}, \quad \text{and hence } H^{-1} = \begin{bmatrix} 2 & 1 \\ -1 & -1 \end{bmatrix},$$

such that

$$H^{-1}KH = \begin{bmatrix} -1 & 0 \\ 0 & -2 \end{bmatrix}.$$

Set

$$Q = (H^{-1})^T H^{-1} = \begin{bmatrix} 5 & 3 \\ 3 & 2 \end{bmatrix}.$$

Define $V(x) = x^TQx$ for $x \in R^2$. It is easy to see

$$\frac{1}{7}|x|^2 \le V(x) \le 7|x|^2.$$

We furthermore compute

$$(LV)(x, y, t) = 2f(x, y, t)^TQx + \sum_{k=1}^{\infty} g_k(y, t)^T Q\, g_k(y, t)$$

$$\le 2x^TK^TQx + 2(0, \sigma_0(y, t))(H^{-1})^T H^{-1}x + 2\sum_{k=1}^{\infty} |\sigma_k(y, t)|^2$$

$$\le -2x^TQx + |(0, \sigma_0(y, t))(H^{-1})^T|^2 + |H^{-1}x|^2 + 2\varepsilon_2|y|^2$$

$$\le -V(x) + 2(\varepsilon_1 + \varepsilon_2)|y|^2 \le -V(x) + 14(\varepsilon_1 + \varepsilon_2)V(y).$$

Therefore, by Corollary 2.3, we conclude that equation (6.1) is exponentially stable in mean square if $\varepsilon_1 + \varepsilon_2 < 1/14$.

Example 6.2 Let $\rho : R_+ \to R_+$ be a continuously differentiable function such that

$$\rho(t) \le t \quad \text{and} \quad \dot\rho(t) \ge 1 \quad \text{for all } t \ge 0.$$

Replacing $t-\tau$ by $\rho(t)$ in (6.1) we have a new stochastic oscillator described by the delay equation

$$\ddot z(t) + 3\dot z(t) + 2z(t) + \sigma_0(z(\rho(t)), \dot z(\rho(t)), t)$$

$$= \sum_{k=1}^{\infty} \sigma_k(z(\rho(t)), \dot z(\rho(t)), t)\, \dot w_k(t) \tag{6.4}$$

on $t \geq 0$ with suitable initial data. Using the arguments of Example 6.1 and Corollary 3.3 we see that equation (6.4) is exponentially stable in mean square if $\varepsilon_1 + \varepsilon_2 < 1/14$.

Example 6.3 Let $\delta : R_+ \to [-\tau, 0]$ be any Borel measurable function. If the time lag τ in equation (6.1) is replaced by $\delta(t)$, then the equation reads as

$$\ddot{z}(t) + 3\dot{z}(t) + 2z(t) + \sigma_0(z(t-\delta(t)), \dot{z}(t-\delta(t)), t)$$

$$= \sum_{k=1}^{\infty} \sigma_k(z(t-\delta(t)), \dot{z}(t-\delta(t)), t)\, \dot{w}_k(t). \tag{6.5}$$

Applying Corollary 3.8 we see that equation (6.5) is exponentially stable in mean square if one of the following conditions holds:

(a) $\varepsilon_1+\varepsilon_2 < 1/14e^{\tau}$,

(b) $1 < \tau$ and $\varepsilon_1+\varepsilon_2 < 1/14\tau e$.

Example 6.4 Even more generally, let us consider a stochastic oscillator described by a stochastic differential functional equation

$$\ddot{z}(t) + 3\dot{z}(t) + 2z(t) + \sigma_0(\hat{z}(t), \hat{\dot{z}}(t), t) = \sum_{k=1}^{\infty} \sigma_k(\hat{z}(t), \hat{\dot{z}}(t), t)\, \dot{w}_k(t) \tag{6.6}$$

on $t \geq 0$ with suitable initial data, where σ_k, $k \geq 0$ are all real-valued functions defined on $C([-\tau, 0]; R^2) \times R_+$. Assume

$$|\sigma_0(u, t)|^2 \leq \varepsilon_1 \int_{-\tau}^{0} \beta(r)\, |u(r)|^2\, dr$$

and

$$\sum_{k=1}^{\infty} |\sigma_k(u, t)|^2 \leq \varepsilon_2 \int_{-\tau}^{0} \beta(r)\, |u(r)|^2\, dr$$

for all $u \in C([-\tau, 0]; R^2)$ and $t \geq 0$, where ε_1 and ε_2 are both positive constants as before and $\beta(.)$ is a Borel measurable function from $[-\tau, 0]$ to R_+ such that

$$\int_{-\tau}^{0} \beta(r)\, dr = 1.$$

Again introduce a new variable $x = (x_1, x_2)^T = (z, \dot{z})^T$ and re-write equation (6.6) as an Itô differential functional equation

$$dx(t) = f(x(t), \hat{x}(t), t)\, dt + \sum_{k=1}^{\infty} g_k(\hat{x}(t), t)\, dw_k(t), \tag{6.7}$$

where

$$f(x, u, t) = Kx + (0, \sigma_0(u, t))^T, \qquad K = \begin{bmatrix} 0 & 1 \\ -2 & -3 \end{bmatrix},$$

$$g_k(u, t) = (0, \sigma_k(u, t))^T$$

for $x \in R^2$, $u \in C([-\tau, 0]; R^2)$ and $t \geq 0$. Define V(x) the same as in Example 6.1 and compute

$$(LV)(x, u, t) = 2f(x, u, t)^T Qx + \sum_{k=1}^{\infty} g_k(u, t)^T Q\, g_k(u, t)$$

$$\leq 2x^T K^T Qx + 2(0, \sigma_0(u, t))(H^{-1})^T H^{-1}x + 2\sum_{k=1}^{\infty} |\sigma_k(u, t)|^2$$

$$\leq -2x^T Qx + |(0, \sigma_0(u, t))(H^{-1})^T|^2 + |H^{-1}x|^2 + 2\varepsilon_2 \int_{-\tau}^{0} \beta(r)\, |u(r)|^2\, dr$$

$$\leq -V(x) + 2(\varepsilon_1 + \varepsilon_2) \int_{-\tau}^{0} \beta(r)\, |u(r)|^2\, dr$$

$$\leq -V(x) + 2(\varepsilon_1 + \varepsilon_2) \int_{-\tau}^{0} \beta(r)\, V(u(r))\, dr.$$

By Corollary 4.5, we therefore conclude that equation (6.6) is exponentially stable in mean square if $\varepsilon_1 + \varepsilon_2 < 1/14$.

Example 6.5 Consider a semi-linear stochastic differential functional equation

$$dx(t) = Kx(t)\, dt + f(\hat{x}(t), t)\, dt + \sum_{k=1}^{\infty} g_k(\hat{x}(t), t)\, dw_k(t) \quad \text{on } t \geq t_o \tag{6.8}$$

with initial data $x(t_o + s) = \xi(s)$, $-\tau \leq s \leq 0$, where K is a $d \times d$ constant matrix and f, g_k are all R^d-valued function defined on $C([-\tau, 0]; R^d) \times R_+$. Assume

(i) K has real eigenvalues $-\lambda_1 < -\lambda_2 < \ldots < -\lambda_d < 0$;

(ii) $|f(u, t)| \leq c_1 \left(\int_{-\tau}^{0} \beta(r)\, |u(r)|^2\, dr \right)^{1/2}$,

$$\text{(iii)}\quad \sum_{k=1}^{\infty} |g_k(u, t)|^2 \le c_2 \int_{-\tau}^{0} \beta(r)\, |u(r)|^2\, dr$$

for all $u \in C([-\tau, 0]; R^d)$ and $t \ge 0$, where c_1 and c_2 are two positive constants and $\beta(.)$ is a Borel measurable function from $[-\tau, 0]$ to R_+ such that

$$\int_{-\tau}^{0} \beta(r)\, dr = 1.$$

Note that there exists an invertible $d \times d$ matrix H such that

$$H^{-1}KH = \text{diag.}(-\lambda_1, -\lambda_2, \ldots, -\lambda_d) := \bar{K}.$$

Define a new stochastic process $\bar{x}(t) = H^{-1}x(t)$. Then $\bar{x}(t)$ satisfies the following stochastic differential functional equation

$$d\bar{x}(t) = \bar{K}\bar{x}(t)\, dt + H^{-1}f(H\hat{\bar{x}}(t), t)\, dt + \sum_{k=1}^{\infty} H^{-1}g_k(H\hat{\bar{x}}(t), t)\, dw_k(t). \tag{6.9}$$

Obviously the exponential stability in mean square of equation (6.8) is equivalent to that of equation (6.9). It is easy to see

$$2\bar{x}^T\bar{K}\bar{x} \le -2\lambda_d|\bar{x}|^2,$$

$$|H^{-1}f(H\bar{u}, t)| \le c_1\, |H^{-1}|\, |H| \left(\int_{-\tau}^{0} \beta(r)\, |\bar{u}(r)|^2\, dr \right)^{1/2},$$

$$\sum_{k=1}^{\infty} |H^{-1}g_k(H\bar{u}, t)|^2 \le c_2\, (|H^{-1}|\, |H|)^2 \int_{-\tau}^{0} \beta(r)\, |\bar{u}(r)|^2\, dr$$

for all $\bar{x} \in R^d$, $\bar{u} \in C([-\tau, 0]; R^d)$ and $t \ge 0$. Hence, by Corollary 4.6, we conclude that equation (6.9), hence the original equation (6.8) is exponentially stable in mean square if $2\lambda_d > |H^{-1}|\, |H| \left[\, 2c_1 + c_2\, |H^{-1}|\, |H| \,\right]$.

Example 6.6 Consider a stochastic differential delay equation

$$dx(t) = \left[\, f(x(t), t) + g_0(x(t-\tau), t) \,\right] dt + \sum_{k=1}^{\infty} g_k(x(t-\tau), t)\, dw_k(t) \qquad \text{on } t \ge t_o \tag{6.10}$$

with initial data $x(t_o + s) = \xi(s)$, $-\tau \le s \le 0$, where f and g_k, $k \ge 0$ are all R^d-valued function defined on $R^d \times R_+$. Assume:

(i) $x^T f(x, t) \le -c_1|x|^2$,

(ii) $|g_0(x, t)| \le c_2|x|$,

(iii) $\sum_{k=1}^{\infty} |g_k(x, t)|^2 \le c_3|x|^2$

for all $x \in R^d$, where c_1–c_3 are three positive constants. Define $V(x) = |x|^4$ and compute, for $(x, y, t) \in R^d \times R^d \times R_+$,

$$(LV)(x, y, t)$$

$$= 4|x|^2 x^T \Big[f(x, t) + g_0(y, t) \Big] + 2\sum_{k=1}^{\infty} \Big[|x|^2 |g_k(y, t)|^2 + 2|x^T g_k(y, t)|^2 \Big]$$

$$\le 4|x|^2 x^T f(x, t) + 4|x|^3 |g_0(y, t)| + 6|x|^2 \sum_{k=1}^{\infty} |g_k(y, t)|^2$$

$$\le -4c_1|x|^4 + 4c_2|x|^3 |y| + 6c_3 |x|^2 |y|^2.$$

Note the elementary inequalities $4|x|^3 |y| \le 3|x|^4 + |y|^4$ and $2|x|^2 |y|^2 \le |x|^4 + |y|^4$. We see

$$(LV)(x, y, t) \le -(4c_1 - 3c_2 - 3c_3)V(x) + (c_2 + 3c_3)V(y).$$

So by Corollary 2.3 we conclude that equation (6.10) is 4th moment exponentially stable if $4c_1 > 4c_2 + 6c_3$. One may note that, by Corollary 2.4, we only require $2c_1 > 2c_2 + c_3$ for the equation to be exponentially stable in mean square.

One may also note that the above conclusions hold if only $t-\tau$ in Equation (6.10) is replace by a continuously differentiable function $\rho(t)$ from R_+ to R satisfying $\rho(t) \le t$ and $\dot{\rho}(t) \ge 1$ for all $t \ge 0$. But if $t-\tau$ is replaced by $t-\delta(t)$, where δ is any Borel measurable function from R_+ to $[-\tau, 0]$, then one can conclude that equation (6.10) is 4th moment exponentially stable provided one of the following two conditions is further satisfied:

(a) $(4c_1 - 3c_2 - 3c_3) \exp\Big[-(4c_1 - 3c_2 - 3c_3)\tau\Big] > c_2 + 3c_3$,

(b) $4c_1 - 3c_2 - 3c_3 \ge 1/\tau$ and $c_2 + 3c_3 < 1/\tau e$.

Example 6.7 Let $F(x, y, t)$, $x, y \in R^2$ be a 2-dimensional continuous C-semimartingale with the local characteristic $(a(x, y, \bar{x}, \bar{y}, t), b(x, y, t), A(t))$. Let $\tau > 0$ and $\delta(.) : R_+ \to [-\tau, 0]$ be any Borel measurable function. Consider a 2-dimensional stochastic differential delay equation driven by F(x, y, t) of the form

$$x(t) = \xi(0) + \int_{t_o}^{t} F(x(s), x(s-\delta(s)), ds) \qquad \text{on } t \geq t_o \tag{6.11}$$

with initial data $x(t_o + s) = \xi(s)$, $-\tau \leq s \leq 0$. We assume

(i) $b(x, y, t) = (-x_1 - x_2^2 \sin t + \varepsilon y_2 \cos x_1, -x_2 + x_1 x_2 \sin t + \varepsilon y_1 \sin x_2)^T$;

(ii) $\text{trace}\, a(x, y, x, y, t) \leq \varepsilon(|x|^2 + |y|^2)$,

(iii) $\tau \leq A(t+\tau) - A(t) \leq 2\tau$,

where ε is a small parameter. Let $V(x) = |x|^3$ and compute

$$(LV)(x, y, t)$$

$$= 3|x|\, x^T b(x, y, t) + \frac{3}{2}\Big[\, |x|\, \text{trace}\, a(x, y, x, y, t) + |x|^{-1} x^T a(x, y, x, y, t) x \,\Big]$$

$$\leq -3|x|^3 + 3\varepsilon|x|^2 |y| + 3\varepsilon|x| (|x|^2 + |y|^2).$$

Note the elementary inequality $3\alpha\beta^2 \leq \alpha^3 + 2\beta^3$ for any $\alpha, \beta \geq 0$. We then see

$$(LV)(x, y, t) \leq -3|x|^3 + \varepsilon(2|x|^3 + |y|^3) + 3\varepsilon|x|^3 + \varepsilon(|x|^3 + 2|y|^3)$$

$$\leq -3(1 - 2\varepsilon)V(x) + 3\varepsilon V(y).$$

Moreover,

$$(HV)(x, y, t) = 9|x|^2 x^T a(x, y, x, y, t)x \leq 9\varepsilon|x|^4 (|x|^2 + |y|^2)$$

$$\leq 9\varepsilon|x|^6 + 3\varepsilon(2|x|^6 + |y|^6) = 15\varepsilon(V(x))^2 + 3\varepsilon(V(y))^2.$$

Hence, one can apply Theorem 3.6 to conclude that the delay equation (6.11) is 3rd moment exponentially stable if

$$3 > 6\varepsilon + 18\varepsilon e^{6\tau(1-2\varepsilon)} + 108\varepsilon\Big[\, 5 + e^{12\tau(1-2\varepsilon)} \,\Big].$$

Example 6.8 Let K be a d × d constant matrix which has negative real part eigenvalues. Let $\tilde{F}(x, y, t)$, $x \in R^d$ and $y \in R^m$, be a d-dimensional continuous C-semimartingale that has the local characteristic $(\tilde{a}(x, y, \bar{x}, \bar{y}, t), \tilde{b}(x, y, t), t)$. Let D be an operator from $C([-\tau, 0]; R^d) \to R^m$ which is Lipschitz continuous and moreover $D(0) = 0$. Consider a semi-linear stochastic differential functional equation of the form

$$x(t) = \xi(0) + \int_{t_o}^{t} Kx(s)\,ds + \int_{t_o}^{t} \tilde{F}(x(s), D(\hat{x}(s)), ds) \tag{6.12}$$

on $t \geq t_o$, with initial data $x(t_o + s) = \xi(s)$, $-\tau \leq s \leq 0$, where $\xi(.)$ is a $C([-\tau, 0]; R^d)$-valued $\mathcal{F}_{t_o}$-measurable random variable. This equation can be regarded as a stochastically perturbed system of a linear ordinary differential equation

$$\dot{y}(t) = Ky(t) \quad \text{on } t \geq t_o. \tag{6.13}$$

Since all the eigenvalues of K have negative real parts, equation (6.13) is exponentially stable. We further assume

$$|\tilde{b}(x, y, t)|^2 \vee \text{trace } \tilde{a}(x, y, x, y, t) \leq \varepsilon(t)\,(|x|^2 + |y|^2),$$

where $\varepsilon(t)$ is a positive real-valued bounded function defined on R_+ such that $\varepsilon(t) \to 0$ as $t \to \infty$. Then, in view of Corollary 5.2, the perturbed system (6.12) is pth moment exponentially stable provided the initial data satisfies $\sup\{E|\xi(s)|^p: -\tau \leq s \leq 0\} < \infty$.

CHAPTER 7

Exponential Stability of Stochastic Differential Equations with Small Time Lag

7.1. INTRODUCTION

To explain what problem we are going to study in this Chapter, let us look a simple one dimensional Itô delay equation

$$dx(t) = -x(t)\,dt + (\alpha x(t) + \beta x(t-\tau))\,dw(t), \quad t \geq 0 \tag{1.1}$$

with suitable initial data on $[-\tau, 0]$, where $w(t)$ is a real-valued Brownian motion and α, β are both constants. It is easy to see from Corollary 6-2.3 that equation (1.1) is exponentially stable in mean square if

$$(|\alpha| + |\beta|)^2 < 2 \tag{1.2}$$

and the stability is independent of τ (i.e. no matter the time lag τ is small or large). But if let $\alpha = 1$ and $\beta = -1$, equation (1.1) reads as

$$dx(t) = -x(t)\,dt + (x(t) - x(t-\tau))\,dw(t), \quad t \geq 0. \tag{1.3}$$

Because (1.2) is not satisfied, we do not know so far whether the equation is exponentially stable in mean square or not. On the other hand, one can regard equation (1.3) as a stochastically perturbed system of an ordinary linear differential equation

$$\dot{x}(t) = -x(t). \tag{1.4}$$

Obviously equation (1.4) is exponentially stable. We observe that when the time lag τ is very small, then $x(t) - x(t-\tau)$ can be expected to be nearly zero, depending on the "reasonable" conjecture that the modulus of continuity of $x(t)$ is like that of Brownian motion. Hence the stochastic perturbation term $(x(t) - x(t-\tau))\,dw(t)$ in equation (1.3) could be so small that it could be absorbed by the stable system (1.4) so that the perturbed system remains stable. In other words, we might expect that equation (1.3) is still exponentially stable provided τ small enough. The aim of this chapter is to prove this fact. This problem was first studied by Mao (1990b).

In order to make our theory more understandable we shall start with the stability problem of linear stochastic differential equations with small time lag in Section 7.2 and then deal with semi-linear ones in Section 7.3. Furthermore we shall discuss the stability of general non-linear stochastic differential equations with small time lag in Section 7.4. Finally several examples will be given in Section 7.5.

It should also be mentioned that in this chapter the driving processes of equations are a sequence of independent Brownian motions, but this is equivalent to that the equations are driven by C-Brownian motions (cf. Kunita (1990)).

7.2. LINEAR EQUATIONS WITH SMALL TIME LAG

In this section we shall let $\{w_k(t)\colon k \geq 1\}$ be a sequence of independent real-valued Brownian motions. Let $\tau > 0$ and $t_o \geq 0$. Let $\xi(.)$ be a $C([-\tau, 0]; R^d)$-valued $\mathcal{F}_{t_o}$-measurable random variable such that $E(\|\xi\|_\tau)^2 < \infty$. Let A_k, $k \geq 0$ be $d \times d$ constant matrices. Consider a stochastic differential equation with small time lag τ of the form

$$dx(t) = A_0 x(t)\,dt + \sum_{k=1}^{\infty} A_k(x(t) - x(t-\tau))\,dw_k(t) \qquad \text{on } t \geq t_o \tag{2.1}$$

with initial data $\hat{x}(t_o) = \xi$. The solution of the equation is denoted by $x(t; t_o, \xi)$.

Theorem 2.1 Assume there exist two positive constants β and λ such that

$$|e^{A_0 t}|^2 \le \beta e^{-\lambda t} \qquad \text{for all } t \ge 0. \tag{2.2}$$

If τ is so small that

$$3\beta\left(e^{-\lambda\tau} |I - e^{-A_0\tau}|^2 + \lambda\tau \right) \sum_{k=1}^{\infty} |A_k|^2 < \lambda, \tag{2.3}$$

where I is the identity matrix, then the delay equation (2.1) is exponentially stable in mean square.

Proof. For the sake of convenience we let

$$\alpha = |I - e^{-A_0\tau}|^2 \qquad \text{and} \qquad K = \sum_{k=1}^{\infty} |A_k|^2.$$

Fix t_o and ξ arbitrarily and write $x(t; t_o, \xi) = x(t)$ simply. By Itô's formula one can easily derive that for $t \ge t_o$

$$e^{-A_0 t} x(t) = e^{-A_0 t_o} x(t_o) + \sum_{k=1}^{\infty} \int_{t_o}^{t} e^{-A_0 s} A_k(x(s) - x(s-\tau))\, dw_k(s). \tag{2.4}$$

Hence $t \ge t_o + \tau$

$$x(t) - x(t-\tau) = e^{A_0(t-t_o)} \left[I - e^{-A_0\tau} \right] x(t_o)$$

$$+ \sum_{k=1}^{\infty} \int_{t_o}^{t-\tau} e^{A_0(t-s)} \left[I - e^{-A_0\tau} \right] A_k(x(s) - x(s-\tau))\, dw_k(s)$$

$$+ \sum_{k=1}^{\infty} \int_{t-\tau}^{t} e^{A_0(t-s)} A_k(x(s) - x(s-\tau))\, dw_k(s).$$

So

$$E|x(t) - x(t-\tau)|^2 \le 3\alpha\beta e^{-\lambda(t-t_o)} E|x(t_o)|^2$$

$$+ 3\alpha\beta K \int_{t_o}^{t-\tau} e^{-\lambda(t-s)} E|x(s) - x(s-\tau)|^2\, ds$$

$$+ 3\beta K \int_{t-\tau}^{t} e^{-\lambda(t-s)} E|x(s) - x(s-\tau)|^2\, ds, \tag{2.5}$$

where hypothesis (2.2) has been used. In view of condition (2.3) one can find a $\gamma \in (0, \lambda)$ such that

$$3\beta K \left[\alpha e^{-(\lambda-\gamma)\tau} + (\lambda - \gamma)\tau \right] < \lambda - \gamma. \tag{2.6}$$

It follows from (2.5) that

$$e^{\gamma t} E|x(t) - x(t-\tau)|^2 \leq 3\alpha\beta e^{\lambda t_o - (\lambda-\gamma)t} E|\xi(0)|^2$$

$$+ 3\alpha\beta K\, e^{-(\lambda-\gamma)t} \int_{t_o}^{t-\tau} e^{\lambda s} E|x(s) - x(s-\tau)|^2\, ds$$

$$+ 3\beta K\, e^{-(\lambda-\gamma)t} \int_{t-\tau}^{t} e^{\lambda s} E|x(s) - x(s-\tau)|^2\, ds.$$

Let $T > t_o + \tau$ be arbitrary. Integrating the both sides of the above inequality from $t_o + \tau$ to T yields

$$\int_{t_o+\tau}^{T} e^{\gamma t} E|x(t) - x(t-\tau)|^2\, dt \leq \frac{1}{\lambda - \gamma} 3\alpha\beta e^{\gamma t_o} E|\xi(0)|^2$$

$$+ 3\alpha\beta K \int_{t_o+\tau}^{T} e^{-(\lambda-\gamma)t} \int_{t_o}^{t-\tau} e^{\lambda s} E|x(s) - x(s-\tau)|^2\, ds\, dt$$

$$+ 3\beta K \int_{t_o+\tau}^{T} e^{-(\lambda-\gamma)t} \int_{t-\tau}^{t} e^{\lambda s} E|x(s) - x(s-\tau)|^2\, ds\, dt. \tag{2.7}$$

But by changing the order of integrations,

$$\int_{t_o+\tau}^{T} e^{-(\lambda-\gamma)t} \int_{t_o}^{t-\tau} e^{\lambda s} E|x(s) - x(s-\tau)|^2\, ds\, dt$$

$$= \int_{t_o}^{T-\tau} e^{\lambda s} E|x(s) - x(s-\tau)|^2 \int_{s+\tau}^{T} e^{-(\lambda-\gamma)t}\, dt\, ds$$

$$\leq \frac{1}{\lambda - \gamma} e^{-(\lambda-\gamma)\tau} \int_{t_o}^{T-\tau} e^{\gamma s} E|x(s) - x(s-\tau)|^2\, ds,$$

and also

$$\int_{t_o+\tau}^{T} e^{-(\lambda-\gamma)t} \int_{t-\tau}^{t} e^{\lambda s} E|x(s) - x(s-\tau)|^2 \, ds \, dt$$

$$\leq \int_{t_o}^{T} e^{\lambda s} E|x(s) - x(s-\tau)|^2 \int_{s}^{s+\tau} e^{-(\lambda-\gamma)t} \, dt \, ds$$

$$\leq \tau \int_{t_o}^{T} e^{\gamma s} E|x(s) - x(s-\tau)|^2 \, ds.$$

Substituting these two inequalities into (2.7) gives

$$\int_{t_o+\tau}^{T} e^{\gamma t} E|x(t) - x(t-\tau)|^2 \, dt \leq \frac{1}{\lambda - \gamma} 3\alpha\beta e^{\gamma t_o} E|\xi(0)|^2$$

$$+ \frac{1}{\lambda - \gamma} 3\alpha\beta K \, e^{-(\lambda-\gamma)\tau} \int_{t_o}^{T-\tau} e^{\gamma s} E|x(s) - x(s-\tau)|^2 \, ds$$

$$+ 3\beta K\tau \int_{t_o}^{T} e^{\gamma s} E|x(s) - x(s-\tau)|^2 \, ds.$$

$$\leq \frac{1}{\lambda - \gamma} 3\alpha\beta e^{\gamma t_o} E|\xi(0)|^2 + C_1 \int_{t_o}^{t_o+\tau} e^{\gamma s} E|x(s) - x(s-\tau)|^2 \, ds$$

$$+ C_1 \int_{t_o+\tau}^{T} e^{\gamma s} E|x(s) - x(s-\tau)|^2 \, ds, \tag{2.8}$$

where

$$C_1 = 3\beta K\Big[\frac{1}{\lambda - \gamma} \alpha e^{-(\lambda-\gamma)\tau} + \tau\Big].$$

Recalling (2.6) we see $C_1 < 1$ and hence it follows from (2.8) that

$$\int_{t_o+\tau}^{T} e^{\gamma t} E|x(t) - x(t-\tau)|^2 \, dt$$

$$\leq (1 - C_1)^{-1} \Big(\frac{1}{\lambda - \gamma} 3\alpha\beta e^{\gamma t_o} E|\xi(0)|^2 + C_1 \int_{t_o}^{t_o+\tau} e^{\gamma s} E|x(s) - x(s-\tau)|^2 \, ds \Big).$$

But it is rather easy to show that there exists a $C_2 > 0$ independent of t_o such that

$$E|x(t)|^2 \leq C_2 \, E\|\xi\|_\tau^2 \qquad \text{for } t_o \leq t \leq t_o + \tau. \tag{2.9}$$

We therefore obtain that

$$\int_{t_o+\tau}^{T} e^{\gamma t} E|x(t) - x(t-\tau)|^2 \, dt \leq C_3 \, e^{\gamma t_o} E\|\xi\|_\tau^2 ,$$

where

$$C_3 \leq (1 - C_1)^{-1} \Big(\frac{1}{\lambda - \gamma} 3\alpha\beta + 2C_1 e^{\gamma\tau} (1 + C_2) \Big).$$

Since T is arbitrary,

$$\int_{t_o+\tau}^{\infty} e^{\gamma t} E|x(t) - x(t-\tau)|^2 \, dt \leq C_3 \, e^{\gamma t_o} E\|\xi\|_\tau^2 . \tag{2.10}$$

On the other hand, it follows from (2.4) that for all $t \geq t_o$

$$x(t) = e^{A_0(t-t_o)} x(t_o) + \sum_{k=1}^{\infty} \int_{t_o}^{t} e^{A_0(t-s)} A_k(x(s) - x(s-\tau)) \, dw_k(s).$$

Thus, by (2.2), (2.9) and (2.10)

$$E|x(t)|^2 \leq 2\beta e^{-\lambda(t-t_o)} E|x(t_o)|^2 + 2\beta K \int_{t_o}^{t} e^{-\lambda(t-s)} E|x(s) - x(s-\tau)|^2 \, ds$$

$$\leq 2\beta e^{-\gamma(t-t_o)} E|x(t_o)|^2 + 2\beta K e^{-\gamma t} \int_{t_o}^{t} e^{\gamma s} E|x(s) - x(s-\tau)|^2 \, ds$$

$$\leq 2\beta e^{-\gamma(t-t_o)} E\|\xi\|_\tau^2 + 2\beta K e^{-\gamma t} (2e^{\gamma\tau}(1 + C_2) + C_3) \, e^{\gamma t_o} E\|\xi\|_\tau^2$$

$$= 2\beta \big[1 + K(2e^{\gamma\tau}(1 + C_2) + C_3) \big] E\|\xi\|_\tau^2 \, e^{-\gamma(t-t_o)}.$$

In other words, the delay equation (2.1) is exponentially stable in mean square. The proof is complete.

The following corollary is somewhat more convenient to be used in applications since it is easier to verify the condition proposed.

Corollary 2.2 Assume (2.2) holds. If τ is so small that

$$3\beta\Big(e^{-\lambda\tau}\,(e^{|A_0|\tau} - 1)^2 + \lambda\tau \Big) \sum_{k=1}^{\infty} |A_k|^2 < \lambda, \tag{2.11}$$

then the delay equation (2.1) is exponentially stable in mean square.

Proof. We need only to check condition (2.3) but this follows from (2.11) and the elementary inequality

$$|I - e^{-A_0\tau}|^2 \le (e^{|A_0|\tau} - 1)^2.$$

7.3. SEMI-LINEAR EQUATIONS WITH SMALL TIME LAG

Let $\{w_k(t): k \ge 1\}$, τ, t_o, $\xi(.)$ and A_0 be the same as in the previous section. Let g_k, $k \ge 1$ be all Borel measurable functions from $R^d \times R_+$ to R^d. Consider a semi-linear stochastic differential equation with small time lag τ of the form

$$dx(t) = A_0x(t)\,dt + \sum_{k=1}^{\infty} g_k(x(t) - x(t-\tau), t)\,dw_k(t) \qquad \text{on } t \ge t_o \tag{3.1}$$

with initial data $\hat{x}(t_o) = \xi$. It is obvious to see from the proof of Theorem 2.1 that the following theorem holds.

Theorem 3.1 Assume there exist three positive constants β, λ and K such that

$$|e^{A_0t}|^2 \le \beta e^{-\lambda t} \qquad \text{for all } t \ge 0 \tag{3.2}$$

and

$$\sum_{k=1}^{\infty} |g_k(x, t)|^2 \le K|x|^2 \qquad \text{for all } (x, t) \in R^d \times R_+. \tag{3.3}$$

If τ is so small that

$$3\beta K\Big(e^{-\lambda\tau}\,|I - e^{-A_0\tau}|^2 + \lambda\tau \Big) < \lambda,$$

then the delay equation (3.1) is exponentially stable in mean square.

We now let ρ be any Borel measurable function from R_+ to $[-\tau, 0]$. Replace the time lag τ in equation (3.1) by ρ(t) then the equation becomes

$$dx(t) = A_0 x(t)\,dt + \sum_{k=1}^{\infty} g_k(x(t) - x(t-\rho(t)), t)\,dw_k(t) \qquad \text{on } t \geq t_o \tag{3.4}$$

with initial data $\hat{x}(t_o) = \xi$. With some modification of the proof of Theorem 2.1 one can show the following more general result but the details are left to the reader.

Theorem 3.2 Assume (3.2) and (3.3) hold. If τ is so small that

$$3\beta K\Big(e^{-\lambda\tau}\Big[\sup_{0 \leq s \leq \tau} |I - e^{-A_0 s}|^2 \Big] + \lambda\tau \Big) < \lambda, \tag{3.5}$$

then equation (3.4) is exponentially stable in mean square. A stronger condition than (3.5) is

$$3\beta K\Big(e^{-\lambda\tau}\,(e^{|A_0|\tau} - 1)^2 + \lambda\tau \Big) < \lambda. \tag{3.6}$$

7.4. NON-LINEAR EQUATIONS WITH SMALL TIME LAG

Let $f : R^d \times R_+ \to R^d$, $F : R^d \times R_+ \to R^d$ and $G_k : R^d \times R_+ \to R^{d\times m}$ $(k \geq 1)$ be all Borel measurable. Consider a d-dimensional stochastic differential delay equation

$$dx(t) = f(x(t), t)\,dt + F(x(t) - x(t-\tau), t)\,dt$$

$$+ \sum_{k=1}^{\infty} G_k(x(t) - x(t-\tau), t)\,dw_k(t) \qquad \text{on } t \geq t_o \tag{4.1}$$

with initial data $\hat{x}(t_o) = \xi$. Assume equation (4.1) has a unique solution on $t \geq t_o$, which is denoted by $x(t; t_o, \xi)$ as before. Also assume $f(0, t) \equiv 0$, $F(0, t) \equiv 0$ and $G_k(0, t) \equiv 0$ so equation (4.1) admits a trivial solution $x(t; t_o, 0) \equiv 0$. Regard equation (4.1) as a stochastically perturbed system of the following ordinary differential equation

$$\dot{y}(t) = f(y(t), t) \quad \text{on } t \geq t_o \text{ with initial data } y(t_o) = y_o \in R^d. \tag{4.2}$$

Denote the solution of this equation by $y(t; t_o, y_o)$.

Theorem 4.1 Assume for all $x, y \in R^d$ and $t \geq 0$,

$$|f(x, t) - f(y, t)|^2 \leq K|x - y|^2 \text{ and } |F(x, t)|^2 \vee \sum_{k=1}^{\infty} |G_k(x, t)|^2 \leq K|x|^2, \tag{4.3}$$

where K is a positive constant. Assume furthermore that equation (4.2) is exponentially stable, that is there exists a pair of positive constants M and λ such that

$$|y(t; t_o, y_o)| \le M|y_o| e^{-\lambda(t-t_o)} \qquad \text{on } t \ge t_o \tag{4.4}$$

for all $t_o \ge 0$ and $y_o \in R^d$. Then there exists a $\tau^* > 0$ such that equation (4.1) is exponentially stable in mean square provided $\tau \le \tau^*$.

Proof. We shall divide the whole proof into three steps.

Step 1. We first claim that if

$$0 < \tau^* \le \min(1, 1/12K), \tag{4.5}$$

then for all t_o and ξ,

$$\int_{t_o+\tau}^{T} E|x(t) - x(t-\tau)|^2\, dt \le (\tau/2) \int_{t_o}^{T} E|x(t)|^2\, dt + \int_{t_o}^{t_o+\tau} E|x(t) - x(t-\tau)|^2\, dt \tag{4.6}$$

whenever $T \ge t_o + \tau$, where we write $x(t; t_o, \xi) = x(t)$ simply. In fact, it is easy to see from (4.1) that if $t \ge t_o + \tau$,

$$E|x(t) - x(t-\tau)|^2 \le 3\tau K \int_{t-\tau}^{t} E|x(s)|^2\, ds + 3(\tau + 1)K \int_{t-\tau}^{t} E|x(s) - x(s-\tau)|^2\, ds,$$

where condition (4.3) has been used of course. Integrating both sides from $t_o + \tau$ to T yields

$$\int_{t_o+\tau}^{T} E|x(t) + x(t-\tau)|^2\, dt$$

$$\le 3\tau K \int_{t_o+\tau}^{T} \int_{t-\tau}^{t} E|x(s)|^2\, ds\, dt + 3(\tau + 1)K \int_{t_o+\tau}^{T} \int_{t-\tau}^{t} E|x(s) - x(s-\tau)|^2\, ds\, dt.$$

But by changing the order of integrations we have

$$\int_{t_o+\tau}^{T} \int_{t-\tau}^{t} E|x(s)|^2\, ds\, dt \le \tau \int_{t_o}^{T} E|x(s)|^2\, ds,$$

$$\int_{t_o+\tau}^{T} \int_{t-\tau}^{t} E|x(s) - x(s-\tau)|^2\, ds\, dt \le \tau \int_{t_o}^{T} E|x(s) - x(s-\tau)|^2\, ds.$$

Therefore

$$\int_{t_o+\tau}^{T} E|x(t) - x(t-\tau)|^2\, dt$$

$$\le 3\tau^2 K \int_{t_o}^{T} E|x(s)|^2\, ds + 3\tau(\tau+1)K \int_{t_o}^{t_o+\tau} E|x(s) - x(s-\tau)|^2\, ds$$

$$+ 3\tau(\tau+1)K \int_{t_o+\tau}^{T} E|x(s) - x(s-\tau)|^2\, ds.$$

In view of (4.5) we obtain (4.6).

Step 2. Let $\delta = 2 + \log(2M^2/0.525)/\lambda$. We claim that if furthermore

$$0 < \tau^* \le e^{-\lambda\delta - 12\delta^2 K}/24(8 + 0.5\delta)(1 + 2\delta)K, \tag{4.7}$$

then for any t_o and ξ,

$$\sup_{t_o-\tau+\delta \le t \le t_o-\tau+2\delta} E|x(t;\, t_o, \xi)|^2 \le e^{-\lambda\delta}\Big[\sup_{t_o-\tau \le r \le t_o-\tau+\delta} E|x(r;\, t_o, \xi)|^2\Big]. \tag{4.8}$$

For convenience we write $x(t;\, t_o, \xi) = x(t)$ and $y(t;\, t_o, \xi(0)) = y(t)$ simply. For $t_o \le t \le t_o+2\delta$, it follows from (4.1) and (4.2) that

$$E|x(t) - y(t)|^2$$

$$\le 6\delta K \int_{t_o}^{t} E|x(s) - y(s)|^2\, ds + 3(1 + 2\delta)K \int_{t_o}^{t} E|x(s) - x(s-\tau)|^2\, ds.$$

By the well-known Gronwall inequality,

$$E|x(t) - y(t)|^2 \le 3(1 + 2\delta)K\, e^{6\delta K(t-t_o)} \int_{t_o}^{t} E|x(s) - x(s-\tau)|^2\, ds$$

$$\le 3(1 + 2\delta)K\, e^{12\delta^2 K} \int_{t_o}^{t} E|x(s) - x(s-\tau)|^2\, ds. \tag{4.9}$$

Note

$$E|x(t)|^2 \le 2E|y(t)|^2 + 2E|x(t) - y(t)|^2.$$

So one can see from (4.4) and (4.9) that if $t_o - \tau + \delta \le t \le t_o - \tau + 2\delta$,

$$E|x(t)|^2 \le 2M^2 E|\xi(0)|^2 e^{-2\lambda(\delta-\tau)}$$

$$+ 6(1 + 2\delta)K\, e^{12\delta^2 K} \int_{t_o}^{t} E|x(s) - x(s-\tau)|^2\, ds. \tag{4.10}$$

But by the definition of δ,

$$2M^2 e^{-\lambda\delta+2\lambda\tau} \le 2M^2 e^{-\lambda\delta+2\lambda} = 0.525. \tag{4.11}$$

Also in view of (4.6),

$$\int_{t_o}^{t} E|x(s) - x(s-\tau)|^2\, ds$$

$$\le 2 \int_{t_o}^{t_o+\tau} E|x(s) - x(s-\tau)|^2\, ds + (\tau/2) \int_{t_o}^{t} E|x(s)|^2\, ds$$

$$\le 4 \int_{t_o}^{t_o+\tau} \left[E|x(s)|^2 + E|x(s-\tau)|^2 \right] ds + (\tau/2) \int_{t_o}^{t_o-\tau+\delta} E|x(s)|^2\, ds$$

$$+ (\tau/2) \int_{t_o-\tau+\delta}^{t} E|x(s)|^2\, ds$$

$$\le \tau(8 + 0.5\delta) \left[\sup_{t_o-\tau\le r\le t_o-\tau+\delta} E|x(t)|^2 \right] + (\tau/2) \int_{t_o-\tau+\delta}^{t} E|x(s)|^2\, ds. \tag{4.12}$$

Substituting (4.11) and (4.12) into (4.10) yields

$$E|x(t)|^2 \le 0.525\, E|\xi(0)|^2\, e^{-\lambda\delta}$$

$$+ 6\tau(8 + 0.5\delta)(1 + 2\delta)K\, e^{12\delta^2 K} \left[\sup_{t_o-\tau\le r\le t_o-\tau+\delta} E|x(t)|^2 \right]$$

$$+ 3\tau(1 + 2\delta)K\, e^{12\delta^2 K} \int_{t_o-\tau+\delta}^{t} E|x(s)|^2\, ds.$$

Recalling (4.7) we see

$$6\tau(8 + 0.5\delta)(1 + 2\delta)K\, e^{12\delta^2 K} \le 0.25e^{-\lambda\delta},$$

$$3\tau(1 + 2\delta)K\, e^{12\delta^2 K} \le 1/4\delta.$$

Consequently

$$E|x(t)|^2 \le 0.775\, e^{-\lambda\delta}\Big[\sup_{t_o-\tau\le r\le t_o-\tau+\delta} E|x(t)|^2\Big]$$

$$+ (1/4\delta) \int_{t_o-\tau+\delta}^{t} E|x(s)|^2\, ds.$$

In view of the Gronwall inequality we obtain

$$E|x(t)|^2 \le 0.775\, e^{-\lambda\delta+1/4}\Big[\sup_{t_o-\tau\le r\le t_o-\tau+\delta} E|x(t)|^2\Big]$$

$$\le e^{-\lambda\delta}\Big[\sup_{t_o-\tau\le r\le t_o-\tau+\delta} E|x(t)|^2\Big] \qquad \text{if } t_o - \tau + \delta \le t \le t_o - \tau + 2\delta.$$

In other words, (4.8) has been proved.

Step 3. Now let $k = 1, 2, \ldots$. Note the solution has the flow property

$$x(t; t_o, \xi) = x(t; t_o + (k-1)\delta, \hat{x}(t_o + (k-1)\delta; t_o, \xi))$$

for any $t \ge t_o + (k-1)\delta$. So by (4.8)

$$\sup\{E|x(t; t_o, \xi)|^2 : t_o - \tau + k\delta \le t \le t_o - \tau + (k+1)\delta\}$$

$$\le e^{-\lambda\delta} \sup\{E|x(t; t_o, \xi)|^2 : t_o - \tau + (k-1)\delta \le t \le t_o - \tau + k\delta\}$$

and then by induction

$$\sup\{E|x(t; t_o, \xi)|^2 : t_o - \tau + k\delta \le t \le t_o - \tau + (k+1)\delta\}$$

$$\le e^{-\lambda k\delta} \sup\{E|x(t; t_o, \xi)|^2 : t_o - \tau \le t \le t_o - \tau + \delta\}. \tag{4.13}$$

On the other hand, it is not difficult to show that there exists a $C > 0$ such that

$$\sup\{E|x(t; t_o, \xi)|^2 : t_o - \tau \le t \le t_o - \tau + \delta\} \le C\, E\|\xi\|_\tau^2 . \tag{4.14}$$

Consequently

$$\sup\{E|x(t;\, t_o, \xi)|^2 :\, t_o - \tau + k\delta \le t \le t_o - \tau + (k+1)\delta\} \le C E\|\xi\|_\tau^2 \; e^{-\lambda k\delta}.$$

Now for any $t > t_o - \tau + \delta$, find a k such that $t_o - \tau + k\delta \le t < t_o - \tau + (k+1)\delta$ and so

$$E|x(t;\, t_o, \xi)|^2 \le C E\|\xi\|_\tau^2 \; e^{-\lambda(t-t_o)}, \tag{4.15}$$

but this holds for any $t_o \le t \le t_o - \tau + \delta$ as well. In other words, we have showed that equation (4.1) is exponentially stable in mean square. The proof is complete.

Remark 4.2 (i) It can be seen from the above proof that one may let

$$\tau^* = \min\{\, 1,\; e^{-\lambda\delta - 12\delta^2 K}/24(8 + 0.5\delta)(1 + 2\delta)K \,\}, \tag{4.16}$$

where $\delta = 2 + \log(2M^2/0.525)/\lambda$. Of course this is not optimal.

(ii) The estimate (4.16) is essentially based on we want to obtain (4.15), i.e. the second moment Lyapunov exponent is not great than $-\lambda$. However, with a little bit modification of the above proof, one can even show that for any $\varepsilon \in (0, 2\lambda)$, there exists a $\tau^*(\varepsilon) > 0$ such that whenever $\tau \le \tau^*(\varepsilon)$ the solution of equation (4.1) satisfies

$$E|x(t;\, t_o, \xi)|^2 \le C E\|\xi\|_\tau^2 \; e^{-(2\lambda-\varepsilon)(t-t_o)}, \quad t \ge t_o \tag{4.17}$$

for all t_o and ξ, where C is a positive constant dependent of only K, M, λ and ε. Moreover, one may let

$$\tau^*(\varepsilon) = \min\{\, 1,\; e^{-(2\lambda-\varepsilon)\delta - 12\delta^2 K}/24(8 + 0.5\delta)(1 + 2\delta)K \,\}, \tag{4.18}$$

where $\delta = \max\{\, 2,\; \left[\, 2\lambda + \log(2M^2/0.525) \,\right]/\varepsilon \,\}$.

In equation (4.1) the time lag is a constant. Nevertheless, the above proof is applicable to the case when the time lag is a variable. More precisely, let ρ be any Borel measurable function from R_+ to $[-\tau, 0]$. Replacing the τ in equation (4.1) we arrive at a stochastic differential delay equation

$$dx(t) = f(x(t), t)\, dt + F(x(t) - x(t-\rho(t)), t)\, dt$$

$$+ \sum_{k=1}^{\infty} G_k(x(t) - x(t-\rho(t)), t)\, dw_k(t) \quad \text{on } t \ge t_o \tag{4.19}$$

with initial data $\hat{x}(t_o) = \xi$. In the same way one can prove

Theorem 4.3 Under assumptions (4.3) and (4.4), there exists a $\tau^* > 0$ such that equation (4.19) is exponentially stable in mean square provided $\tau \le \tau^*$. Moreover, Remark 4.2 is also valid in this case.

An interesting result is described as follows.

Corollary 4.4 In addition to assumptions (4.3) and (4.4), assume $\rho(t) \to 0$ as $t \to \infty$. Then equation (4.19) is exponentially stable in mean square. Moreover, the second moment Lyapunov exponent of the solution is not greater than -2λ.

Proof. Let $\varepsilon \in (0, 2\lambda)$ be arbitrary. Since $\rho(t) \to 0$ as $t \to \infty$, by Theorem 2.4 with Remark 2.3, we can find a sufficiently large T such that for all $t_o \ge T$ and ξ,

$$E|x(t; t_o, \xi)|^2 \le C_1 E\|\xi\|_\tau^2 \, e^{-(2\lambda-\varepsilon)(t-t_o)} \qquad \text{on } t \ge t_o,$$

where C_1 is a positive constant and may let $C_1 \ge 1$. But it is not difficult to show that there exists a $C_2 \ge 1$ such that for all $0 \le t_o \le T$ and ξ,

$$E|x(t; t_o, \xi)|^2 \le C_2 \, E\|\xi\|_\tau^2 \qquad \text{on } t_o \le t \le T.$$

Hence, for any $0 \le t_o \le T$ and ξ,

$$E|x(t; t_o, \xi)|^2 \le C_2 \, e^{2\lambda T} \Big[\sup_{-\tau \le s \le 0} E|\xi(s)|^2 \Big] e^{-(2\lambda-\varepsilon)(t-t_o)}$$

if $t_o \le t \le T$, and

$$E|x(t; t_o, \xi)|^2 = E|x(t; T, \hat{x}(T; t_o, \xi))|^2$$

$$\le C_1 \Big[\sup_{-\tau \le s \le 0} E|x(T+s; t_o, \xi)|^2 \Big] e^{-(2\lambda-\varepsilon)(t-T)}$$

$$\le C_1 C_2 \, e^{2\lambda T} \Big[\sup_{-\tau \le s \le 0} E|\xi(s)|^2 \Big] e^{-(2\lambda-\varepsilon)(t-t_o)}$$

if $t > T$. In other words, we have already showed that for all $t_o \ge 0$ and ξ,

$$E|x(t; t_o, \xi)|^2 \le C_1 C_2 \, e^{2\lambda T} \Big[\sup_{-\tau \le s \le 0} E|\xi(s)|^2 \Big] e^{-(2\lambda-\varepsilon)(t-t_o)} \qquad \text{on } t \ge t_o.$$

So equation (4.19) is exponentially stable in mean square. Moreover, the above inequality yields

$$\limsup_{t \to \infty} \frac{1}{t} \log\Big[E|x(t; t_o, \xi)|^2 \Big] \le -(2\gamma-\varepsilon).$$

Since ε is arbitrary, the second moment Lyapunov exponent of the solution is not greater than -2λ. The proof is complete.

7.5. FURTHER REMARKS AND EXAMPLES

In this section we first state a few more general results which can be proved by employing the technique used in the previous section and then give a number of examples for illustration.

First of all, one may have already noted that in the previous section the unperturbed system is described by a deterministic ordinary differential equation $\dot{y}(t) = f(y(t), t)$. But this setting is just to make the theory more understandable. As a matter of fact, the method used above can be adopted to deal with more general cases, e.g. the unperturbed system itself is described by a stochastic differential equation

$$dy(t) = f(y(t), t)\, dt + \sum_{k=1}^{\infty} g_k(y(t), t)\, dw_k(t) \quad \text{on } t \geq t_o \tag{5.1}$$

with initial data $y(t_o) = y_o$, where y_o is an $\mathcal{F}_{t_o}$-measurable R^d-valued random variable and g_k, $k \geq 1$ are Borel measurable functions from $R^d \times R_+$ to R^d such that $g_k(0, t) \equiv 0$. Suppose its perturbed system is described by a stochastic delay equation

$$dx(t) = f(x(t), t)\, dt + \sum_{k=1}^{\infty} g_k(x(t), t)\, dw_k(t)$$

$$+ F(x(t) - x(t-\tau), t)\, dt + \sum_{k=1}^{\infty} G_k(x(t) - x(t-\tau), t)\, dw_k(t) \quad \text{on } t \geq t_o \tag{5.2}$$

with initial data $\hat{x}(t_o) = \xi$. In the same way as the proof of Theorem 4.1, one can prove the following result.

Theorem 5.1 Assume for all $x, y \in R^d$ and $t \geq 0$,

$$|f(x, t) - f(y, t)|^2 \vee \sum_{k=1}^{\infty} |g_k(x, t) - g(y, t)|^2 \leq K|x - y|^2 \tag{5.3}$$

and

$$|F(x, t)|^2 \vee \sum_{k=1}^{\infty} |G_k(x, t)|^2 \leq K|x|^2, \tag{5.4}$$

where K is a positive constant. Assume furthermore that equation (5.1) is exponentially stable in mean square, that is there exists a pair of positive constants M and λ such that

$$E|y(t; t_o, y_o)|^2 \leq ME|y_o|^2 e^{-\lambda(t-t_o)} \quad \text{on } t \geq t_o \tag{5.5}$$

for all $t_o \geq 0$ and $\mathcal{F}_{t_o}$-measurable R^d-valued random variables y_o with $E|y_o|^2 < \infty$. Then there exists a $\tau^* > 0$ such that the perturbed equation (5.2) is also exponentially stable in mean square provided $\tau \leq \tau^*$.

Secondly, so far we have always dealt with exponentially stability in mean square. Nevertheless, we should, in the same spirit, expect the same result holds for pth moment exponentially stability. For example, one can slightly modify the proof of Theorem 4.1 to prove the following even more general result.

Theorem 5.2 Let (5.3) and (5.4) hold and let $p \geq 2$. Assume that equation (5.1) is pth moment exponentially stable. Then there exists a $\tau^* > 0$ such that the perturbed equation (5.2) is also pth moment exponentially stable provided $\tau \leq \tau^*$.

Thirdly it should be pointed out that the constant time lag τ in equation (5.2) may replaced by a variable $\rho(t)$ without changing the results above as we have seen in the previous section.

In the remainder of this section we are going to discuss a few examples for illustration.

Example 5.1 Let us first return back to the example described in Section 7.1, i.e. the stochastic differential delay equation

$$dx(t) = -x(t)\,dt + (x(t) - x(t-\tau))\,dw(t). \tag{5.6}$$

It has been pointed out that the theory of the previous chapter is not applicable to this equation. But in view of Theorem 2.1, one can now conclude that this equation is exponential stable in mean square if τ is so small that

$$3\Big[\,|e^{-\tau} - 1|^2 + 2\tau\,\Big] < 2.$$

But $|e^{-\tau} - 1| \leq \tau$. So it is enough that $3(\tau^2 + 2\tau) < 2$, or $\tau < (\sqrt{60} - 6)/6$.

Example 5.2 Consider a stochastic oscillator described by a stochastic delay equation

$$\ddot{z}(t) + 3\dot{z}(t) + 2z(t) = \sum_{k=1}^{\infty} \sigma_k(z(t) - z(t-\tau), \dot{z}(t) - \dot{z}(t-\tau), t)\,\dot{w}_k(t) \tag{5.7}$$

on $t \geq 0$ with suitable initial data, where σ_k, $k \geq 1$ are all real-valued functions defined on $R^2 \times R_+$. Introduce a new variable $x = (x_1, x_2)^T = (z, \dot{z})^T$ and then equation (5.7) can be read as an Itô delay equation

$$dx(t) = A_0 x(t)\,dt + \sum_{k=1}^{\infty} g_k(x(t) - x(t-\tau), t)\,dw_k(t), \tag{5.8}$$

where

$$A_0 = \begin{bmatrix} 0 & 1 \\ -2 & -3 \end{bmatrix}, \qquad g_k(x, t) = (0, \sigma_k(x, t))^T$$

for $x \in R^2$. Assume

$$\sum_{k=1}^{\infty} |\sigma_k(x, t)|^2 \le |y|^2 \quad \text{for } x \in R^2 \text{ and } t \ge 0. \tag{5.9}$$

It is rather easy to obtain

$$H = \begin{bmatrix} 1 & 1 \\ -1 & -2 \end{bmatrix}, \quad \text{and hence } H^{-1} = \begin{bmatrix} 2 & 1 \\ -1 & -1 \end{bmatrix},$$

such that

$$H^{-1}A_0H = B := \begin{bmatrix} -1 & 0 \\ 0 & -2 \end{bmatrix}.$$

Note

$$e^{A_0t} = e^{HBH^{-1}t} = He^{Bt}H^{-1} = H\,\text{diag.}(e^{-t}, e^{-2t})\,H^{-1}.$$

It is then easy to see

$$|e^{A_0t}|^2 \le 10e^{-2t} \qquad \text{for all } t \ge 0.$$

In view of Theorem 3.1, if τ is so small that

$$30\left(e^{-2\tau}\,|I - e^{-A_0\tau}|^2 + 2\tau \right) < 2, \tag{5.10}$$

then the delay equation (3.1) is exponentially stable in mean square. Note furthermore

$$e^{-A_0\tau} = e^{H(-B\tau)H^{-1}} = He^{-B\tau}H^{-1} = H\,\text{diag.}(e^{\tau}, e^{2\tau})\,H^{-1}.$$

So

$$I - e^{-A_0\tau} \le e^{2\tau}\left(e^{-2\tau}I - H\,\text{diag.}(e^{-\tau}, 1)\,H^{-1}\right)$$

$$= e^{2\tau}\begin{bmatrix}(1 - e^{-\tau})^2, & 1 - e^{-\tau} \\ 2(1 - e^{-\tau}), & e^{-2\tau} + e^{-\tau} + 2\end{bmatrix}.$$

Using $1 - e^{-\tau} \le \tau$ we then see

$$e^{-2\tau}\,|I - e^{-A_0\tau}|^2 \le e^{2\tau}\,(\tau^4 + 13\tau^2).$$

So for (5.10) to be satisfied, it is enough to require

$$\tau\left(e^{2\tau}(\tau^3 + 13\tau) + 2\right) < 1/15,$$

or even simply $\tau < 1/45$.

Example 5.3 We finally consider a stochastic delay equation

$$dx(t) = f(x(t))\,dt + \sum_{k=1}^{\infty} g_k(x(t), t)\,dw_k(t)$$

$$+ F(x(t) - x(t-\rho(t)), t)\,dt + \sum_{k=1}^{\infty} G_k(x(t) - x(t-\rho(t)), t)\,dw_k(t) \qquad (5.11)$$

on $t \ge t_o$ with initial data $\hat{x}(t_o) = \xi$. Here:

(i) $f(x) = (-2x_1 + x_2 \sin x_1,\ -2x_2 + x_1 e^{-|x_2|})^T$.

(ii) F, g_k and G_k $(k \ge 1)$ are all Borel measurable functions from $R^2 \times R_+$ to R^2 such that for all $(x, t) \in R^2 \times R_+$ and some $K > 0$,

$$\sum_{k=1}^{\infty} |g_k(x, t)|^2 \le |x|^2 \quad \text{and} \quad |F(x, t)|^2 \vee \sum_{k=1}^{\infty} |G_k(x, t)|^2 \le K|x|^2.$$

(iii) $\rho : R_+ \to [-\tau, 0]$ such that $\rho(t) \to 0$ as $t \to \infty$.

In order to employ the theory established above, we introduce the unperturbed system

$$dy(t) = f(y(t))\,dt + \sum_{k=1}^{\infty} g_k(y(t), t)\,dw_k(t). \qquad (5.11)$$

It is easy to see that $L|y|^2 \le -|y|^2$, where the operator L is defined in Chapter 4. So in view of Theorem 4-3.2, equation (5.11) is exponentially stable in mean square. Hence, by Theorem 5.1 and the proof of Corollary 4.4, we can conclude that equation (5.10) is also exponentially stable in mean square.

CHAPTER 8

Exponential Stability of Large-Scale Stochastic Differential Delay Systems

8.1. INTRODUCTION

Numerous problems in control theory and dynamical systems as well as applications to engineering, economy, biology lead to the study of exponential stability of a large-scale system described, for instance, by an ordinary differential equation

$$\dot{x} = f(x, t) \tag{1.1}$$

($x \in R^d$ and d is large). In theory, there exist several methods in dealing with the exponential stability. However, enormous calculations involved in the large-scale case make the methods difficult to be used in practice. For example, it is known that the exponential stability of a linear differential equation $\dot{x} = Ax$ is equivalent to that all the eigenvalues of the $d \times d$ matrix A have negative real parts. Unfortunately it takes ages to calculate all the eigenvalues when d is large even with the help of modern computers.

One of the effective methods developed for large-scale systems is the decomposition technique which has been studied by many authors, for example, Michel, Miller and Tang (1978), Siljak (1977), Sontag (1980), Vidyasagar (1980). The idea is as follows: Decompose the system into several subsystems (with a necessary rearrangement of the order of the variables)

$$\dot{x}_i = f_i(x_1, ..., x_m, t), \quad i = 1, ..., m, \tag{1.2}$$

where $x_i \in R^{d_i}$ and $\sum d_i = d$, and introduce the isolated subsystems

$$\dot{y}_i = f_i(0, y_i, 0, t). \tag{1.3}$$

Under certain hypotheses added on the interconnected terms

$$f_i(x_1, ..., x_m, t) - f_i(0, y_i, 0, t)$$

it is shown that the large-scale system (1.1) is exponentially stable if each of the isolated subsystems (1.3) is exponentially stable. In particular, if the large-scale system can be decomposed into a hierarchical form, i.e.

$$\dot{x}_i = f_i(x_1, ..., x_i, t), \quad i = 1, ..., m, \tag{1.4}$$

and the f_i's are all Lipschitz continuous, then it has been showed that the large-scale system (1.4) is exponentially stable if and only if each of the corresponding isolated subsystems is exponentially stable.

On the other hand, so far little is known on exponential stability of large-scale stochastic delay systems. This is the aim of this chapter: to extend the above decomposition technique to large-scale stochastic delay systems. We shall first in Section 8.2 give the new definition of the exponential stability in $L^2(\Omega; C([-\tau, 0]; R^d))$ and discuss the relationships between the various exponential stabilities e.g. almost surely, in mean square and in $L^2(\Omega; C([-\tau, 0]; R^d))$. Sections 8.3 and 8.4 is then devoted to the discussion of the exponential stability in $L^2(\Omega; C([-\tau, 0]; R^d))$ for large-scale stochastic delay systems in hierarchical form and general large-scale stochastic delay system respectively. Applications to linear large-scale stochastic delay systems and others will be finally discussed in Section 8.5 for illustration.

As in the last chapter, the driving processes of the equations studied in this chapter are a sequence of independent Brownian motions, and this is equivalent to that the equations are driven by C-Brownian motions (cf. Kunita (1990)).

8.2. DEFINITIONS AND RELATIONSHIPS BETWEEN VARIOUS EXPONENTIAL STABILITIES

In this chapter we shall always let $\{w_k(t): k \geq 1\}$ be a sequence of independent real-valued Brownian motions. Let $\tau > 0$ and $t_o \geq 0$. Let $L^2(\Omega, \mathcal{F}_{t_o}; C([-\tau, 0]; R^d))$ denote all $\mathcal{F}_{t_o}$-measurable $C([-\tau, 0]; R^d)$-valued random variables ξ such that $E(\|\xi\|_\tau)^2 < \infty$. Write $L^2(\Omega; C([-\tau, 0]; R^d))$ for $L^2(\Omega, \mathcal{F}; C([-\tau, 0]; R^d))$ simply.

Let f, g_k $(k \geq 1)$ are all Borel measurable functions from $C([-\tau, 0]; R^d) \times R_+$ to R^d. Consider a stochastic delay equation

$$dx(t) = f(\hat{x}(t), t)\, dt + \sum_{k=1}^{\infty} g_k(\hat{x}(t), t)\, dw_k(t) \quad \text{on } t \geq t_o \tag{2.1}$$

with initial data $\hat{x}(t_o) = \xi \in L^2(\Omega, \mathcal{F}_{t_o}; C([-\tau, 0]; R^d))$. Assume that the equation has a unique solution which is denoted by $x(t; t_o, \xi)$. Also assume that $f(0, t) = g_k(0, t) \equiv 0$ so the equation has a trivial solution $x(t; t_o, 0) \equiv 0$.

Definition 2.1 The stochastic delay equation (2.1) is said to be *exponentially stable in $L^2(\Omega; C([-\tau, 0]; R^d))$* if there exist two positive constants λ and M such that

$$E\|\hat{x}(t; t_o, \xi)\|_\tau^2 \leq ME\|\xi\|_\tau^2\, e^{-\lambda(t-t_o)} \quad \text{on } t \geq t_o \tag{2.2}$$

for all $t_o \geq 0$ and $\xi \in L^2(\Omega, \mathcal{F}_{t_o}; C([-\tau, 0]; R^d))$.

Let us now discuss the relationships between various exponential stabilities e.g. almost surely, in mean square and in $L^2(\Omega; C([-\tau, 0]; R^d))$. First of all, we would like to point out that the exponential stability in $L^2(\Omega; C([-\tau, 0]; R^d))$ implies the almost surely exponential stability. As a matter of fact we have the following proposition.

Proposition 2.1 (2.2) implies

$$\limsup_{t\to\infty} \frac{1}{t} \log|x(t; t_o, \xi)| \leq -\lambda/2 \quad \text{a.s.} \tag{2.3}$$

In other words, if equation (2.1) is exponentially stable in $L^2(\Omega; C([-\tau, 0]; R^d))$, it is also almost surely exponentially stable.

Proof. Let $\varepsilon \in (0, \lambda)$ arbitrarily and let $n = 1, 2, \ldots$. Then (2.2) gives

$$P\Big(\omega: \|\hat{x}(n\tau + t_o; t_o, \xi)\|_\tau > e^{-(\lambda-\varepsilon)n\tau/2}\Big)$$

$$\leq e^{(\lambda-\varepsilon)n\tau}\, E\|\hat{x}(n\tau + t_o; t_o, \xi)\|_\tau^2 \leq ME\|\xi\|_\tau^2\, e^{-\varepsilon n\tau}.$$

In view of the well-known Borel-Cantelli lemma, we see that for almost all $\omega \in \Omega$,

$$\|\hat{x}(n\tau + t_o; t_o, \xi)\|_\tau \le e^{-(\lambda-\varepsilon)n\tau/2} \tag{2.4}$$

holds for all but finitely many n. Hence there exists $n_o(\omega)$, for all $\omega \in \Omega$ excluding a P-null set, for which (2.4) holds whenever $n \ge n_o(\omega)$. Consequently,

$$\frac{1}{t}\log|x(t; t_o, \xi)| \le \frac{1}{(n-1)\tau + t_o}\left(-(\lambda-\varepsilon)n\tau/2\right)$$

if $(n-1)\tau + t_o \le t \le n\tau + t_o$, $n \ge n_o$ almost surely. Therefore

$$\limsup_{t\to\infty} \frac{1}{t}\log|x(t; t_o, \xi)| \le -(\lambda - \varepsilon)/2 \qquad \text{a.s.}$$

and then (2.3) follows since ε is arbitrary. The proof is complete.

We then compare the exponential stability in $L^2(\Omega; C([-\tau, 0]; R^d))$ with the exponential stability in mean square (see Definition 6-2.1). Obviously the exponential stability in $L^2(\Omega; C([-\tau, 0]; R^d))$ implies the exponential stability in mean square. Although we do not know whether the exponential stability in mean square implies the exponential stability in $L^2(\Omega; C([-\tau, 0]; R^d))$ in general, we shall show it does in many important and interesting cases. For instance, if the coefficients of equation (2.1) are Lipschitz continuous (in fact we only need a weaker condition as shown below) and τ is small enough, then both of the stabilities are equivalent. This is described as follows:

Proposition 2.2 Assume there exists a $K > 0$ such that

$$|f(u, t)|^2 \vee \sum_{k=1}^{\infty} |g_k(u, t)|^2 \le K\|u\|_\tau^2 \tag{2.5}$$

for all $u \in C([-\tau, 0]; R^d)$ and $t \ge 0$. If τ is so small that

$$3K(4 + \tau)\tau < 1, \tag{2.6}$$

then both of the exponential stability in $L^2(\Omega; C([-\tau, 0]; R^d))$ and the exponential stability in mean square are equivalent.

Proof. We need only prove that the exponential stability in mean square implies the exponential stability in $L^2(\Omega; C([-\tau, 0]; R^d))$ so we assume there exist two positive constants λ and M such that

$$E|\hat{x}(t; t_o, \xi)|^2 \le ME\|\xi\|_\tau^2 e^{-\lambda(t-t_o)} \qquad \text{on } t \ge t_o \tag{2.7}$$

for all $t_o \geq 0$ and $\xi \in L^2(\Omega, \mathcal{F}_{t_o}; C([-\tau, 0]; R^d))$. From (2.6) one can find an $\varepsilon \in (0, \lambda)$ such that

$$3K(4+\tau)\tau\, e^{\varepsilon\tau} < 1. \tag{2.8}$$

We shall show that there exists a $C > 0$ such that

$$E\|\hat{x}(t; t_o, \xi)\|_\tau^2 \leq CE\|\xi\|_\tau^2\, e^{-\varepsilon(t-t_o)} \quad \text{on } t \geq t_o. \tag{2.9}$$

We fix t_o and ξ arbitrarily and write $x(t; t_o, \xi) = x(t)$ simply. Using Doob's martingale inequality, Hölder's inequality and hypotheses (2.5) and (2.7) we can easily derive that if $t \geq t_o + \tau$,

$$E\|\hat{x}(t)\|_\tau^2 \leq 3E|x(t-\tau)|^2 + 3K(4+\tau)\int_{t-\tau}^{t} E\|\hat{x}(r)\|_\tau^2\, dr$$

$$\leq 3ME\|\xi\|_\tau^2\, e^{-\lambda(t-\tau-t_o)} + 3K(4+\tau)\int_{t-\tau}^{t} E\|\hat{x}(r)\|_\tau^2\, dr. \tag{2.10}$$

On the other hand, if $t_o \leq t \leq t_o + \tau$,

$$E\|\hat{x}(t)\|_\tau^2 \leq E\|\xi\|_\tau^2 + E\Big(\sup_{t_o<r<t} |x(r)|^2\Big)$$

$$\leq 4ME\|\xi\|_\tau^2 + 3K(4+\tau)\int_{t_o}^{t} E\|\hat{x}(r)\|_\tau^2\, dr. \tag{2.11}$$

Combining (2.10) and (2.11) we see that for all $t \geq t_o$,

$$E\|\hat{x}(t)\|_\tau^2 \leq 4ME\|\xi\|_\tau^2\, e^{-\lambda(t-\tau-t_o)} + 3K(4+\tau)\int_{(t-\tau)\vee t_o}^{t} E\|\hat{x}(r)\|_\tau^2\, dr. \tag{2.12}$$

Therefore, for any $T \geq t_o$,

$$\int_{t_o}^{T} e^{\varepsilon t} E\|\hat{x}(t)\|_\tau^2\, dt$$

$$\leq 4ME\|\xi\|_\tau^2\, e^{\lambda(\tau+t_o)}\int_{t_o}^{T} e^{-(\lambda-\varepsilon)t}\, dt + 3K(4+\tau)\int_{t_o}^{T} e^{\varepsilon t}\int_{(t-\tau)\vee t_o}^{t} E\|\hat{x}(r)\|_\tau^2\, dr\, dt$$

$$\leq \frac{4M}{\lambda-\varepsilon} E\|\xi\|_\tau^2\, e^{\lambda\tau+\varepsilon t_o} + 3K(4+\tau)\int_{t_o}^{T}\Big(\int_{r}^{(r+\tau)\wedge T} e^{\varepsilon t} dt\Big) E\|\hat{x}(r)\|_\tau^2\, dr$$

$$\leq \frac{4M}{\lambda - \varepsilon} E\|\xi\|_\tau^2 e^{\lambda\tau+\varepsilon t_o} + 3K(4+\tau)\tau e^{\varepsilon\tau} \int_{t_o}^{T} e^{\varepsilon r} E\|\hat{x}(r)\|_\tau^2 \, dr.$$

This implies

$$\int_{t_o}^{T} e^{\varepsilon t} E\|\hat{x}(t)\|_\tau^2 \, dt \leq C_1 E\|\xi\|_\tau^2 e^{\varepsilon t_o} \qquad \text{for all } T \geq t_o, \tag{2.13}$$

where $C_1 = 4Me^{\lambda\tau}\left[(\lambda - \varepsilon)(1 - 3K(4+\tau)\tau e^{\varepsilon\tau}) \right]^{-1}$. Finally (2.12) and (2.13) yield that for $t \geq t_o$,

$$e^{\varepsilon t} E\|\hat{x}(t)\|_\tau^2 \leq 4ME\|\xi\|_\tau^2 e^{\lambda\tau+\varepsilon t_o} + 3K(4+\tau) e^{\varepsilon t} \int_{(t-\tau)\vee t_o}^{t} E\|\hat{x}(r)\|_\tau^2 \, dr$$

$$\leq 4ME\|\xi\|_\tau^2 e^{\lambda\tau+\varepsilon t_o} + 3K(4+\tau) e^{\varepsilon\tau} \int_{(t-\tau)\vee t_o}^{t} e^{\varepsilon r} E\|\hat{x}(r)\|_\tau^2 \, dr$$

$$\leq \left(4Me^{\lambda\tau} + 3K(4+\tau)e^{\varepsilon\tau}C_1\right) E\|\xi\|_\tau^2 e^{\varepsilon t_o}. \tag{2.14}$$

This is (2.9) by putting $C = 4Me^{\lambda\tau} + 3K(4+\tau)e^{\varepsilon\tau}C_1$. The proof is complete.

Another important case for both of the stabilities to be equivalent is a stochastic differential equation with variable time lags of the form:

$$dx(t) = F(x(\tau_1(t)), \ldots, x(\tau_n(t)), t) \, dt$$

$$+ \sum_{k=1}^{\infty} G_k(x(\tau_1(t)), \ldots, x(\tau_n(t)), t) \, dw_k(t) \qquad \text{on } t \geq t_o \tag{2.15}$$

with initial data $\hat{x}(t_o) = \xi$. Here τ_i, $1 \leq i \leq n$ are all real-valued functions defined on R_+ such that $t-\tau \leq \tau_i(t) \leq t$ for all $t \geq 0$, and F, G_k are all Borel measurable functions from $R^{d\times n} \times R_+$ to R^d. For this delay equation we have the following proposition which gives a weaker condition than that of Proposition 2.2.

Proposition 2.3 Assume there exists a $K > 0$ such that

$$|F(v_1, \ldots, v_n, t)|^2 \vee \sum_{k=1}^{\infty} |G(v_1, \ldots, v_n, t)|^2 \leq K \sum_{i=1}^{n} |v_i|^2 \tag{2.16}$$

for all $v_i \in R^d$ and $t \geq 0$. Then, for equation (2.15), both of the exponential stability in $L^2(\Omega; C([-\tau, 0]; R^d))$ and the exponential stability in mean square are equivalent.

Proof. Again we need only to prove that the exponential stability in mean square implies the exponential stability in $L^2(\Omega; C([-\tau, 0]; R^d))$ so we assume (2.7) holds. Fix any t_o and ξ, and write $x(t; t_o, \xi) = x(t)$ simply. Then we can derive that, for any $t \geq t_o+2\tau$,

$$E\|\hat{x}(t)\|_\tau^2 \leq 3E|x(t-\tau)|^2 + 3(4+\tau)K \int_{t-\tau}^{t} \Big(\sum_{i=1}^{n} E|x(\tau_i(s)|^2 \Big)\, ds$$

$$\leq 3ME\|\xi\|_\tau^2\, e^{-\lambda(t-\tau-t_o)} + 3(4+\tau)K \int_{t-\tau}^{t} \Big(\sum_{i=1}^{n} ME\|\xi\|_\tau^2\, e^{-\lambda(\tau_i(s)-t_o)} \Big)\, ds$$

$$\leq (\, 3Me^{\lambda\tau} + 3(4+\tau)KMn\tau e^{2\tau}\,)\, E\|\xi\|_\tau^2\, e^{-\lambda(t-t_o)}.$$

On the other hand, it is easy to show that there exists a $C_1 > 0$ such that

$$E\|\hat{x}(t)\|_\tau^2 \leq C_1 E\|\xi\|_\tau^2 \qquad \text{whenever } t_o \leq t \leq t_o + 2\tau.$$

Setting $C = C_1 e^{2\tau} \vee (3Me^{\lambda\tau} + 3(4+\tau)KMn\tau e^{2\tau})$ we therefore have

$$E\|\hat{x}(t)\|_\tau^2 \leq CE\|\xi\|_\tau^2\, e^{-\lambda(t-t_o)} \qquad \text{for all } t \geq t_o$$

as required. The proof is complete.

8.3. LARGE-SCALE STOCHASTIC DELAY SYSTEMS IN HIERARCHICAL FORM

In this section we shall consider a large-scale stochastic delay system in hierarchical form, i.e. the system which consists of several subsystems and each subsystem interacts only with "lower" subsystems but not with "higher" subsystems. In other words we shall study a large-scale stochastic delay system of the form

$$\begin{aligned} dx_i(t) &= f_i(\hat{x}_1(t), \ldots, \hat{x}_i(t), t)\, dt \\ &\quad + \sum_{k=1}^{\infty} g_{ki}(\hat{x}_1(t), \ldots, \hat{x}_i(t), t)\, dw_k(t) \qquad \text{on } t \geq t_o, \\ \hat{x}_i(t_o) &= \xi_i, \quad i = 1, \ldots, m. \end{aligned} \tag{3.1}$$

Here:

$x_i \in R^{d_i}$, $m \geq 2$, $\sum d_i = d$,

$f_i, g_{ki} : C([-\tau, 0]; R^{d_1+\ldots+d_i}) \times R_+ \to R^{d_i}$,

$\xi_i \in L^2(\Omega, \mathcal{F}_{t_o}; C([-\tau, 0]; R^{d_i}))$.

Set $x = (x_1, \ldots, x_m)$ and $\xi = (\xi_1, \ldots, \xi_m)$. Assume the equation has a unique global solution and we denote the solution by $x(t; t_o, \xi)$. Assume also that $f_i(0, t) \equiv 0$ and $g_{ki}(0, t) \equiv 0$ for all i, k. Therefore the equation admits a trivial solution $x(t; t_o, 0) \equiv 0$. Introduce the isolated subsystems

$$dy_i(t) = f_i(0, \hat{y}_i(t), t)\, dt + \sum_{k=1}^{\infty} g_{ki}(0, \hat{y}_i(t), t)\, dw_k(t) \quad \text{on } t \geq t_o, \tag{3.2}$$

$$\hat{y}_i(t_o) = \xi_i.$$

The solution of each subsystem is denoted by $y_i(t; t_o, \xi_i)$. We propose a standing hypothesis:

(H3.1) There exists a $K > 0$ such that

$$|f_i(u_1, \ldots, u_i, t) - f_i(0, v_i, t)|^2 \vee \sum_{k=1}^{\infty} |g_{ki}(u_1, \ldots, u_i, t) - g_{ki}(0, v_i, t)|^2$$

$$\leq K\Big(\sum_{j<i} \|u_j\|_\tau^2 + \|u_i - v_i\|_\tau^2\Big)$$

for all $u_j, v_j \in C([-\tau, 0]; R^{d_j})$, $1 \leq j \leq m$ and $2 \leq i \leq m$.

Theorem 3.1 Under hypothesis (H3.1), the large-scale delay system (3.1) is exponentially stable in $L^2(\Omega; C([-\tau, 0]; R^d))$ if and only if each of the isolated delay subsystems (3.2) is exponentially stable in $L^2(\Omega; C([-\tau, 0]; R^{d_i}))$.

Proof. "only if": Assume (3.1) is exponentially stable in $L^2(\Omega; C([-\tau, 0]; R^d))$. By definition, there exist two positive constants M and λ such that

$$E\|\hat{x}(t; t_o, \xi)\|_\tau^2 \leq ME\|\xi\|_\tau^2\, e^{-\lambda(t-t_o)} \quad \text{on } t \geq t_o$$

for all $t_o \geq 0$ and $\xi \in L^2(\Omega, \mathcal{F}_{t_o}; C([-\tau, 0]; R^d))$. Let us now fix any $t_o \geq 0$ and $\xi_i \in L^2(\Omega, \mathcal{F}_{t_o}; C([-\tau, 0]; R^{d_i})$. Set $\xi = (0, \xi_i, 0)$ and it is easy to see that

$$y_i(t; t_o, \xi_i) = x_i(t; t_o, \xi).$$

Therefore

$$E\|\hat{y}_i(t; t_o, \xi_i)\|_\tau^2 = E\|\hat{x}_i(t; t_o, \xi)\|_\tau^2 \leq E\|\hat{x}(t; t_o, \xi)\|_\tau^2$$

$$\leq ME\|\xi\|_\tau^2\, e^{-\lambda(t-t_o)} = ME\|\xi_i\|_\tau^2\, e^{-\lambda(t-t_o)}.$$

In other words, each of the isolated subsystems (3.2) is exponentially stable in $L^2(\Omega; C([-\tau, 0]; R^{d_i}))$.

"if": We give the proof for the case m = 2; the general case follows by induction. The hypotheses in this case are that both of the equations

$$dy_1(t) = f_1(\hat{y}_1, t)\,dt + \sum_{k=1}^{\infty} g_{k1}(\hat{y}_1, t)\,dw_k(t)$$

and

$$dy_2(t) = f_2(0, \hat{y}_2, t)\,dt + \sum_{k=1}^{\infty} g_{k2}(0, \hat{y}_2, t)\,dw_k(t)$$

are exponentially stable in $L^2(\Omega; C([-\tau, 0]; R^{d_1}))$ and $L^2(\Omega; C([-\tau, 0]; R^{d_2}))$ respectively. Hence one can find a pair of positive constants M and λ such that

$$E\|\hat{y}_i(t; t_o, \xi_i)\|_\tau^2 \le ME\|\xi_i\|_\tau^2\, e^{-\lambda(t-t_o)} \quad \text{on } t \ge t_o, \ i = 1, 2, \tag{3.3}$$

for all $t_o \ge 0$ and $\xi_i \in L^2(\Omega, \mathcal{F}_{t_o}; C([-\tau, 0]; R^{d_i}))$. We need to show that the equation

$$dx_1(t) = f_1(\hat{x}_1, t)\,dt + \sum_{k=1}^{\infty} g_{k1}(\hat{x}_1, t)\,dw_k(t),$$

$$dx_2(t) = f_2(\hat{x}_1, \hat{x}_2, t)\,dt + \sum_{k=1}^{\infty} g_{k2}(\hat{x}_1, \hat{x}_2, t)\,dw_k(t),$$

$$(\hat{x}_1(t_o), \hat{x}_2(t_o)) = \xi = (\xi_1, \xi_2)$$

is exponentially stable in $L^2(\Omega; C([-\tau, 0]; R^d))$. Note $x_1(t; t_o, \xi) = y_1(t; t_o, \xi_1)$. We therefore need only to show that there exist two positive constants M_1 and λ_1 such that

$$E\|\hat{x}_2(t; t_o, \xi)\|_\tau^2 \le M_1 E\|\xi\|_\tau^2\, e^{-\lambda_1(t-t_o)} \quad \text{on } t \ge t_o.$$

In fact we shall show that for any $\varepsilon \in (0, \lambda/2)$ there exists a $C > 0$ such that

$$E\|\hat{x}_2(t; t_o, \xi)\|_\tau^2 \le C\, E\|\xi\|_\tau^2\, e^{-(\lambda-2\varepsilon)(t-t_o)} \quad \text{on } t \ge t_o \tag{3.4}$$

for all $t_o \ge 0$ and $\xi \in L^2(\Omega, \mathcal{F}_{t_o}; C([-\tau, 0]; R^d))$. Fix such t_o and ξ arbitrarily and write $x(t; t_o, \xi) = x(t) = (x_1(t), x_2(t))$ simply. Let $\varepsilon \in (0, \lambda/2)$ and define

$$\delta = \tau \vee \frac{1}{\varepsilon}\log(2M),$$

$$\rho = \delta \vee \frac{1}{\varepsilon}\Big(8K\delta(2+\delta) - \log\big[\lambda/8KM(2+\delta)\big] \Big). \tag{3.5}$$

We first claim that for any $\sigma \ge \rho$,

$$E\|\hat{x}_2(t)\|_\tau^2 \le E\|\hat{x}_2(t_o+\sigma)\|_\tau^2 e^{-(\lambda-\varepsilon)\delta} + E\|\xi_1\|_\tau^2 e^{-(\lambda-\varepsilon)\sigma} \tag{3.6}$$

if $t_o+\sigma+\delta \le t \le t_o+\sigma+2\delta$. In fact, if $t_o+\sigma \le t \le t_o+\sigma+2\delta$, we define

$$y_2(t) = y_2(t; t_o+\sigma, \hat{x}_2(t_o+\sigma)).$$

Then by Doob's martingale inequality, Hölder's inequality, the standing hypothesis (H3.1) and inequality (3.3) we can derive that

$$E\Big(\sup_{t_o+\sigma\le s\le t} |x_2(s)-y_2(s)|^2\Big)$$

$$\le 2K(4+2\delta)\int_{t_o+\sigma}^{t} E\|\hat{x}_1(s)\|_\tau^2\,ds + 2K(4+2\delta)\int_{t_o+\sigma}^{t} E\|\hat{x}_2(s)-\hat{y}_2(s)\|_\tau^2\,ds$$

$$\le 4KM(2+\delta)E\|\xi_1\|_\tau^2 \int_{t_o+\sigma}^{t} e^{-\lambda(s-t_o)}\,ds$$

$$+\,4K(2+\delta)\int_{t_o+\sigma}^{t} E\Big(\sup_{t_o+\sigma\le s\le r}|x_2(s)-y_2(s)|^2\Big)\,dr$$

$$\le \big[4KM(2+\delta)/\lambda\big]\,E\|\xi_1\|_\tau^2\, e^{-\lambda\sigma}$$

$$+\,4K(2+\delta)\int_{t_o+\sigma}^{t} E\Big(\sup_{t_o+\sigma\le s\le r}|x_2(s)-y_2(s)|^2\Big)\,dr.$$

The well-known Gronwall inequality gives

$$E\Big(\sup_{t_o+\sigma\le s\le t}|x_2(s)-y_2(s)|^2\Big) \le \big[4KM(2+\delta)/\lambda\big]\,E\|\xi_1\|_\tau^2\, e^{-\lambda\sigma+4K(2+\delta)(t-t_o-\sigma)}.$$

In particular, we have

$$E\Big(\sup_{t_o+\sigma\le s\le t_o+\sigma+2\delta}|x_2(s)-y_2(s)|^2\Big)$$

$$\le \big[4KM(2+\delta)/\lambda\big]\,E\|\xi_1\|_\tau^2\, e^{-\lambda\sigma+8K\delta(2+\delta)} \le \frac{1}{2}E\|\xi_1\|_\tau^2\, e^{-(\lambda-\varepsilon)\sigma},$$

where (3.5) has been used. Hence, if $t_o+\sigma+\delta \le t \le t_o+\sigma+2\delta$,

$$E\|\hat{x}_2(t)\|_\tau^2 \le 2E\|\hat{y}_2(t)\|_\tau^2 + 2E\|\hat{x}_2(t)-\hat{y}_2(t)\|_\tau^2$$

$$\le 2ME\|\hat{x}_2(t_o+\sigma)\|_\tau^2 e^{-\lambda(t-t_o-\sigma)} + E\|\xi_1\|_\tau^2 e^{-(\lambda-\varepsilon)\sigma}$$

$$\le 2ME\|\hat{x}_2(t_o+\sigma)\|_\tau^2 e^{-\lambda\delta} + E\|\xi_1\|_\tau^2 e^{-(\lambda-\varepsilon)\sigma}$$

$$\le E\|\hat{x}_2(t_o+\sigma)\|_\tau^2 e^{-(\lambda-\varepsilon)\delta} + E\|\xi_1\|_\tau^2 e^{-(\lambda-\varepsilon)\sigma}$$

which is the required (3.6), where (3.5) has been used again. We now let $n = 1, 2, \dots$. Taking $\sigma = \rho + (n-1)\delta$ in (3.6) we obtain that, if $t_o + \rho + n\delta \le t \le t_o + \rho + (n+1)\delta$,

$$E\|\hat{x}_2(t)\|_\tau^2 \le E\|\hat{x}_2(t_o+\rho+(n-1)\delta)\|_\tau^2 e^{-(\lambda-\varepsilon)\delta} + E\|\xi_1\|_\tau^2 e^{-(\lambda-\varepsilon)(\rho+(n-1)\delta)},$$

and then by induction

$$E\|\hat{x}_2(t)\|_\tau^2 \le E\|\hat{x}_2(t_o+\rho)\|_\tau^2 e^{-(\lambda-\varepsilon)n\delta} + nE\|\xi_1\|_\tau^2 e^{-(\lambda-\varepsilon)(\rho+(n-1)\delta)}$$

$$\le \left(E\|x_2(t_o+\rho)\|_\tau^2 + \frac{1}{\varepsilon\delta e} E\|\xi_1\|_\tau^2 \right) e^{-(\lambda-2\varepsilon)n\delta}, \tag{3.7}$$

where we have used $\rho \ge \delta$ and an elementary inequality

$$n\, e^{-\varepsilon n\delta} \le \frac{1}{\varepsilon\delta e} \qquad \text{for all } n \ge 1.$$

On the other hand, it is easy to show that there exists a positive constant C_1 which depends only on K, ρ and δ such that

$$E\|\hat{x}_2(t)\|_\tau^2 \le C_1 E\|\xi\|_\tau^2 \qquad \text{on } t_o \le t \le t_o+\rho+\delta. \tag{3.8}$$

Therefore, if $t > t_o + \rho + \delta$, we choose n such that $t_o + \rho + n\delta \le t \le t_o + \rho + (n+1)\delta$, and then (3.7) and (3.8) gives

$$E\|\hat{x}_2(t)\|_\tau^2 \le \left(C_1 + \frac{1}{\varepsilon\delta e}\right) E\|\xi\|_\tau^2 e^{-(\lambda-2\varepsilon)n\delta}$$

$$\le \left(C_1 + \frac{1}{\varepsilon\delta e}\right) E\|\xi\|_\tau^2 e^{-(\lambda-2\varepsilon)(t-t_o-\rho-\delta)}$$

$$\le CE\|\xi\|_\tau^2 e^{-(\lambda-2\varepsilon)(t-t_o)}, \tag{3.9}$$

where

$$C = \left(C_1 + \frac{1}{\varepsilon\delta e}\right) e^{(\lambda-2\varepsilon)(\rho+\delta)}.$$

However, (3.9) also holds for $t_o \le t \le t_o + \rho + \delta$ obviously. In other words, (3.4) has been proved and the proof of the theorem is complete.

8.4. GENERAL LARGE-SCALE STOCHASTIC DELAY SYSTEMS

In this section we shall consider a general large-scale stochastic differential delay system

$$dx_i(t) = f_i(\hat{x}_1(t), ..., \hat{x}_m(t), t)\, dt$$

$$+ \sum_{k=1}^{\infty} g_{ki}(\hat{x}_1(t), ..., \hat{x}_m(t), t)\, dw_k(t) \qquad \text{on } t \geq t_o, \tag{4.1}$$

$$\hat{x}_i(t_o) = \xi_i, \qquad i = 1, ..., m.$$

Here:

$$x_i \in R^{d_i}, \quad m \geq 2, \quad \sum d_i = d,$$
$$f_i, g_{ki} : C([-\tau, 0]; R^d) \times R_+ \to R^{d_i},$$
$$\xi_i \in L^2(\Omega, \mathcal{F}_{t_o}; C([-\tau, 0]; R^{d_i})).$$

Set $x = (x_1, ..., x_m)$ and $\xi = (\xi_1, ..., \xi_m)$ as before. Assume the equation has a unique global solution which is denoted by $x(t; t_o, \xi)$. Assume also that $f_i(0, t) \equiv 0$ and $g_{ki}(0, t) \equiv 0$ for all i, k. Therefore the equation admits a trivial solution $x(t; t_o, 0) \equiv 0$. In order to study the exponential stability of this large-scale delay system, we introduce the corresponding isolated subsystems

$$dy_i(t) = f_i(0, \hat{y}_i(t), 0, t)\, dt + \sum_{k=1}^{\infty} g_{ki}(0, \hat{y}_i(t), 0, t)\, dw_k(t) \qquad \text{on } t \geq t_o, \tag{4.2}$$

$$\hat{y}_i(t_o) = \xi_i.$$

The solution of each subsystem is denoted by $y_i(t; t_o, \xi_i)$. In the previous section we proposed hypothesis (H3.1) under which the large-scale delay system in hierarchical form is exponentially stable in $L^2(\Omega; C([-\tau, 0]; R^d))$ if and only if each of the corresponding isolated delay systems is exponentially stable in $L^2(\Omega; C([-\tau, 0]; R^{d_i}))$. However for the general large-scale delay system (4.1) we have not yet obtained such a nice condition under which the stability of (4.1) is equivalent to the stability of its subsystems. Nevertheless we obtain a sufficient condition. So let us propose a standing hypothesis:

(H4.1) There exist two constants $K > 0$ and $\theta > 0$ such that

$$|f_i(u_1, ..., u_m, t) - f_i(0, v_i, 0, t)|^2 \vee \sum_{k=1}^{\infty} |g_{ki}(u_1, ..., u_m, t) - g_{ki}(0, v_i, 0, t)|^2$$

$$\leq K\|u_i - v_i\|_\tau^2 + \theta \sum_{j\neq i} \|u_j\|_\tau^2$$

for all $u_j, v_j \in C([-\tau, 0]; R^{d_j})$, $1 \leq j, i \leq m$.

Theorem 4.1 Let (H4.1) hold. Assume each of the isolated delay systems (4.2) is exponentially stable in $L^2(\Omega; C([-\tau, 0]; R^{d_i}))$, that is there exists a pair of positive constants M and λ such that

$$E\|\hat{y}_i(t; t_o, \xi_i)\|_\tau^2 \leq ME\|\xi_i\|_\tau^2 e^{-\lambda(t-t_o)} \quad \text{on } t \geq t_o \tag{4.3}$$

for all $t_o \geq 0$, $\xi_i \in L^2(\Omega, \mathcal{F}_{t_o}; C([-\tau, 0]; R^{d_i}))$, and $1 \leq i \leq m$. Then there exists a positive θ^*, which depends on K, M, λ only, such that if $\theta < \theta^*$ then the large-scale delay system (4.1) is exponentially stable in $L^2(\Omega; C([-\tau, 0]; R^d))$.

Proof. Write equation (4.1) as

$$dx_i(t) = \Big(f_i(0, \hat{x}_i(t), 0, t) + f_i(\hat{x}_1(t), ..., \hat{x}_m(t), t) - f_i(\hat{x}_1(t), ..., \hat{x}_m(t), t) \Big) dt$$

$$+ \sum_{k=1}^{\infty} \Big(g_{ki}(0, \hat{x}_i(t), 0, t) + g_{ki}(\hat{x}_1(t), ..., \hat{x}_m(t), t) - g_{ki}(0, \hat{x}_i(t), 0, t) \Big) dw_k(t),$$

or equivalently

$$dx(t) = \Big(f(\hat{x}(t), t) + F(\hat{x}(t), t)\Big) dt + \sum_{k=1}^{\infty} \Big(g_k(\hat{x}(t), t) + G_k(\hat{x}(t), t)\Big) dw_k(t), \tag{4.4}$$

where

$$f(\hat{x}, t) = (f_1(0, \hat{x}_1, 0, t), ..., f_m(0, \hat{x}_m, 0, t))^T,$$
$$g_k(\hat{x}, t) = (g_{k1}(0, \hat{x}_1, 0, t), ..., g_{km}(0, \hat{x}_m, 0, t))^T,$$
$$F(\hat{x}, t) = (f_1(\hat{x}, t) - f_1(0, \hat{x}_1, 0, t), \; ..., \; f_m(\hat{x}, t) - f_m(0, \hat{x}_m, 0, t))^T,$$
$$G_k(\hat{x}, t) = (g_{k1}(\hat{x}, t) - g_{k1}(0, \hat{x}_1, 0, t), \; ..., \; g_{km}(\hat{x}, t) - g_{km}(0, \hat{x}_m, 0, t))^T.$$

Now regard equation (4.4) as a stochastically perturbed equation of the following delay equation

$$dx(t) = f(\hat{x}(t), t)\, dt + \sum_{k=1}^{\infty} g_k(\hat{x}(t), t)\, dw_k(t). \tag{4.5}$$

Obviously equation (4.5) is equivalent to the isolated subsystems (4.2). So as assumed equation (4.5) is exponentially stable in $L^2(\Omega; C([-\tau, 0]; R^d))$. All we need to do is to show that its stochastically perturbed system (4.4) is also exponentially stable in

$L^2(\Omega; C([-\tau, 0]; R^d))$. This is can be done in a similar way as the proof of Theorem 6-5.1 but the details are left to the reader. The proof is complete.

The following useful corollary is now immediate.

Corollary 4.2 Assume for all $u_j, v_j \in C([-\tau, 0]; R^{d_j})$, $1 \le j, i \le m$,

$$|f_i(u_1, ..., u_m, t) - f_i(0, v_i, 0, t)|^2 \vee \sum_{k=1}^{\infty} |g_{ki}(u_1, ..., u_m, t) - g_{ki}(0, v_i, 0, t)|^2$$

$$\le K\|u_i - v_i\|_\tau^2 + \theta(t) \sum_{j \ne i} \|u_j\|_\tau^2,$$

where $K > 0$ and $\theta : R_+ \to R_+$ with property $\theta(t) \to 0$ as $t \to \infty$. If each of the isolated delay subsystems (4.2) is exponentially stable in $L^2(\Omega; C([-\tau, 0]; R^{d_i}))$, then the large-scale delay system (4.1) is exponentially stable in $L^2(\Omega; C([-\tau, 0]; R^d))$.

8.5. APPLICATIONS

In this section we shall apply the theory established in the previous sections as well as that in the previous chapters to study the exponential stability for some special but important large-scale stochastic delay systems.

Let us start with a linear large-scale stochastic delay system in hierarchical form: For $i = 1, 2, ..., m$,

$$dx_i(t) = A_i x_i(t)\,dt + \sum_{k=1}^{\infty} \sum_{j=1}^{i} \left(B_{kij} x_j(t) + C_{kij} x_j(t-\tau)\right) dw_k(t) \quad \text{on } t \ge t_o, \tag{5.1}$$

$$\hat{x}_i(t_o) = \xi_i \in L^2(\Omega, \mathcal{F}_{t_o}; C([-\tau, 0]; R^{d_i})),$$

where $A_i \in R^{d_i \times d_i}$, $B_{kij} \in R^{d_i \times d_j}$ and $C_{kij} \in R^{d_i \times d_j}$. Perhaps it should be pointed out that the theory developed below can obviously be extended to more general linear cases, for instance, a system with several constant time lags, but here we restrict ourselves to the above linear system just to make the theory more understandable and to reduce the complication of notations. Note the corresponding isolated subsystems are as follows:

$$dy_i(t) = A_i\, y_i(t)\,dt + \sum_{k=1}^{\infty} \left(B_{kii} y_i(t) + C_{kii} y_i(t-\tau)\right) dw_k(t) \quad \text{on } t \ge t_o, \tag{5.2}$$

$$\hat{y}_i(t_o) = \xi_i \in L^2(\Omega, \mathcal{F}_{t_o}; C([-\tau, 0]; R^{d_i})).$$

One can see from Theorem 3.1 that the linear delay system (5.1) is exponentially stable in $L^2(\Omega; C([-\tau, 0]; R^d))$ if and only if each of subsystems (5.2) is exponentially stable in $L^2(\Omega; C([-\tau, 0]; R^{d_i}))$. On the other hand, in view of Proposition 2.3, for such linear delay subsystems the exponential stability in $L^2(\Omega; C([-\tau, 0]; R^{d_i}))$ is equivalent to the exponential stability in mean square. We therefore can apply the results established in Chapter 6, e.g. Corollary 2.4, to obtain the following theorem.

Theorem 5.1 Assume that for each $i = 1, ..., m$, there exists a constant λ_i such that

$$2x_i^T A_i x_i \le -\lambda_i |x_i|^2 \quad \text{for all } x_i \in R^{d_i} \tag{5.3}$$

and

$$2\sum_{k=1}^{\infty}\left(|B_{kii}|^2 + |C_{kii}|^2 \right) < \lambda_i. \tag{5.4}$$

Then each of subsystems (5.2) is exponentially stable in $L^2(\Omega; C([-\tau, 0]; R^{d_i}))$, and therefore the linear large-scale delay system (5.1) is also exponentially stable in $L^2(\Omega; C([-\tau, 0]; R^d))$.

We now observe that each of the isolated subsystems (5.2) can be regarded as a stochastically perturbed system of the corresponding linear ordinary differential equation

$$\dot{y}_i(t) = A_i\, y_i(t), \tag{5.5}$$

and condition (5.3) ensures equation (5.5) is exponentially stable. Theorem 5.1 shows that the stochastically perturbed system (5.2) remains exponentially stable provided the stochastic perturbation is small enough; the perturbation is measured by (5.4) and is independent of τ. Let us now consider a special case of system (5.1) with $C_{kii} = -B_{kii}$ but the norms of B_{kii} may be so large that (5.4) is no longer satisfied and hence the result above can not be used. However, note that the isolated subsystems now become

$$dy_i(t) = A_i\, y_i(t)\, dt + \sum_{k=1}^{\infty} B_{kii}(y_i(t) - y_i(t-\tau))\, dw_k(t) \quad \text{on } t \ge t_o, \tag{5.6}$$

$$\hat{y}_i(t_o) = \xi_i \in L^2(\Omega, \mathcal{F}_{t_o}; C([-\tau, 0]; R^{d_i})).$$

and observe the stochastic perturbation is dependent of the difference $y_i(t) - y_i(t-\tau)$ which can be nearly zero if τ is very small. The exponential stability of such delay systems has been studied in Chapter 7, where it has been showed that if τ is small enough the stochastic perturbation could be so small that it could absorbed by the stable system (5.5) so that system (5.6) remains stable, that is system (5.6) is still exponentially stable provided τ small enough (see Theorem 7-2.1). So the following theorem is now immediate.

Theorem 5.2 Let $C_{kii} = -B_{iiq}$ for all $1 \le i \le m$, $k \ge 1$ in the large-scale stochastic delay system (5.1). Assume for each i = 1, 2, ..., m, there exist two positive constants β_i and λ_i such that

$$|e^{A_i t}|^2 \le \beta_i e^{-\lambda_i t} \quad \text{for all } t \ge 0. \tag{5.7}$$

If τ is so small that

$$3\beta_i\left(e^{-\lambda_i \tau} |I - e^{-A_i \tau}|^2 + \lambda_i \tau \right) \sum_{k=1}^{\infty} |B_{kii}|^2 < \lambda_i, \quad 1 \le i \le m, \tag{5.8}$$

where I is the identity matrix, then the large-scale delay system (5.1) is exponentially stable in $L^2(\Omega; C([-\tau, 0]; R^d))$.

Let us now turn to non-linear large-scale stochastic delay systems. First consider a system in hierarchical form: For i = 1, 2, ..., m

$$dx_i(t) = F_i(x_1(t), x_1(t-\tau), ..., x_i(t), x_i(t-\tau), t)\, dt$$

$$+ \sum_{k=1}^{\infty} G_{ki}(x_1(t), x_1(t-\tau), ..., x_i(t), x_i(t-\tau), t)\, dw_k(t) \quad \text{on } t \ge t_o, \tag{5.9}$$

$$\hat{x}_i(t_o) = \xi_i \in L^2(\Omega, \mathcal{F}_{t_o}; C([-\tau, 0]; R^{d_i})),$$

where F_i, G_{ki} are all Borel measurable functions from $R^{2(d_1+...+d_i)} \times R_+ \to R^{d_i}$ such that $F_i(0, t) = G_{ki}(0, t) \equiv 0$. The corresponding isolated subsystems are:

$$dy_i(t) = F_i(0, x_i(t), x_i(t-\tau), t)\, dt$$

$$+ \sum_{k=1}^{\infty} G_{ki}(0, x_i(t), x_i(t-\tau), t)\, dw_k(t) \quad \text{on } t \ge t_o, \tag{5.10}$$

$$\hat{y}_i(t_o) = \xi_i.$$

Theorem 5.3 Assume for each i = 1, 2, ..., m, there exist positive constants $c_{i1} - c_{i4}$ and K such that

(i) $2x_i^T F_i(0, x_i, 0, t) \le -c_{i1}|x_i|^2$,

(ii) $|F_i(x_1, \bar{x}_1, \dots, x_i, \bar{x}_i, t) - F_i(0, y_i, \bar{y}_i, t)|^2$

$$\leq K \sum_{j<i} (|x_j|^2 + |\bar{x}_j|^2) + K|x_i - y_i|^2 + c_{i2}|\bar{x}_i - \bar{y}_i|^2,$$

(iii) $\sum_{k=1}^{\infty} |G_{ki}(x_1, \bar{x}_1, \dots, x_i, \bar{x}_i, t) - G_{ki}(0, y_i, \bar{y}_i, t)|^2$

$$\leq K \sum_{j<i} (|x_j|^2 + |\bar{x}_j|^2) + c_{i3}|x_i - y_i|^2 + c_{i4}|\bar{x}_i - \bar{y}_i|^2,$$

for all $x_j, \bar{x}_j, y_j, \bar{y}_j \in R^{d_j}$, $1 \leq j \leq i$. If

$$c_{i1} > 2\sqrt{c_{i2}} + c_{i3} + c_{i4}, \qquad 1 \leq i \leq m,$$

then the large-scale system (5.9) is exponentially stable in $L^2(\Omega; C([-\tau, 0]; R^d))$.

Proof. Define

$$f_i(u_1, \dots, u_i, t) = F_i(u_1(-\tau), u_1(0), \dots, u_i(-\tau), u_i(0), t),$$
$$g_{ki}(u_1, \dots, u_i, t) = G_{ki}(u_1(-\tau), u_1(0), \dots, u_i(-\tau), u_i(0), t)$$

for all $u_j, v_j \in C([-\tau, 0]; R^{d_j})$, $1 \leq j \leq i \leq m$, $k \geq 1$. Then equation (5.9) can be written as equation (3.1). Obviously the standing hypothesis (H3.1) is satisfied so in view of Theorem 3.1, to show the large-scale system (5.9) is exponentially stable in $L^2(\Omega; C([-\tau, 0]; R^d))$ we need only to check each of the isolated subsystems (5.10) is exponentially stable in $L^2(\Omega; C([-\tau, 0]; R^{d_i}))$ or equivalently in mean square (see Proposition 2.3). But we see from conditions (ii) and (iii) that for each $i = 1, 2, \dots, m$,

$$|F_i(0, x_i, \bar{x}_i, t) - F_i(0, x_i, 0, t)| \leq \sqrt{c_{i2}}\, |\bar{x}_i|,$$

and

$$\sum_{k=1}^{\infty} |G_{ki}(0, x_i, \bar{x}_i, t)|^2 \leq c_{i3}|x_i|^2 + c_{i4}|\bar{x}_i|^2,$$

for all $x_j, \bar{x}_j \in R^{d_j}$, $1 \leq j \leq i$. By Corollary 2.4, each of the isolated subsystems (5.10) is exponentially stable in mean square indeed. The proof is complete.

We finally consider a more general large-scale delay system of the form: For $i = 1, 2, \dots, m$,

$$dx_i(t) = F_i(x_1(t), x_1(t-\tau), \dots, x_m(t), x_m(t-\tau), t)\, dt$$

$$+\sum_{k=1}^{\infty} G_{ki}(x_1(t), x_1(t-\tau), \ldots, x_m(t), x_m(t-\tau), t)\, dw_k(t) \qquad \text{on } t \geq t_o, \tag{5.11}$$

$$\hat{x}_i(t_o) = \xi_i \in L^2(\Omega, \mathcal{F}_{t_o}; C([-\tau, 0]; R^{d_i})),$$

where F_i, G_{ki} are all Borel measurable functions from $R^{2d} \times R_+ \to R^{d_i}$ such that $F_i(0, t) = G_{ki}(0, t) \equiv 0$. The corresponding isolated subsystems are:

$$dy_i(t) = F_i(0, x_i(t), x_i(t-\tau), 0, t)\, dt$$

$$+\sum_{k=1}^{\infty} G_{ki}(0, x_i(t), x_i(t-\tau), 0, t)\, dw_k(t) \qquad \text{on } t \geq t_o, \tag{5.12}$$

$$\hat{y}_i(t_o) = \xi_i.$$

It is easy to see the following useful result holds.

Theorem 5.4 Let $K > 0$ and $\theta : R_+ \to R_+$ be a Borel measurable function such that $\theta(t) \to 0$ as $t \to 0$. Assume for each $i = 1, 2, \ldots, m$, there exist positive constants $c_{i1} - c_{i4}$ such that

(i) $2x_i^T F_i(0, x_i, 0, t) \leq -c_{i1}|x_i|^2$,

(ii) $|F_i(x_1, \bar{x}_1, \ldots, x_m, \bar{x}_m, t) - F_i(0, y_i, \bar{y}_i, 0, t)|^2$

$$\leq \theta(t) \sum_{j \neq i} (|x_j|^2 + |\bar{x}_j|^2) + K|x_i - y_i|^2 + c_{i2}|\bar{x}_i - \bar{y}_i|^2,$$

(iii) $\sum_{k=1}^{\infty} |G_{ki}(x_1, \bar{x}_1, \ldots, x_m, \bar{x}_m, t) - G_{ki}(0, y_i, \bar{y}_i, 0, t)|^2$

$$\leq \theta(t) \sum_{j \neq i} (|x_j|^2 + |\bar{x}_j|^2) + c_{i3}|x_i - y_i|^2 + c_{i4}|\bar{x}_i - \bar{y}_i|^2,$$

for all $x_j, \bar{x}_j, y_j, \bar{y}_j \in R^{d_j}$, $1 \leq j \leq m$. If

$$c_{i1} > 2\sqrt{c_{i2}} + c_{i3} + c_{i4}, \qquad 1 \leq i \leq m,$$

then the large-scale system (5.11) is exponentially stable in $L^2(\Omega; C([-\tau, 0]; R^d))$.

References

Akinyele, O. (1984), On Gronwall-Bellman-Bihari-type integral inequalities in several variables with retardation, J. Math. Anal. Appl. 104, pp1-26.

Altman, M. (1973), Inverse differentiability, contractors and equations in Banach spaces, Studia Math. 46, pp1-15.

Altman, M. (1977), Contractors and Contractor Directions, Theory and Applications, Dekker, New York.

Anderson, W. J. (1972), Local behaviour of solutions of stochastic integral equations, Tran. Amer. Math. Soci. 164, pp309-321.

Arnold, L. (1974), Stochastic Differential Equations: Theory and Applications, John Wiley and Sons.

Arnold, L. and Crauel, H. (1991), Random dynamical system, Lecture Notes in Mathematics 1486, pp1-22.

Arnold, L. and Kliemann, W. (1987), On unique ergodicity for degenerate diffusions, Stochastics 21, pp41-61.

Arnold, L., Oeljeklaus, E. and Pardoux, E. (1984), Almost sure and moment stability for linear Itô equations, Lecture Note in Math. 1186, pp129-159.

Baxendale, P. (1984), Brownian motions in the diffeomorphism group I, Compositio Mathematica 53, pp19-50.

Beckenbach, E. F. and Bellman, R. (1961), Inequalities, Springer-Verlag, Berlin and Heidelberg.

Bell, D. R. and Mohammed, S. E. A. (1989), On the solution of stochastic differential equations via small delays, Stochastics and Stocahstics Reports 29, pp293-299.

Bellman, R. (1953), Stability Theory of Differential Equations, McGraw-Hill, New York.

Bellman, R. and Cooke, K. L. (1963), Differential-Difference Equations, Academic Press, New York.

Bertram, J. E. and Sarachik, P. E. (1959), On the stability of systems with random parameters, Trans. IRE-PGCT 5.

Bihari, I. (1957), A generalization of a lemma of Bellman and its application to uniqueness problem of differential equations, Acta Math. Acad. Sci. Hungar. 7, pp71-94.

Bondge, B. K. and Pachpatte, B. G. (1979a), On Wendroff type integral inequalities in n independent variables, Chinese J. Math. 7, pp37-46.

Bondge, B. K. and Pachpatte, B. G. (1979b), On nonlinear integral inequalities of the Wendroff type, J. Math. Anal. Appl. 70, pp161-169.

Bucy, H. J. (1965), Stability and positive supermartingales, J. Differential Equation 1, pp151-155.

Carmona, R. and Nulart, D. (1990), Nonlinear Stochastic Integrators, Equations and Flows, Stochastics Monographs, London New York, Gordon and Breach.

Carverhill, A. (1985), Flows of stochastic dynamical systems: ergodic theory, Stochastics 14, pp273-317.

Chappell, M. (1984), Bounds for average Lyapunov exponents of gradient stochastic systems, Lecture Notes Math. 1186, pp292-307.

Coddington, R. F. and Levinson, N. (1955), Theory of Ordinary Differential Equations, McGraw-Hill, New York, Toronto, London.

Corduneanu, A. (1988), Nonlinear integral inequalities in n-independent variables, An. Strint. Univ. Al. I. Giza Iasi Sect. I a Mat. (N. S.) 31(3), pp281-288.

Courrège, P. (1963), Intégeales stochastiques et martingales de carré integrable, Sem. Th. Potentials 7, Inst. Henri Poincaré, Paris.

Crauel, H. (1984), Lyapunov numbers of Markov solutions of linear stochastic systems, Stochastics 14, pp11-28.

Curtain, R. F. (ed.) (1972), Stability of Stochastic Dynamical Systems, Lecture Notes Math. 294, Berlin-Heidelberg-New York, Springer.

Curtain, R. F. and Pritchard, A. J. (1978), Infinite Dimensional Linear System Theory,

Lecture Notes in Control and Information Sciences 8, Springer-Verlag, Berlin.

Dannan, F. M. (1985), Integral inequalities of Gronwall-Bellman-Bihari type and asymptotic behavior of certain second order nonlinear differential equations, J. Math. Anal. Appl. 108, pp151-164.

Da Prato, G. and J. Zabczyk (1992), Stochastic Equations in Infinite Dimensions, Cambridge University Press.

Dellacherie, C. and Meyer, P. A. (1980), Probabilités et Potentiels, 2e édition, Chapitres V-VIII, Hermann.

Doléans-Dade, C. and Meyer, P. A. (1977), Equations Différentilles Stochastiques, Sém. Probab. XI, Lect. Notes in Math. 581.

Doob, J. L. (1953), Stochastic Processes, John Wiley, New York.

Dubins, L. E. and Schwartz, G. (1965), On continuous martingales, Proc. Nat. Acad. Sci. USA 53, pp913-916.

Dynkin, E. B. (1965), Markov Processes, Vol. I, Berlin-Heideberg-New York.

Elliott, R. J. (1982), Stochastic Calculus and Applications, Springer-Verlag.

Elworthy, K. D. (1982), Stochastic Differential Equations on Manifolds, London Math. Society, Lect. Notes Series 70, C. U. P.

Emery, E. (1978), Stabilité des solutions des équations différentielles stochastiques, application aux intégrales multiplicatives stochastiques, Z. W. Verw. Geb. 41(3).

Fisk, D. L. (1963), Quasi-martingales and stochastic integrals, Tech. Reports 1, Dept. Math. Michigan State Univ.

Friedman, A. (1975), Stochastic Differential Equations and Applications, New York, Vol. 1 and Vol. 2.

Gal'cuk, L. I. and Davis, M. H. A. (1982), A note on a comparison theorem for equations with different diffusions, Stochastics 6, pp147-149.

Halmos, P. R. (1974), Measure Theory, Springer-Verlag, New York, 1974.

Has'minskii, R. Z. (1962), On the stability of the trajectory of Markov processes, J. Appl. Math. and Mech. (PMM) 26, pp1554-1565.

Has'minskii, R. Z. (1967), Necessary and sufficient conditions for the asymptotic stability of linear stochastic systems, Theory Probability Appl. 12, pp144-147.

Has'minskii, R. Z. (1969), Stochastic Stability of Differential Equations, Alphen: Sijtjoff and Noordhoff 1980 (translation of the Russian edition, Moscow, Nauka 1969).

Ichikawa, A. (1982), Stability of semilinear stochastic evolution equations, J. Math. Anal. Appl. 90, pp12-44.

Ikeda, N. and Watanabe, S. (1977), A comparison theorem for solutions of stochastic differential equations and its applications, Osaka J. Math. 14, pp619-633.

Itô, K. (1942), Differential equations determining Markov processes, Zenkoku Shijo Suguku Danwakai 1077, pp1352-1400.

Jacod, J. (1979), Calcul Stochastique et Problèmes de Martingales, Lect. Notes in

Math. 714, Springer-Verlag.
Kaneko, T. and Nakao, S. (1988), A note on approximation for stochastic differential equations, Séminaire de Probabilités 22, Lect. Notes Math. 1321, pp525-533.
Karatzas, I. and Shreve, S. E. (1988), Brownian Motion and Stochastic Calculus, Springer-Verlag.
Kolmogorov, A. (1931), Uber analytische methoden in der wahrcheinlichkeitsrechnung, Math. Ann. 104, pp415-458.
Kunita, H. (1986), Lectures on Stochastic Flows and Applications, Tata Institute of Fundamental Research, Springer-Verlag.
Kunita, H. (1990), Stochastic Flows and Stochastic Differential Equations, Cambridge University Press.
Kunita, H. and Watanabe, S. (1976), On square integrable martingales, Negoya Math. J. 30, pp209-245.
Kuo, H.-H. (1979), On integral contractors, J. Integral Equations 1, pp35-46.
Kushner, H. J. (1965), On the construction of stochastic Lyapunov functions, IEEE Trans. Automatic Control AC-10, pp477-478.
Kushner, H. J. (1967), Stochastic Stability and Control, Academic Press.
Kushner, H. J. (1968), On the stability of processes defined by stochastic differential-difference equations, J. Differential Equations 4, pp424-443.
Ladde, G. S. (1974), Differential inequalities and stochastic functional differential equations, J. Math. Phys. 15, pp738-743.
Ladde, G. S. (1981), Stochastic stability analysis of model ecosystems with time-delay, in "Differential Equations and Applications in Biology, Epidemics, and Population Problems" edited by S. N. Busenberg and K. Cook, Academic Press, pp215-228.
Ladde, G. S. and Lakshmikantham, V. (1980), Random Differential Inequalities, Academic Press.
Lakshmikantham, V. (1973), A variation of constants formula and Bellman-Gronwall-Reid inequalities, J. Math. Anal. Appl. 41, pp199-204.
Lakshmikantham, V., Leeda, S. and Martynyuk, A. A. (1989), Stability Analysis of Nonlinear Systems, Marcel Dekker, Inc.
Lee, A. C. H. and Padgett, W. J. (1977a), Random contractors and the solution of random nonlinear equations, Nonlinear Analysis: Theory, Methods, Appl. 1, pp175-185.
Lee, A. C. H. and Padgett, W. J. (1977b), On random nonlinear contractors, Math. System Theory 11, pp77-84.
Le Jan, Y. (1982), Flots de diffusions dans R^d, C. R. Acad. Sci., Paris, Ser. I 294 (1982), pp697-699.
Le Jan, Y. and Watanabe, S. (1984), Stochastic flows of diffeomorphisms, Stochastic Analysis, Proceedings of the Taniguchi Conference, Kyoto 1982, Amsterdam

New York Oxford, North Holland.

Li, C. W. and Mao, X. R. (1986a), Delay stochastic differential equations, J. East China Inst. Chem. Tech. 6, pp779-792.

Li, C. W. and Mao, X. R. (1986b), Global asymptotic stability of solutions of stochastic differential equations, J. East China Inst. Chem. Tech. 4, pp531-536.

Li, C. W. and Mao, X. R. (1987), Stochastic boundedness of solutions of Itô's equations, J. East China Inst. Chem. Tech.3, pp383-388.

Lyapunov, A. M. (1892), Probleme general de la stabilite du mouvement, Comm. Soc. Math. Kharkov 2, pp265-272

Mao, X. R. (1989a), Lebesgue-Stieltjes integral inequalities and stochastic stabilities, Quarterly J. Math. Oxford (2), 40, pp301-311.

Mao, X. R. (1989b), Existence and uniqueness of the solutions of delay stochastic integral equations, Stochastic Analysis and Applications 7(1), pp59-74.

Mao, X. R. (1989c), Lebesgue-Stieltjes integral inequalities of the Wendroof type in n-independent variables, Chinese J. Math. 37(1), pp29-50.

Mao, X. R. (1989d), Exponential stability of delay Itô's equations, Proceedings of ICCON'89 IEEE International Conference on Control and Applications, Israel, RA-6-5, 1-5.

Mao, X. R. (1989e), Lebesgue-Stieltjes integral inequalities of the Greene Type in n-independent variables, Chinese J. Math. 17(4), pp295-305.

Mao, X. R. (1989f), Exponential stability for nonlinear stochastic differential equations with respect to semimartingales, Stochastics and Stochastics Reports 28, pp343-355.

Mao, X. R. (1990a), Eventual asymptotic stability for stochastic differential equations with respect to semimartingales, Quarterly J. Math. Oxford (2), 41, pp71-77.

Mao, X. R. (1990b), Exponential stability in mean square for stochastic differential equations, Stochastic Analysis and Applications 8(1), pp91-103.

Mao, X. R. (1990c), Lyapunov functions and almost sure exponential stability of stochastic differential equations based on semimartingales with spatial parameters, SIAM Journal on Control and Optimization 28(6), pp1481-1490.

Mao, X. R. (1990d), Exponential stability for stochastic differential equations with respect to semimartingales, Stochastic Processes and Their Applications 35, pp267-277.

Mao, X. R. (1990e), A note on global solution to stochastic differential equations based on semimartingales with spatial parameters, Journal of Theoretical Probability 4(1), pp161-167.

Mao, X. R. (1990f), Lebesgue-Stieltjes integral inequalities in several variables with retardation, Proceedings Indian Academy Sciences (Math. Sciences) 100(3), pp231-243.

Mao, X. R. (1991a), Almost sure exponential stability for delay stochastic differential equations with respect to semimartingales, Stochastic Analysis and Applications 9(2), pp177-194.

Mao, X. R. (1991b), Exponential stability for stochastic differential delay equations in Hilbert spaces, Quarterly Journal of Math. Oxford (2), 42, pp77-85.21

Mao, X. R. (1991c), A note on comparison theorems for stochastic differential equations with respect to semimartingales, Stochastics and Stochastics Reports 37, pp49-59.

Mao, X. R. (1991d), Eventual uniform asymptotic stability for stochastic differential equations based on semimartingales with spatial parameters, Journal of Theoretical Probability 4(2), pp441-456.

Mao, X. R. (1991e), Approximate solutions for a class of delay stochastic differential equations, Stochastics and Stochastics Reports 35, pp111-123.

Mao, X. R. (1991f), Stability of Stochastic Differential Equations with Respect to Semimartingales, Pitman Research Notes in Mathematics Series 251, Longman Scientific and Technical.

Mao, X. R. (1991g), Lyapunov functions and almost sure exponential stability, Proceedings of Lyapunov Exponents-Oberwolfach 1990, Lect. Notes Math. 1486, pp165-177.

Mao, X. R. (1991h), Approximate solutions for a class of stochastic evolution equations with variable delays, Numer. Funct. Anal and Optimiz. 12(5&6), pp525-533.

Mao, X. R. (1992a), Polynomial stability for perturbed stochastic differential equations with respect to semimartingales, Stochastic Processes and Their Applications 41 pp101-116.

Mao, X. R. (1992b), Almost sure polynomial stability for a class of stochastic differential equations, Quarterly J. Math. Oxford (2) 43, pp339-348.

Mao, X. R. (1992c), Lyapunov functions and almost sure polynomial stability, in "Stochastics and Quantum Mechanics" edited by A. Truman and I. M. Davis, World Scientific, pp189-204.

Mao, X. R. (1992d), Almost sure asymptotic bounds for a class of stochastic differential equations, Stochastics and Stochastics Reports 41, pp57-69.

Mao, X. R. (1992e), Solutions of stochastic differential-functional equations via bounded stochastic integral contractors, J. Theoretical Probability 5(3), pp487-502.

Mao, X. R. (1992f), Exponential stability of large-scale stochastic differential equations, Systems and Control Letters 19, pp71-81.

Mao, X. R. (1992g), Robustness of stability of nonlinear systems with stochastic delay perturbations, Systems and Control Letters 19, pp391-400.

Mao, X. R. (1993a), Almost sure exponential stability for a class of stochastic differential equations with applications to stochastic flows, Stochastic Analysis and Applications 11(1), pp77-95.

Mao, X. R. (1993b), Necessary and sufficient condition for exponential stability of large-scale stochastic delay systems in hierarchical form, in "Differential Equations, Dynamical Systems, and Control Science" edited by Elworthy et al., Marcel Dekker, Inc.

Mao, X. R. (1994a), Stochastic stabilization and destabilization, to appear in Systems and Control Letters.

Mao, X. R. (1994b), Approximate solutions of stochastic differential equations with pathwise uniqueness, to appear in Stochastic Analysis and Applications.

Mao, X. R. and Markus, L. (1991), Energy bounds for nonlinear dissipative stochastic differential equations with respect to semimartingales, Stochastics and Stochastics Reports 37, pp1-14.

Mao, X. R. and Markus, L. (1993), Wave equations with stochastic boundary conditions, J. Math. Anal. Appl. 177 (2) (1993), 315-341.

Markus, L. (1986), Global Lorentz geometry and relativistic Brownian motion, in "From Local Times to Global Geometry, Control and Physics" Edited by K. D. Elworthy, Pitman Research Notes in Mathematics 150, pp273-286.

Markus, L. and Weerasinghe, A. (1988), Stochastic oscillators, J. Differential Equations 71(2), pp288-314.

McKean, H. P. (1969), Stochastic Integrals, Academic Press, 1969.

Mcshane, E. J. (1974), Stochastic Calculus and Stochastic Models, Academic Press.

Métivier, M. (1982), Semimartingales, Walter de Gruyter, Berlin-New York.

Meyer, P. A. (1967), Integrales stochastiques, I-IV Semi. de Prob., Lect. Notes in Math. 39, pp72-162.

Meyer, P. A. (1972), Martingales and stochastic integrals I, Lect. Notes in Math. 284, Springer.

Michel, A. N., Miller, R. K. and Tang, W. (1978), Lyapunov stability of interconnected systems: decomposition into strongly connected subsystems, IEEE Trans. Circuits Syst., CAS-25, 799-809.

Mitrinović, D. S. (1970), Analytic Inequalities, Springer-Verlag.

Mizel, V. J. and Trutzer, V. (1984), Stochastic hereditary equations: existence and asymptotic stability, J. Integral Equations 7, pp1-72.

Mohammed, S-E. A. (1984), Stochastic Functional Differential Equations, Longman Scientific and Technical.

Mohammed, S-E. A. (1986), Stability of linear delay equations under a small noise, Proceedings of Edinburgh Mathematical Society 29, pp233-254.

Mohammed, S-E. A. (1990), The Lyapunov spectrum and stable manifolds for stochastic linear delay equations, Stochastics and Stochastics Reports 29, pp89-131.

Mohammed, S-E. A., Scheutzow, M. and Weizsächer, H. V. (1986), Hyperbolic state space decomposition for a linear stochastic delay equations, SIAM J. on Control and Optimization 24(3), pp543-551.

Natanson, I. P. (1964), Theory of Functions of Real Variables, Vol. 1, Frederick Unger Pub. Co., New York.

Nualart, D. and Pardoux, E. (1987), Stochastic calculus with anticipating integrands, Probab. Theory and Rel. Fields 78, pp535-581.

Nualart, D. and Zakai, M. (1986), Generalized stochastic integrals and the Malliavin calculus, Probab. Theory and Rel. Fields 73(1986), pp225-280.

Oseledec, V. I. (1968), A multiplicative ergodic theorem: Laypunov characteristic numbers for dynamical systems, Trans. Moscow Math. Soc. 19, 197-231.

Pachpatte, B. G. (1979), On Wendroff type partial integral inequalities, Tamkang J. Math. 10, pp141-150.

Padgett, W. J. (1983), The method of random contractors and its applications to random nonlinear equations, Prob. Anal. and Related Topics (ed. A. T. Bharucha-Reid), Vol.3, Academic Press, New York, 1983.

Padgett, W. J. and Rao, A. N. V. (1979), Solution of a stochastic integral equation using integral contractors, Inform. and Control 41, pp56-66.

Pardoux, E. and Wihstutz, V. (1988), Lyapunov exponent and rotation number of two-dimensional stochastic systems with small diffusion, SIAM J. Applied Math. 48, pp442-457.

Pinsky, M. A. and Wihstutz, V. (1988), Lyapunov exponents of nilponent Itô systems, Stochastics 25, pp43-57.

Raghavendra, V. and Rao, M. R. M. (1974), On the stability of differential systems with respect to impulsive perturbations, J. Math. Anal. Appl. 48, pp515-526.

Rao, A. N. V. and Padgett, W. J. (1982), On the solution of a class of stochastic integral systems, J. Integral Equations 4, pp145-162.

Rao, A, N. V. and Tsokos, C. P. (1975), On the existence, uniqueness, and stability behavior of a random solution to a nonlinear perturbed stochastic integro-differential equation, Inform. and Control 27, pp61-74.

Rogers, L. C. G. and Williams, D. (1987), Diffusions, Markov Processes and Martingales, Vol.2, John Wiley and Sons.

Sagirow, P. (1970), Stochastic Methods in the Dynamics of Satellites, Lecture Notes, Udline, CISM.

Senitman, A. S. (1982), Martingales dépendant d'un paramètre: une formule d'Itô, Z. W. Verw. Geb. 68, pp41-70.

Siljak, D. D. (1977), Large-Scale Dynamic Systems: Stability and Structure, Amsterdam, the Netherlands: North-Holland.

Skorokhod, A. V. (1965), Studies in the Theory of Random Processes, Addison-Wesley.

Sontag, E. D. (1980), Further facts about input to state stabilization, IEEE Trans. Automat. Contr., AC-35(4), 473-476.

Tanaka, H. and Hasegawa, M. (1964), Stochastic differential equations, Vol. 19 of

Seminar on Probability (in Japanese).

Thandapani, E. and Agarwal, R. P. (1982), On some new inequalities in n-independent variables, J. Math. Anal. Appl. 86, pp542-561.

Truman, A. (1986), An introduction to the stochastic mechanics of stationary states with applications, in "From Local Times to Global Geometry, Control and Physics" edited by K. D. Elworthy, Pitman Research Notes in Math. 150, pp329-344.

Tsokos, C. P. and Padgett, W. J. (1974), Random Integral Equations with Applications to Life Sciences and Engineering, Academic Press.

Vidyasagar, M. (1980), Decomposition techniques for large-scale systems with non-additive interactions: stability and stabilizablity, IEEE Trans. Automat. Contr., AC-25(4), pp773-779.

Wang, Z. K. (1986), Theory of Stochastic Processes, China Science Press.

Watanabe, S. and Yamada, T. (1971), On the uniqueness of solutions of stochastic differential equations II, J. Math. Kyoto University 11, pp553-563.

Wu, R. Q. and Mao, X. R. (1983), Existence and uniqueness of the solutions of stochastic differential equations, Stochastics 11, pp19-32.

Wu, R. Q. and Mao, X. R.(1986), On Comparison theorems for a kind of integral equations with respect to semimartingales, J. Engineering Math. 3(1), pp1-10.

Wu, R. Q. and Mao, X. R. (1988), A kind of stochastic differential games, Mathematical Statistics and Applied Probability 3(1), pp99-111.

Yamada, T. (1981), On the successive approximation of solutions of stochastic differential equations, J. Math. Kyoto University 21, pp501-515.

Yamada, T. and Watanabe, S. (1971), On the uniqueness of solutions of stochastic differential equations, J. Math. Kyoto University 11, pp155-167.

Yan, J. A. (1981), Introduction to Martingales and Stochastic Integrals, Shanghai Science and Technology Press.

Yeh, C. C. (1980), Wendroff's inequality in n independent variables, J. Math. Anal. Appl. 78 (1980), pp78-87.

Yeh, C. C. (1982a), On some integral inequalities in n-independent variables and their applications, J. Math. Anal. Appl. 86, pp387-410.

Yeh, C. C. (1982b), Bellman-Bihari inequality in several variables, J. Math. Anal. Appl. 87, pp311-32.

Yeh, C. C. (1985a), Discrete inequalities of the Gronwall-Bellman type in n-independent variables, J. Math. Anal. Appl. 105, pp322-332.

Yeh, C. C. (1985b), Discrete inequalities of the Gronwall-Bellman type in n-independent variables, II, J. Math. Anal. Appl. 106, pp282-285.

Yeh, C. C. and Shih, M. H. (1982), The Gronwall-Bellman inequality in several variables, J. Math. Anal. Appl. 86, pp157-167.

Young, E. C. (1973), Gronwall's inequality in n-independent variables, Proc. Amer.

Math. Soc. 41, pp241-144.
Zabczyk, J. (1979), On stability of infinite dimensional stochastic systems, Probability Theory, Z. Ciesislski (Ed.), Banach Center Publications, 5, Warswa, pp273-281.
Zabczyk, J. (1985), Linear stochastic systems in Hilbert spaces: structural properties and limit bahaviour, Banach Center Publications, 14, Warswa.
Zahariev, A. I. and Bainov, D. D. (1981), A note on Bellman-Gronwall's inequality, J. Math. Anal. Appl. 80, pp147-149.

Index

Adapted process 2
Almost surely asymptotic
  bound 76, 136

Bihari's inequality 25
  Generalized ~ 25
  Stochastic ~ 31
Bounded stochastic integral
  contractor 104, 115
Brownian flow 89
Brownian motion 7
Burkholder-Davis-Gundy's
  inequality 7

Cadlag process 2
Carathéodory's approximation 124
Cauchy-Maruyama's
  approximation 54, 55, 117
Class $B^{m,\delta}$ 14, 17, 34
Class $(B^{m,\delta}, B^{m',\delta'})$ 17, 34
Class $\tilde{C}^m$ 13
Class $\tilde{C}^{m,\delta}$ 13
Continuous
  local martingale part 6, 7
  martingale part 5
  process 2
C-Brownian motion 89
C^m-local martingale 13
C^m-process 12
C^m-semimartingale 16

$C^{m,\delta}$-local martingale 13
$C^{m,\delta}$-process 12
$C^{m,\delta}$-semimartingale 16

Decomposition 5, 7
Dominated convergence theorem 11
Doob's martingale inequality 4
Doob's stopping theorem 4

Explosion time 95
Exponential martingale inequality 6
Exponential stability
  Almost sure ~ 148, 184
  pth moment ~ 163, 224
  in mean square 164, 224
  in $L^p(\Omega; C([-\tau, 0]; R^d))$ 279

Finite variation process 3
Forward Brownian motion 89
Forward stochastic flow 89

Gronwall's inequality 25
  Generalized ~ 24
  Stochastic ~ 31

Increasing process 3
Indistinguishable processes 3
Infinitesimal covariance 90
Infinitesimal mean 90
Integrable process 3
Itô's formula 11
  Generalized ~ 21
Itô's integral 18, 20

Joint local characteristic 15
Joint optional quadratic
  variation 5, 6, 8

Lare-scale system 277
  in hierrachial form 283

Left continuous process 2
Left limit function 24
Lévy's theorem 7
Linear growth condition 38, 98
Lipschitz condition 94
Local characteristic 14, 16
Local Lipschitz condition 94
Local martingale 6
Local process of finite variation 6
Local solution 95
Locally square integrable process 6
Lyapunov exponent 172
  Almost sure ~ 148, 184
  pth moment ~ 163, 224
L^p-estimate 72, 100, 132

Martingale 3
Maximal solution 95
Measurable process 3
Meyer process 5
Modification 2
Multiplier function 28

Nonlinear integrator 17

Optional process 3
Optional quadratic variation 5, 6, 7
Orthogonal martingales 5

Predictable process 3
Predictable quadratic variation 5, 6
Prelocally integrable process 6
Progressive process 3
Progressively measurable 3
Purely discontinuous
  local martingale part 6
  martingale 5
  martingale part 5

Quadratic variation 5, 6

Regular contractor 104, 115
Right continuous process 2

Semimartingale 7, 11
Solution 35, 94
Square integrable martingale 4
Stochastic
  destabilization 167
  differential delay equation 93
  differential equation 33
  flow of homomorphisms 89
  integral 8, 10, 17
  integral inequality 31
  interval 3
  oscillator 176
  process 2
  stabilization 167
Stochastic integral convergence theorem 9
Stochastically closed 104
Stopped process 6
Stopping time 3
Submultiplicative function 28

Time change formula 23

Uniformly integrable martingale 4
Usual conditions 2

Version 3

Wendroff inequality 26
  Generalized ~ 26